Quantum Finance

Raymond S. T. Lee

Quantum Finance

Intelligent Forecast and Trading Systems

 Springer

Raymond S. T. Lee
Division of Science and Technology
Beijing Normal University-Hong Kong
Baptist University United International
College (UIC)
Zhuhai, Guangdong, China

ISBN 978-981-32-9798-2   ISBN 978-981-32-9796-8   (eBook)
https://doi.org/10.1007/978-981-32-9796-8

This Springer imprint is published by the registered company Springer Nature Singapore Pte Ltd.
The registered company address is: 152 Beach Road, #21-01/04 Gateway East, Singapore 189721,
Singapore

This book is dedicated to all readers and students in UIC taking Quantum Finance course; your enthusiasm for learning new concepts and seeking knowledge prompts me to write this book.

Preface

Motivation of This Book

Everything happens for a reason. Anything that might be happened with a cause.
These two sentences seem to be very simple, but they are major motivations to drive the advances in science and technology of human history.

However, modern cosmology tells us that we live in a *multiverse*, the world we all experience is just one of the many *realities*. Quantum theory tells us that all fundamental particles in the subatomic world coexist in a particle–wave duality state. Heisenberg's uncertainty principle even tells us that we can never 100% correctly measure both the position and momentum of any quantum particle simultaneously. Chaos theory shapes the world as complex systems with all predictions are highly sensitive to its initial conditions; only a tiny observation error will lead to a totally different future. Fuzzy logic tells us to accept this as nature of reality; all we can do is to model the world as a collection of *fuzzy linguistics and events*.

Does it mean that we cannot predict the future?

The answer is definitely ... NO.

The motivation of this book—quantum finance, a cross-discipline of financial engineering with the integration of quantum field theory, classical finance theory, computer science, and AI technology—is to find a scientific path to resolve this *grant puzzle*.

Quest for the Knowledge of Financial Prediction

This book is also about *an over 30-year journey of knowledge seeking on financial prediction*.

The idea of using quantum field theory and AI to solve financial problems (now called *quantum finance*) emerged when I was an undergraduate physics student at the Hong Kong University (HKU) in 1986 studying quantum mechanics and general relativity courses. At that time, I was deeply impressed by all great minds and innovative ideas for the interpretation of the world in two totally different perspectives: the subatomic world of quantum mechanics versus the cosmic world of general relativity. Since then, I strongly believe that both the Schrödinger equation and Feynman's path integrals hold the keys to unlock the *dynamics of everything*—to truly observe the past and predict the future. They also hold the keys to resolve two major forecasting challenges in human history: weather forecast in meteorology and worldwide financial prediction in finance.

After graduation from HKU in 1989, I joined the Hong Kong Observatory of the Government of Hong Kong Special Administrative Region and worked as experimental office on weather prediction and in the building of computational weather prediction model known as numerical weather prediction (NWP) system using Fortran. Although at that time it was still based on Newton's laws of motion, fluid dynamics, and thermodynamics equations to predict weather, it gave me an opportunity to realize two important facts: (1) In order to ultimately unlock the mystery of prediction using Schrödinger equation by computer system, we need to convert (model) the Schrödinger equation wisely into other formats such that it can be easily computed by digital computers using numerical computation techniques; (2) Only Schrödinger equation and Feynman's path integral are not adequate to predict the future. In Chaps. 3–5, these two mathematical models are the keys to model the quantum dynamics of quantum financial particles (QFPs) and to evaluate the quantum price levels (QPLs), but they are not the ideal tools to predict future. In order to predict the financial market wisely and effectively, we need some other *intelligent* tools. Such motivation precipitated me to pursue M.Sc. and Ph.D. studies at Hong Kong Polytechnic University in 1994 and 1997, respectively.

During this period, I have learned and explored all different kinds of artificial intelligence (AI) technologies including artificial neural network (ANN), fuzzy logic (FL), genetic algorithms (GAs), and chaos and fractals theories. Among them, I realized in order to solve this *grant puzzle*, one must look for four *puzzle pieces* first:

(1) All these AI tools and technologies are not only closely related but also each of them has its own advantages and shortcomings. If we can integrate these AI tools and techniques wisely, we might have an opportunity to solve more complex and fundamental problems, worldwide financial prediction—for instance;

(2) All these AI tools and techniques are in fact *different faces of the same coin*; how to integrate them wisely and effectively should not only focus on the methodology but rather the fundamental structure of the AI tools themselves;

(3) How to model quantum finance effectively is another important issue. In Chap. 3, we will see that traditional Feynman's path integral although can provide a feasible solution to model quantum dynamics of financial markets, the computational complexity of this method hinders us to apply it for real-time financial market prediction. In other words, we need a totally new idea to model

quantum finance effectively, not only at the analytical level but also at the numerical computation level;

(4) Even though if we can model financial markets using quantum finance effectively, how to integrate such model with AI is another problem. In order to integrate quantum finance model and AI technologies nicely to implement worldwide financial prediction and intelligent quantum trading system, we need a *common ground* to link up these two worlds. Again, not at the methodology level, but rather at the *core* model structure itself.

Not until 2004, I proposed the Lee-oscillator—a chaotic time-discrete neural oscillator with profound transient-chaotic property, which gave us some new hopes to resolve part of this puzzle. The interesting thing is that, when the Lee-oscillators are used to replace the neurons in any time series neural networks, such as traditional feedforward backpropagation networks (FFBPNs), they would convert the neural networks into powerful chaotic neural oscillatory networks (CNONs) with promising prediction and machine learning capabilities. In Chap. 9, such chaotic neural oscillators can be served as the *missing link* to integrate all different kinds of AI technologies effortlessly, which sheds light on the resolution of the first two puzzle pieces to model complex time series prediction problems such as real-time worldwide financial forecast. After that, it leaves us with the last two puzzle pieces.

In 2007, Dasgupta et al. published their paper titled as *Simple systematics in the energy eigenvalues of quantum anharmonic oscillators* in the *Journal of Physics A* (Chap. 3) which explained how to model Schrödinger equation as quantum anharmonic oscillators (QAHOs). More importantly, this paper nicely described how to resolve a special type of QAHO so-called λx^{2m} AHO in which at the same time I had derived the quantum finance Schrödinger equation (QFSE) as an AHO of order 4 (i.e., m = 2) quantum dynamic model (Chap. 4). By the integration of this λx^4 AHO mathematical model and the finite difference method (FDM) learnt for the implementation of the NWP model in Hong Kong Observatory years ago, I can finally construct a numerical computation model with the integration of AHO (to model quantum financial particles) and CNON (to model chaotic time series prediction) into the so-called quantum finance forecast system (QFFS).

In order to prove the applicability of this totally new AI-fintech invention in real-world financial markets, during 2012–2017, I joined Leanda Investment Group (one of the major commodity investment and trading companies in China) as the chief analyst and group CTO for the implementation and national promotion of QFFS in China. With the success of the implementation of QFFS in commodity product forecast for over 5 years in China and Hong Kong, in Mar 2017, I set up the quantum finance forecast center (aka QFFC, website QFFC.org), a nonprofit making, AI-fintech R&D, and worldwide financial forecast center aims at the R&D and the provision of an open platform for worldwide traders and individual investors to acquire free worldwide real-time financial prediction. QFFS now provides daily (weekly) financial forecasts for over 120 worldwide financial products including 9 major cryptocurrencies, 17 worldwide financial indices, 84 forex, and 19 major commodities. As of March 2019, QFFC has over 10,000+

registered worldwide members which consists of worldwide professional traders, quants, and independent investors from major fund houses and financial institutions using the free quantum finance daily forecast services.

Organization of This Book

This book is the collection of my past 20 years (1999–2019) of R&D works and practical implementation of quantum finance and related AI technology including chaotic neural networks, fuzzy logics and genetic algorithms (GAs) in financial market modeling, time series financial prediction, and intelligent agent-based quantum trading techniques and methodologies.

This book consists of two parts: Part I describes the basic concept and theory of quantum finance and related AI technology. Part II describes the current and ongoing R&D projects for the application of quantum finance technology on intelligent real-time financial prediction and intelligent agent-based quantum trading systems.

This book is organized as follows:

- Chapter 1—Introduction to Quantum Finance
  This chapter introduces the origin of quantum finance and the two different perspectives to *see* the world. It investigates the wave–particle duality and Heisenberg's uncertainty principle in quantum mechanics and how these phenomena are related to quantum finance, which leads to our study on the philosophy of quantum finance. After that, it introduces the major components of quantum finance and the quantum finance four-tier concentric sphere model.
- Chapter 2—Quantum Field Theory for Quantum Finance
  To avoid complex mathematical derivation, the chapter mainly focuses on the basic concept and physical meaning of how quantum field theory can be applied to financial engineering for the modeling of worldwide financial markets.
- Chapter 3—An Overview of Quantum Finance Models
  This chapter investigates two main branches of quantum finance models—the Feynman's path integral model and the quantum anharmonic oscillator model. It also explores the future of quantum finance which relates to intelligent finance systems' development.
- Chapter 4—Quantum Finance Theory
  This chapter starts with a basic concept of quantum finance theory—quantum anharmonic oscillator model. It focuses on the complete mathematical derivation of the author's original work of quantum finance Schrödinger equation (QFSE). It also investigates the quantum dynamics in financial markets, the notion of quantum price levels (QPLs) and their relationship with quantum finance energy levels (QFELs).
- Chapter 5—Quantum Price Levels—Basic Theory and Numerical Computational Technique

This chapter focuses on detailed mathematical and numerical derivations of quantum price levels to solve the quantum finance Schrödinger equation (QFSE) effectively using numerical computational method—the *core* of quantum finance in financial market modeling. First, it presents the basic concept of quantum price levels (QPLs) and its relationship with quantum finance energy levels (QFELs) in QFSE. Second, it shows how to interpret the quantum price return wave function in terms of probability density function (*pdf*) using finite difference method (FDM). Third, it explores how to solve QFSE using numerical computational technique and describes the computer algorithm to determine all QFEL and QPL. This chapter will be ended with the first quantum finance computing workshop for QPL evaluation on worldwide financial products using metatrader query language (MQL) in metatrader (MT) platform.

- Chapter 6—Quantum Trading and Hedging Strategies
- This chapter focuses on the practical use of quantum finance in quantum trading and hedging operations. First, it starts with the discourse of seven major trading and hedging techniques—a collection of the author 20+ years of trading experience in stock, commodity, and forex trading with numerous technical trading courses conducted in China and Hong Kong for past 10 years. Second, it presents the main concept on how to make use of quantum finance, especially on quantum price levels (QPLs), quantum forecasts to design quantum trading and hedging strategies.
- Chapter 7—AI Powerful Tools in Quantum Finance
  This chapter introduces three major AI tools in quantum finance: artificial neural networks (ANNs) on machine learning and time series prediction, fuzzy logics (FL) on fuzzy and inexact financial modeling, and genetic algorithms (GAs) on trading strategy optimization. More importantly, it shows how these AI tools are integrated with quantum finance technology to implement quantum finance forecast and intelligent agent-based program trading systems.
- Chapter 8—Chaos and Fractals in Quantum Finance
  This chapter introduces two vital contemporary finance engineering theories: chaos and fractals. It explains the duality behavior of financial markets modeled by these two theories, and more importantly, on how they are related to quantum finance for financial engineering in contemporary financial institutions.
- Chapter 9—Chaotic Neural Networks in Quantum Finance
  This chapter presents basic theory of chaotic neural oscillators and neural networks with the author's original work on Lee-oscillator, how it can be adopted with quantum field signals (QFSs) into quantum finance oscillators (QFOs) and converted into chaotic neural networks for financial prediction.
- Chapter 10—Quantum Price Levels for Worldwide Financial Products
  As Part II's first chapter on quantum finance applications, this chapter discusses R&D project to apply quantum finance technology to calculate the quantum price levels (QPLs) for over 129 worldwide financial products including: forex (84), cryptocurrency (9), major commodities (17), and worldwide financial indices (17).

- Chapter 11—Time Series Chaotic Neural Oscillatory Networks for Financial Prediction
  This chapter discusses R&D project on how quantum price levels (QPLs) can be integrated with chaotic neural oscillatory network learnt in Chap. 9 to time series chaotic neural oscillatory networks (TSCNONs) for financial prediction.
- Chapter 12—Chaotic Type-2 Transient-Fuzzy Deep Neuro-Oscillatory Network (CT2TFDNN) for Worldwide Financial Prediction
  This chapter discusses the author's latest R&D for the integration of chaotic type-2 transient-fuzzy logic with chaotic neural networks for the implementation of chaotic type-2 transient-fuzzy deep neuro-oscillatory network (CT2TFDNN) for worldwide financial prediction.
- Chapter 13—Quantum Trader—A Multiagent-based Quantum Financial Forecast and Trading System
  This chapter discusses the latest research project to apply multiagent-based real-time financial prediction and trading system using quantum finance technology.
- Chapter 14—Future Trends in Quantum Finance
  This chapter concludes all concepts and theories in quantum finance, their relations to contemporary AI technology on intelligent financial systems. It also discusses the latest research on financial engineering and the trend of quantum finance.

Readers of This Book

This book is both a textbook and professional book tailored for

- Undergraduate and master-degree students for various courses on quantum finance/computational finance/AI/fintech/machine learning;
- Researchers and scientists working in the field of financial engineering and intelligent systems, AI, computational finance, econophysics, data science, and computational intelligence;
- Professional traders/quants/independent investors who would like to learn the basic concept and theory of quantum finance, more importantly on how to use this innovative AI-fintech technology to implement intelligent financial forecast and program trading systems;
- Lecturers and tutors who would like to teach students quantum finance with related AI technologies, and to conduct labs to teach students on how to design and implement intelligent financial forecast and trading systems.

How to Use This Book?

This book can be served as both textbook and research monograph for various undergraduate and postgraduate courses on quantum finance and related AI/AI-fintech courses.

Basically, Part I (Chaps. 1–9) covers all basic concepts and theory about quantum finance and related AI technology can be conducted as a self-contained undergraduate/postgraduate course on quantum finance.

Part II (Chaps. 10–14) covers all advance studies and applications of quantum finance, which can be adopted as the selected topics on various advanced AI/fintech courses for postgraduate and research studies.

In UIC, starting from early 2019, the author launched a new free elective course, namely, "Quantum Finance" for Year 3/4 undergraduate students of different disciplines including computer science, financial mathematics, data science, and statistics. In terms of course syllabus, this course covers Chaps. 1–9 in the 14-week study, and Chaps. 10 and 11 as workshops and term projects for the implementation of QPL-based quantum finance forecast systems. Besides, students and course instructors can use MQL workshops provided by QFFC.org enabling students to learn MQL programming and how to develop simple program trading programs.

For advanced studies such as postgraduate AI courses or M.Phil./Ph.D. studies, Chaps. 11–13 provide advanced topics on how to combine quantum finance theory with contemporary type-2 fuzzy logic, chaotic neural oscillators, and reinforcement learning agent theories for further R&D on quantum finance and related studies.

Last but not least, for quantum finance application developers such as quants and financial analysts make use of the QFSDK provided by QFFC.org to develop quantum finance applications. They can directly read Chaps. 10 and 11 first as technical reference for QPLs' calculation to implement quantum finance forecast system for their own interesting financial products and revert to Part I to learn the basic concepts and theories.

Companion Website—Quantum Finance Forecast Center (QFFC.org)

Quantum finance forecast center (QFFC.org) is a nonprofit making, AI-fintech R&D and worldwide financial forecast center that aims at R&D and the provision of an open platform for worldwide traders and individual investors to acquire free worldwide financial products' quantum finance forecast.

As a companion website, QFFC provides a web portal for readers and worldwide quantum finance system developers to learn and develop quantum finance systems with the following:

- Interactive quantum finance computing workshops (QFCWs) provide step-by-step online tutorials and programming workshops for readers to apply the knowledge learnt from this book.
- Quantum finance software development kit (QFSDK) is a C-library fully integrated with MT platform which consists of all related quantum finance and AI tools and functions for quantum finance developers to build their own quantum finance forecast and intelligent agent-based trading and hedging systems. In other words, this book can be also used as the guidebook and main reference to learn the basic concept and theory of quantum finance technology.

Zhuhai, China   Raymond S. T. Lee

Acknowledgements

While it only required around 6 months to write this book, the whole journey of quantum finance—from seeking concepts and knowledge to the actual implementation—took me almost 30 years, started when I was still an undergraduate at HKU since 1986 studying quantum mechanics and general relativity courses to currently teach quantum finance to my fellow students at UIC.

I would like to express my gratitude to the following people for their support and assistance.

To my wife Iris for her patience, encouragement, and understanding, especially during my time spent on research and writing in the past 30 years.

To Ms. Celine Cheng, executive editor of Springer Nature and her editorial team members including Jane Li and Haiying Li, for their support, valuable comments, and advice.

To Prof. James Liu, my M.Sc. and Ph.D. mentor, for his support to led me to the *sacred land* of artificial intelligence.

To Jim He, my TA, and all my lovely research students in UIC including Turing Qiu, Henry Liang, Alex Yuan, and Peter Chen, for their help and assistance for the development of Quantum Finance Software Development Kit and online tutorials, labs, and workshops.

To Prof. Tang Tao, President of UIC, for the provision of excellent environment for the research, teaching, and the writing of this book.

To Prof. Hua Xiong Huang, Dean of Division of Science and Technology of UIC, and Prof. Weifeng Su, Program Director of Computer Science and Technology of UIC, for their support for the opening of new course in quantum finance in UIC.

To all fellow colleagues in UIC, for their support and fruitful advice in course design.

To UIC, for the prominent support on research grant # R201948; and for the provision of an excellent environment, facilities for system development, testing, and evaluations.

Zhuhai, China
June 2019

Raymond S. T. Lee

Contents

About the Author

Dr. Raymond S. T. Lee founder of quantum finance forecast center with over 20 years of IT consultancy, R&D experiences in AI, chaotic neural networks, intelligent fintech system, quantum finance, and intelligent e-commerce systems, had successfully commercialized his AI-fintech invention at business sectors in China and Hong Kong.

He attained his B.Sc. (Physics) from Hong Kong University in 1989, and M.Sc. (Information Technology) and Ph.D. (Computer Science) from Hong Kong Polytechnic University in 1997 and 2000, respectively.

After graduation from Hong Kong University, he joined the Hong Kong Observatory of the Government of the Hong Kong Special Administrative Region as a meteorological scientist on weather forecasting and developing numerical weather forecast system from 1989 to 1993.

From academic perspective, he had worked at the Department of Computing of Hong Kong Polytechnic University (HKPolyU) as Lecturer, and promoted as Assistant Professor in 2000 and Associate Professor in 2005, respectively. During this time, he had published over 90+ publications and author of six textbooks and monographs covering the fields at AI, chaotic neural networks, AI-based fintech systems, intelligent agent technology, chaotic cryptosystems, ontological agents, neural oscillators, biometrics, weather simulation, and forecasting systems.

From commercial perspective, he was invited to join Leanda Investment Group in China (2012–2017) as Group CTO and Chief Analyst to implement his self-invented and patented AI-fintech invention—quantum finance forecast system on major commodities in China for 1,000+ investors.

In March 2017, he set up the quantum finance forecast center (aka QFFC) (http://qffc.org), a nonprofit making, AI-fintech R&D, and worldwide financial forecast center, that aims at the R&D and provision of a wholly FOC and open platform for worldwide traders and individual investors to acquire free knowledge of worldwide 129 financial product forecasts based on the state-of-the-art AI, chaotic neural networks and quantum finance technologies. At present, QFFC has over 10,000+ registered worldwide members, which consists of worldwide professional traders, quants, and independent investors from major fund houses and financial institutions using the free quantum finance daily forecasting services.

Upon the completion of the fully automated quantum finance forecast system, he joined Beijing Normal University-Hong Kong Baptist University United International College (UIC) in China as Associate Professor in 2018 to further his R&D works on quantum finance and to contribute his knowledge on quantum finance, AI-fintech, chaotic neural networks, and related intelligent forecast systems to the fellow students and the community in China.

Abbreviations

ANN	Artificial Neural Networks
CBTF	Chaotic Bifurcation Transfer Function
CDNONRS	Chaotic Deep Neuro-oscillatory Network with Retrograde Signaling
CNON	Chaotic Neural Oscillatory Network
CT1-FNON	Chaotic T1 Fuzzy Neuro-oscillatory Network
CT2TFDNN	Chaotic T2 Transient-Fuzzy Deep Neuro-oscillatory Network with Retrograde Signaling
CT2TFL	Chaotic T2 Transient-Fuzzy Logic
CT2TFLS	Chaotic T2 Transient-Fuzzy Logic System
CT2TFMF	Chaotic T2 Transient-Fuzzy Membership Function
CTU	Chaotic Transfer Unit
DNN	Deep Neural Network
EC System	Evolutionary Computing System
FFBPN	Feedforward Backpropagation Networks
FFS	Fuzzy Financial Signals
FFSCM	Fuzzy Financial Signals Crossover Module
FFSFEM	Fuzzy Financial Signals Fitness Evaluation Module
FFSMM	Fuzzy Financial Signals Mutation Module
FFSPGM	Fuzzy Financial Signals Population Generation Module
FFSV	Fuzzy Financial Signal Vector
FLS	Fuzzy Logic System
FOU	Footprint Of Uncertainty
FSGM	Financial Signal Generator Module
GA	Genetic Algorithms
HFAPT	High-Frequency Algorithmic Program Trading
IT2FLS	Interval Type-2 Fuzzy Logic System
IT2-FNN	Interval Type-2 Fuzzy-neuro Network
LORS	Lee-oscillator with Retrograde Signaling
MQL	MetaQuotes Language

MT4	MetaTrader4
NGPSM	New Generation Population Selection Module
PCA	Principal Component Analysis
QAOM	Quantum Anharmonic Oscillatory Model
QFFS	Quantum Finance Forecast Centre
QPL	Quantum Price Level
RMSE	Root Mean Square Error
STR	State Transition Region
SVM	Support Vector Machine
T2FL	Type-2 Fuzzy Logic
T2FLS	Type-2 Fuzzy Logic System
T2FMF	Type-2 Fuzzy Membership Function
TOP10-FFSSM	TOP-10 Fitness FFS Selection Module

Part I
Quantum Finance Theory

Chapter 1
Introduction to Quantum Finance

*As far as the laws of mathematics refer to reality, they are not
certain; and as far as they are certain, they do not refer to reality.*
Albert Einstein (1879–1955)

What is the world of reality?
Is there any physical laws governing our world?
Does everything happen simply by chance or predetermined?

Professor Albert Einstein's quotation gave us an exceedingly admirable description
of the true nature as regards *reality*. The fact is that this fundamental question remains
unsolved throughout the evolution of human civilization. Does it mean that we can't
do anything about it? The answer is definitely *no*.

Human beings by nature are exceptionally adequate with one trait in countless
years—live in our predefined models. Even though we never know the real aspect
of the *world of reality*, our minds can unconsciously create the *model of reality*
and make us *believe* that this is the real world (reality) we live in. In science, we
follow the same logic. To study any phenomenon that occurs in the real world (for
instance, *weather*), we build a so-called *model* to represent it. By observing and
studying object and matter dynamics within this model, we try to formulate some
laws of dynamics and/or equations to generalize their activities to find out some
patterns or even to predict what will happen in the future. In finance, we follow the
same footsteps. Quantum finance, as a new age of finance technology that tackles
the ultimate problem of finance—the quantum world of finance—is no exception.

In this chapter, we will begin with the origin of quantum finance—the way we
percept and model our physical world, from Newtonian's classical physical world to
Heisenberg's quantum world of subatomic particles. Next, we will study a unique
feature of the quantum world, the wave–particle duality phenomena of quantum

© Springer Nature Singapore Pte Ltd. 2020
R. S. T. Lee, *Quantum Finance*,
https://doi.org/10.1007/978-981-32-9796-8_1

particles, and how such phenomena are related to quantum finance. After that, we will study quantum field theory and the birth of quantum finance followed by the Heisenberg's uncertainty principle in quantum mechanics, and how it relates to two important theories in AI—chaos theory and fuzzy logic. Lastly, we will explore major components of quantum finance—the four-tier concentric sphere quantum finance model.

1.1 A Tale of Two Worlds—From the World of General Relativity to the World of Quantum Mechanics

From ancient Greeks to the dawn of science, philosophers, mathematicians, physicists, psychologists, and natural scientists over centuries strive to better understand the *world of reality* we live in. We observe all sorts of phenomena, ranging from weather changes in our atmosphere to comets and supernova of the universe. Based on these observations, we discover some common patterns and dynamics. Above these common patterns and dynamics, we build certain physical and mathematical models, equations, and laws of dynamics to make sense of them and to generalize them to what we called *knowledge*. If we are fortunate to exert our sagacity, we can derive from these laws of dynamics and mathematical models not only to extract the patterns that had already occurred but also have a chance to *glimpse* the future.

Finance, especially the worldwide financial markets, is believed to be the phenomena that resulted in *high-level intellectual activities* and *the collective behaviors* of worldwide investors and market regulators. Different from classical finance which focuses on the statistical models and analysis of financial markets and market patterns, quantum finance applies the latest research of quantum field theory and quantum mechanics to study the *actual dynamics* of quantum particles (so-called *quantum financial particles*, QFPs) in every financial market. By exploring these quantum financial dynamic activities, we would have a better chance not only to observe the financial patterns but also to predict their future movements.

To study the actual dynamics of financial particles, why do we use quantum model, but not others?

To answer this question, we need to explore a more fundamental question: What are the laws of dynamics governing the world of reality we inhibit?

Thanks to the prominent minds and scientific discovery in the past centuries, contemporary physics and cosmology tell us that the world of reality can be defined by three physical models: world of classical mechanics discovered by Isaac Newton (1643–1727), cosmic world of general relativity discovered by Albert Einstein (1879–1955), and the subatomic world of quantum mechanics discovered by Werner Heisenberg (1901–1976), together with the three distinctive *laws and equations* to govern these worlds as shown in Fig. 1.1.

| Einstein's
Cosmic world
General Relativity
E = mc2 | Newton's
Physical world
Classical Mechanics
F = ma | Heisenberg's
Subatomic world
Quantum Mechanics
$\hat{H}|\Psi\rangle = E|\Psi\rangle$ |

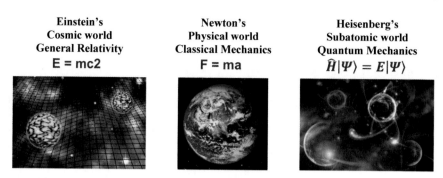

Fig. 1.1 Laws of physical worlds (Tuchong 2019a, b and c)

1.1.1 Newton's Physical World of Classical Mechanics

Sir Isaac Newton (1643–1727) was an English mathematician, physicist, astronomer, and theologian. His book *Mathematical Principles of Natural Philosophy* (Brackenridge 1999) published in 1687 laid out the foundations of classical mechanics, together with his remarkable works on optics and calculus. These three great contributions bestowed on him as one of the most important prominent minds and influential scientists in the human history (Pourciau 2006; Stewart 2004).

In Newton's model, the world of reality is called the *physical world*—the world we live and experience physically. All matter and object motions in this *physical world* are strictly governed by his three laws of dynamics known as *Newton's laws of motion*. Causes and effects (so-called *causation*) of these matter and object motions must follow these laws agreeably in this world almost without any exceptions. Yet, modern cosmology and quantum mechanics tell us that it is not always the case. The fact is that the laws of physics only hold under two conditions:

(1) Objects and matters of *normal* size, what we called *macroscopic objects*; and
(2) Object motions at *normal* speed, as compared with the speed of light ($c = 3 \times 10^8 \text{ms}^{-1}$).

1.1.2 Einstein's Cosmic World of General Relativity

General relativity (GR, also known as the *general theory of relativity*) is the geometric theory of gravitation coined by Professor Albert Einstein (1879–1955) at the presentation of his work *The Field Equations of Gravitation* at the Prussian Academy of Science in November 1915 (Einstein 1915), and *The Foundation of the General Theory of Relativity published by Annalen der Physik in 1916* (Einstein 1916), which set the cornerstone of general relativity along with quantum mechanics as the two pillars of modern physics (Woodhouse 2007). He is best known to the general public for his groundbreaking mass–energy equation $E = mc^2$, which has been considered as *"the world's most famous equation"* after Newton's second law of motion $F = ma$.

Different from centuries of beliefs that *time* as a special one-way, linear, and non-reversible quantity, the groundbreaking idea of general relativity is the consideration of *time* as a dimension identical to the three dimensions of *space* we are known of. General relativity generalizes Einstein's special relativity and Newton's law of universal gravitation, providing a unified description of gravity as a geometric property of space and time, known as *spacetime*—a critical mass of modern cosmology (Woodhouse 2007; Böhmer 2016; Carroll 2016).

In Einstein's world of general relativity, *space*, *time*, and *matters* are interconnected to each other. Even his early work on special relativity already introduced the notion of *time dilation*—in which *time* is not an absolute constant quantity but can vary according to the observer's so-called *frame of reference* (Chaturvedi et al. 2006). In layman's term, to describe matters and objects with normal mass and velocity (i.e., much slower than the speed of light), Newton's laws of motion are good enough to do the job. But, when matters and objects with normal mass move close to the speed of light, it is the time to apply special relativity. In case that matters and objects are extremely massive, such as planets and black holes in which even light and time can be distorted by their massive gravitational fields as visual interpretation is shown in Fig. 1.1a, general relativity should be applied. In fact, general relativity is the basis of current cosmological models of a consistently expanding universe and *Big Bang Theory*.

1.1.3   Heisenberg's Subatomic World of Quantum Mechanics

Professor Werner Karl Heisenberg (1901–1976) was a German theoretical physicist and one of the key figures of quantum mechanics. He coined the term quantum mechanics in his influential paper *Quantum theoretical re-interpretation of kinematic and mechanical relations* published at *Zeitschrift für Physik* in September 1925 (Heisenberg 1925). Quantum mechanics (QM) and quantum field theory describe nature at the smallest scales of energy levels of atoms and subatomic particles. Differing from Newtonian classical physics, quantum mechanics (Norsen 2017) has five characteristics:

1. Objects and matters in this quantum world (or what we called *subatomic world*) are called quantum particles, and their dynamics are governed by the famous *Schrödinger equation*.
2. *Quantization*—a unique phenomenon in quantum mechanics—describes energy, momentum, angular momentum, and related quantities in this subatomic world as *discrete states* and *energy levels*.
3. *Wave–particle duality* states that objects and matters in this world have characteristics of both particles and waves simultaneously.

4. *Heisenberg's uncertainty principle* states that there are *natural* limits of precision for *paired quantity* of every quantum particle, for example, *position* and *momentum* of any quantum particle.
5. *Quantum entanglement*—a unique phenomenon in quantum mechanics—states that multiple quantum particles are linked (interacted) together in a way such that the measurement of one particle's quantum state determines the possible quantum states of the other particles, even they are separated by very far distance (Sen 2017).

As a groundbreaking theory to describe all fundamental particles and their events, quantum mechanics believes to be a perfect tool to model any fundamental phenomena in the physical world such as finance, for example. Figure 1.1c shows a visual interpretation of Heisenberg's subatomic world of quantum mechanics.

1.2 Two Sides of the Same Coin—Wave–Particle Duality in Quantum Finance

1.2.1 Wave–Particle Duality and Double-Slit Experiment

Wave–particle duality is the fundamental concept in quantum mechanics which states that every quantum particle (or so-called *subatomic particle*) coexists in two different states, *waves* and *particles*. In classical physics, we all learnt that *waves* and *particles* are two totally different entities and notions governed by two distinct physical laws: Newton's laws of motion govern all motions of physical objects and matters, whereas Maxwell's wave equations govern all the physical dynamics of waves. As we will explain the details in the coming chapters, Schrödinger equation is the foundation of quantum mechanics that provides a unique way to visualize how wave and particle dynamics are related and coexist from the mathematical perspective in this quantum world.

Can we visualize wave–particle duality in the physical world?

The answer is yes: The famous *double-slit experiment* for photon particles.

Figure 1.2 shows the experiment setup of double-slit experiment.

The setup of double-slit experiment in quantum mechanics borrows from Professor Thomas Young's famous double-slit experiment of interference phenomena in classical wave theory. But this time, laser beam was used as the light source to emit photons in the experiment as shown. Same as the classical double-slit experiments, a wall with two slits of openings was set between the light source and the screen. The whole experiment consists of two parts: part 1 experiment used photoelectric detector to detect the motion of every photon emission reflecting its full path to the destination (screen) and part 2 experiment was the same setup except that the detector was switched off.

Although the two setups were basically the same, yet interesting effects occurred. Part 1's experiment that used photoelectric detector showed that particle motions of

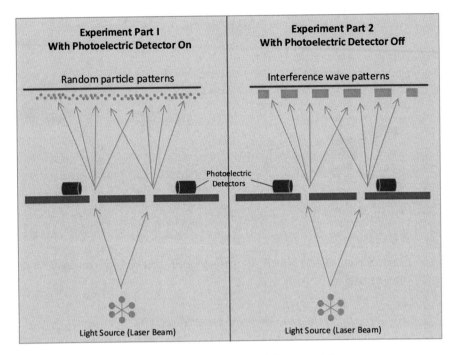

Fig. 1.2 Experimental setup of double-slit experiment of photon particles

light photon projection were randomly distributed as every photon path is a particle motion-independent event. Part 2's experiment that did not use photoelectric detector showed that alterative bright-and-dark bands are similar in any typical wave interference phenomena. What do they mean? What we watch (observe) can really affect the experimental results? How can light particles (photon) be sometimes *particle-like* and sometimes *wave-like*?

Quantum mechanics provides a simple but extraordinary explanation: first, all quantum particles coexist in two totally different states: *wave* and *particle* states, we called it wave–particle duality. Second, how they behave all depend on how we observe (detect) their motions (phenomena). In other words, if we use photoelectric detector to detect light, it will behave as photon particles, that is, photoelectric detector can only detect photons. On the other hand, if we don't use any photoelectric detector, it will behave as its usual state, that is, light waves with typical interference phenomena, which means that the observer oneself (or the detector itself) will affect the experimental results; it is rather against the traditional beliefs in science in which the observer should not be part of the experiment. Is it true?

Despite the notion that was extraordinary among science community in the early twentieth century, it was not a novel concept in philosophy. Professor Immanuel Kant's (1724–1804) remarkable work *Critique of Pure Reason* published in 1781 (Kant 2008) stated that: *the world of reality we percept is just the model reality shaped by the senses and mind.* In layman's term, it can be interpreted as wearing a pair of colored spectacles to see the world. Different colors of the world naturally depend on the color of the spectacles we wear. As mentioned at the beginning of this chapter, humans are good at creating models to reflect everything we see and percept in our minds, and the observation results we obtain naturally depend on the model we create.

In fact, the major significance of the wave–particle duality is that all behaviors of light and matter can be explained through a single equation which relates quantum particle's wave dynamics with its particle dynamics—the Schrödinger equation. This ability to describe reality in the form of waves is the core of quantum mechanics and is also the quintessence of this book.

1.2.2 Wave–Particle Duality in Finance

In financial markets, can we observe any wave–particle duality?

The answer is definitely *"yes"* and *"always"*.

Experienced financial analyst or trader will tell us that financial markets appear with both *particle* and *wave* properties constantly.

The two pillars of modern financial analytical tools, technical analysis and chart analysis, are very good examples of such wave–particle duality.

Technical analysis (Murphy 1999) describes the financial markets as physical motion of price (or index); we analyze the price movement in respect of moving averages, momentums, and acceleration, just like any physical objects we encounter in the physical world, whereas chart analysis shapes the financial markets as financial waves and patterns. From that, it creates and defines various major reversal and trend patterns such as golden ratio patterns, Fibonacci patterns, and the famous Elliott wave patterns shown in Fig. 1.3 that gave credence to affect the market price trends, trading patterns, and individual reasonings for investors to decide the pre-eminent time to trigger the buy or sell decisions (Borden 2018; Brown 2012; Bulkowski 2005).

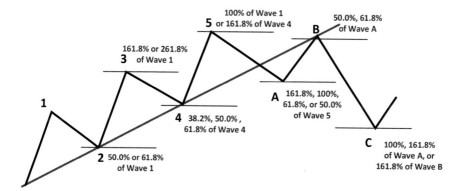

Fig. 1.3 Elliot waves chart principle in financial analysis

In other words, in the financial world at any particular instance, when we can observe (or measure) the price (or index value) of any financial product such as the Dow Jones index (DJI), we treat this as a *particle*, and use technical analytical tools to study and analyze it. At the same time, we can also observe not only on price but also the entire financial time series as waves and patterns; use chart analysis to discover the embedded waves and patterns.

In quantum finance, can we model financial market using quantum mechanics and quantum field theory?

1.3   Quantum Field Theory and the Birth of Quantum Finance

Quantum field theory is a unified theoretical framework to describe motion and dynamics of fundamental particles. It consists of

- Maxwell field theory,
- Special relativity, and
- Quantum mechanics.

Quantum field theory regards all fundamental particles as excited states (quanta) of their underlying fields (quantum fields).

In quantum field theory, atoms absorb and emit electromagnetic radiation, as tiny oscillators with their energies can only take on a series of discrete values—quantum anharmonic oscillators (QAHOs).

This process of restricting energies to discrete values is called *quantization*.

Their interactions and dynamics can be visually represented by the famous Feynman diagrams as shown in Fig. 1.4.

Quantum finance (QF) is a newly developed interdisciplinary subject applying quantum mechanics (QM) and quantum field theory (QFF)—econophysics. The first

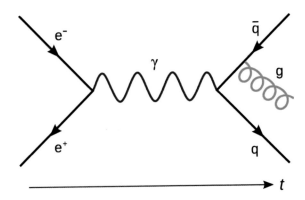

Fig. 1.4 Feynman diagram

published work *An Introduction to Econophysics—Correlations and Complexity in Finance* was written by Mantegna and Stanley in (1999).

Modern econophysics view this as an application of Brownian motion—fundamental phenomenon of statistical physics.

During the past decades, various methods and theories had been proposed for stock price/returns analysis, option pricing, and portfolio analysis.

In this book, we will study how to combine quantum mechanics and quantum field theories including Feynman's path integrals, Hamiltonians, quantum wave functions, quantum oscillators with contemporary AI tools such as fuzzy logics, genetic algorithms, and chaos and fractals to model and interpret worldwide financial markets for financial forecast and quantum trading.

1.4 Heisenberg's Uncertainty Principle in Finance and Modern AI Technology

In 1927, German physicist Professor Werner Heisenberg stated that the more precisely the *position* of some particle is determined, the less precisely its *momentum*— the famous *Heisenberg's uncertainty principle* (Gribbin 1984).

$$\sigma_x \sigma_p \geq \frac{\hbar}{2} \tag{1.1}$$

where $\hbar$ is the reduced Planck constant.

Unlike classical physics, uncertainty principle states that all events occur in quantum states and will (only will) *collapse* into a particular reality when the observer takes the measurement (or observe the event).

This phenomenon is associated with the famous *Schrödinger's cat paradox* (Gribbin 1984; Monroe et al. 1996).

1.4.1 Schrödinger's Cat Paradox

- In this paradox as shown in Fig. 1.5, a cat, a flask of poison, and a radioactive source are placed in a sealed box.
- If an internal monitor (e.g., GM counter) detects radioactivity (i.e., a single atom decaying), the flask is shattered and releases poison which kills the cat.
- The interpretation of quantum mechanics implies that after a while, the cat is simultaneously alive and dead. Yet, when one looks in the box, the cat is either alive or dead (Bhaumik 2017).
- This poses the question of when exactly quantum superposition ends and reality collapses into one possibility or the other (Stamper-Kurn et al. 2016).

Fig. 1.5 Schrödinger's cat paradox in quantum mechanics

1.4.2 Uncertainty Principle in Finance

One may wonder, does uncertainty principle (phenomena) occur in financial markets?

The answer is: *always*.

Any experienced traders can tell us that there is no 100% sure in financial market. Anything can happen in every second. The only thing we are 100% sure and confirm is the time when we look at the market, which is similar to what Einstein said in his famous quote:

> I like to think that the moon is there even if I am not looking at it
>
> Albert Einstein (1879–1955)

In other words, without *actually* observing the market, anything can occur.

If that is the case, how can we model it by using the technology we have? Or, is it merely a fancy thought?

In fact, there are two technologies that provide an excellent analog to uncertainty principle in modern AI (Russell and Norvig 2015).

They are *fuzzy logics* and *chaos theory*.

Fuzzy logics (FLs): In the world of fuzzy logic, everything is uncertain. Fuzzy logic (Zadeh and Aliev 2019) provides an easy-to-implement solution to model multiple attributes that can be occurred at the same time the so-called fuzzy variables (FVs).

Chaos theory (CT): In the world of chaos theory, everything is a complex system, so-called *deterministic chaos* (Schuste 2005) (Fig. 1.6). Chaos theory provides a framework and mathematical model to simulate highly chaotic or random-like phenomena, such as weather patterns and financial markets (Lee 2005).

Fig. 1.6 Deterministic chaos in chaos theory and butterfly effect

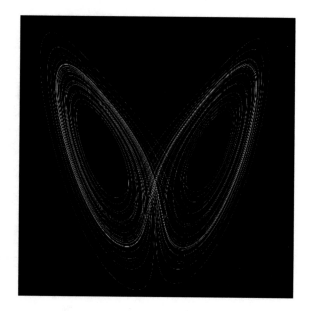

We will study these two fascinating and useful topics in Chap. 7—AI powerful tools in quantum finance.

1.5 Basic Components of Quantum Finance

In this book, we will explore how quantum mechanics and quantum field theory can be used to model financial markets. More importantly, we will study how modern AI technologies such as artificial neural networks, fuzzy logics, genetic algorithms, chaos theory, and fractals can be integrated with quantum finance model to design and implement intelligent real-time financial forecast and quantum trading systems.

Figure 1.7 shows the *concentric sphere model of quantum finance*.

1. First Tier—Energy Field Layer

   - The core,
   - Provides quantum field—quantum price field (QPF).

2. Second Tier—Chaotic Neural Network Layer

   - Provides the neural dynamics in quantum finance,
   - Supports neural oscillators and chaotic neural networks.

3. Third Tier—AI-fintech Layer

   Provides AI-finance technology tools,

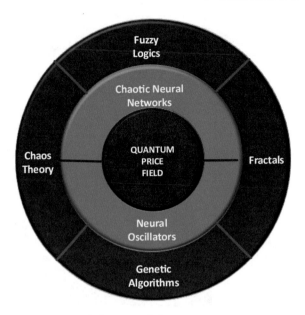

Fig. 1.7 Concentric sphere model of quantum finance

Supports fuzzy logics (FLs), genetic algorithms (GAs), chaos theory (CT), fractals, support vector machine (SVM), etc.

4. Fourth Tier—Application Layer

- Quantum price levels (QPLs),
- Short-term price prediction,
- Long-term trend prediction, and
- Intelligent multiagent-based trading systems.

1.6 Conclusion

In this chapter, we introduced the foundation of quantum finance with exploration of basic dynamics laws of the world we live in from three different perspectives, ranging from macroscopic world of general relativity to microscopic world of quantum mechanics. We also reviewed key properties and features of quantum mechanics including quantization, wave–particle duality, and Heisenberg's uncertainty principle. More importantly, we studied how these unique and extraordinary quantum phenomena are related and can be observed in real-world financial markets, which constituted the motivation in developing quantum finance model in the following chapters.

In the last section, we introduced quantum finance four-tier concentric sphere model, the central framework to integrate quantum finance system core model with modern AI technology as the neural network, AI-fintech, and application tiers which constituted Part I of this book—quantum finance theory.

In Part II, we will study AI-fintech application including the implementation of quantum price level (QPL) for worldwide financial products in Chap. 10, design and implementation of worldwide 120+ quantum finance forecast system in Chap. 11, quantum finance prediction using *chaotic transient-fuzzy deep network* in Chap. 12, and the implementation of intelligent agent-based *quantum trading system* in Chap. 13.

Problems

1.1   What is quantum finance? State and explain briefly the major difference between quantum finance and traditional finance theory.

1.2   What is the four-tier concentric sphere model of quantum finance? How can it relate to contemporary AI technology?

1.3   What is wave–particle duality phenomenon in quantum theory? Give an example of wave–particle duality phenomenon in financial market and explain how it works.

1.4   State and explain the three major components of quantum field theory and discuss how they are related to quantum finance.

1.5   State and explain the difference between quantum mechanics and quantum field theory.

1.6   What are the basic limitations of Newtonian classical mechanics?

1.7   Why the understanding of nature of time is critical in modern finance?

1.8   State and explain the five characteristics of quantum mechanics, and how they are related to quantum finance.

1.9   State and explain the double-slit experiment in quantum mechanics, and its analog to the financial market.

1.10 State and explain how Kant's theory provides a probable explanation to wave–particle duality in quantum theory, and hence quantum finance.

1.11 What are technical analysis and chart analysis in finance engineering? Discuss and explain how these analytical techniques are related to wave–particle duality in quantum finance.

1.12 State and explain the Schrödinger's cat paradox in quantum mechanics. So, at any moment, is the cat alive or die? why? and how it relates to the financial markets?

1.13 What are Elliot waves in finance? Discuss and explain how Elliot waves can be interpreted by quantum finance.

1.14 If everything is uncertain according to uncertainty principle, how can we predict the future such as financial prediction?

1.15 What are the roles of fuzzy logics and chaos theory in quantum finance?

References

Bhaumik (2017) M. L. Is Schrödinger's Cat Alive? *Quanta* 6(1): 70–80.

Böhmer (2016) C. G. *Introduction To General Relativity And Cosmology*, WSPC.

Borden, C. (2018) Fibonacci Trading: How to Master the Time and Price Advantage. UK: McGraw-Hill Education.

Brackenridge, J. B. (1999) *ISAAC NEWTON. the Principia: Mathematical Principles of Natural Philosophy*, 3rd Edition (1726). Newly Translated by I. Bernard Cohen and Anne Whitman. Berkeley: University of California Press.

Brown, C. (2012) Elliot Wave Principle: Elementary Concepts, Wave Patterns, and Practice Exercises. New York: Bloomberg Press.

Bulkowski, T. N. (2005) *Encyclopedia of Chart Patterns*. UK: Wiley.

Carroll (2016) *Spacetime And Geometry: An Introduction To General Relativity*. Pearson Education India.

Chaturvedi, S. et al. (2006) Space, time and relativity. *Resonance* 11(7): 14–29.

Einstein, A. (1915) Die Feldgleichungen der Gravitation. *Sitzungsberichte der Preussischen Akademie der Wissenschaften zu* Berlin: 844–847.

Einstein, A. (1916) The Foundation of the General Theory of Relativity. *Annalen der Physik*. 354 (7): 769.

Gribbin, J. (1984) *In Search of Schrödinger's Cat: Quantum Physics and Reality*. Bantam Books; Reprint edition.

Heisenberg, W. (1925) Über quantentheoretische Umdeutung kinematischer und mechanischer Beziehungen. *Zeitschrift für Physik*. 33 (1): 879–893.

Kant, I. (2008) *Critique of Pure Reason (Translated by M. Weigelt & M. Muller)*. Penguin Classics.

Lee, R. S. T. () *Advanced Paradigms in Artificial Intelligence From Neural Oscillators, Chaos Theory to Chaotic Neural Networks*. Advanced Knowledge International, Australia, 2005.

Mantegna, R. N. and Stanley, H. E. (1999) *Introduction to Econophysics: Correlations and Complexity in Finance*, Cambridge University Press.

Monroe et al, C. (1996) A "Schrödinger Cat" Superposition State of an Atom. *Science* 272(5265): 1131–1136.

Murphy, J. J. (1999) Technical Analysis of the Financial Markets: A Comprehensive Guide to Trading Methods and Application. New York: New York Institute of Finance.

Norsen, T. (2017) *Foundations of Quantum Mechanics: An Exploration of the Physical Meaning of Quantum Theory (Undergraduate Lecture Notes in Physics)*. Springer.

Pourciau, B. (2006) Newton's Interpretation of Newton's Second Law. *Archive for History of Exact Sciences* 60(2): 157–207.

Russell, S. and Norvig, P. (2015) *Artificial Intelligence: A Modern Approach,* 3rd edition. Pearson Education India.

Schuste, H. G. and Just, W. (2005) *Deterministic Chaos: An Introduction*. Wiley-VCH.

Sen, A. (2017) Quantum Entanglement and its Applications. *Current Science* 112(7): 1361.

Stamper-Kurn, D. M. et al. (2016) Verifying quantum superpositions at metre-scales. *Nature* 537, (7618): E1-E2.

Stewart, I. G. (2004) The principia: mathematical principles of natural philosophy. *Studies in History and Philosophy of Science* Part A 35(3): 665–667.

Tuchong (2019a) Einstein's Cosmic World of General Relativity. http://stock.tuchong.com/image?imageId=289742135997235449. Accessed 22 Aug 2019.

Tuchong (2019b) Newton's Physical World of Classical Mechanics. http://stock.tuchong.com/image?imageId=217614207574867979. Accessed 21 Aug 2019.

Tuchong (2019c) Heisenberg's Subatomic World of Quantum Mechanics. http://stock.tuchong.com/image?imageId=423495114220503185. Accessed 22 Aug 2019.

Woodhouse, N. M. J. (2007) *General Relativity (Springer Undergraduate Mathematics Series)*. Springer.

Zadeh, L. A. and Aliev, R. A. (2019) Fuzzy Logic Theory and Applications: Part I and Part II. World Scientific Publishing Company.

Chapter 2
Quantum Field Theory for Quantum Finance

Anyone who is not shocked by quantum theory has not understood it.
Niels Bohr (1885–1962)

What is quantum field theory?
How does it relate to quantum mechanics?
How does it relate to quantum finance?

The quote by Professor Niels Bohr, renowned Danish physicist and Nobel prizewinner for Physics in 1922, is exceptionally accurate. Anyone studies in quantum theory will soon discover that models, basic concepts, and the related phenomena associated with quantum theory are completely different from what we have learnt in classical physics. It involves many riveting topics and phenomena that have never been considered or even thought about from the world of classical physics. In fact, it shows us an entirely new perspective to interpret the world of reality we live in.

As mentioned in Chap. 1, the major component in quantum theory (QT)—quantum field theory—is originated from quantum mechanics (QM), which is an integrated theoretical physics theory with the combination of classical field theory, quantum mechanics, and special relativity.

In this chapter, we will first begin with a brief history of quantum mechanics.

Second, we will explore four unique features and phenomena in quantum mechanics which include wave–particle duality, Heisenberg's uncertainty principle, quantization of energy, and quantum entanglement.

© Springer Nature Singapore Pte Ltd. 2020
R. S. T. Lee, *Quantum Finance*,
https://doi.org/10.1007/978-981-32-9796-8_2

Lastly, we will study four main topics in quantum field theory: classical field versus quantum field, Feynman's path integral, and quantum oscillators which are related to our study of quantum finance theory.

In Chaps. 4 and 5, basic understanding of quantum mechanics and quantum field theory are vital not only to realize the motivation in acquiring quantum theory to model the dynamics of financial markets but also to set out a concrete foundation for further study and research on quantum finance theory with related theoretical and mathematical models.

2.1 Quantum Mechanics—The Basics

Quantum mechanics (QM), known as quantum physics or quantum theory, is a theoretical physics coined by Professor Werner Heisenberg in 1925 (Heisenberg 1925) which describes the nature and dynamics of matters at the smallest scales including atoms and subatomic particles.

Prior QM, the whole world of physics (so-called *physical world*), is governed by two distinctive laws of physics:

1. Newton's law of motion governs nature, dynamics, and motions of all objects (macroscopic objects);
2. Maxwell's electromagnetic wave equation governs nature, properties, and the propagation of electromagnetic waves within the material world.

Quantum mechanics differs from classical physics in four aspects:

1. Wave–particle duality—objects have characteristics of both particles and waves;
2. Quantization—all energy, momentum, angular momentum, and other quantities of a bound system are restricted to discrete values;
3. Heisenberg's uncertainty principle—limits to the precision with which quantities can be measured;
4. Quantum entanglement—instantaneous interactions between subatomic particles even a great distance.

See Fig. 2.1.

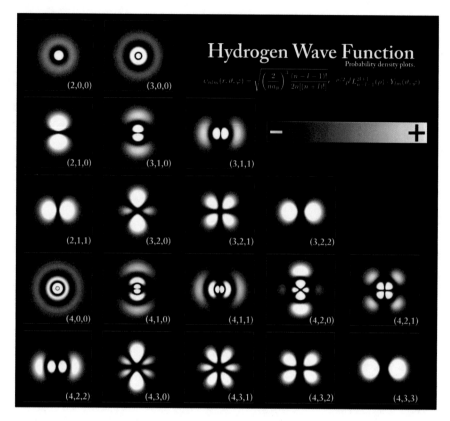

Fig. 2.1 Wave functions of the electron in a hydrogen atom at different energy levels (PoorLeno 2019). *Note* Quantum mechanics cannot predict the exact location of a particle in space, only the probability of finding it at different locations. The brighter areas represent a higher probability of finding the electron

2.1.1 A Brief History Quantum Mechanics

Table 2.1 shows a snapshot of distinguished figures in the course of quantum mechanics since 1800s.

2.1.2 Quantum Mechanics—Wave–Particle Duality Phenomena

Prior quantum mechanics (QM), waves (governed by Maxwell's electromagnetic wave equation) and particles motion (governed by Newton's laws of motion) are two totally different realms in the physical world:

Table 2.1 A brief history of quantum mechanics

Professor Michael Faraday (1791–1867)
A British scientist who contributed to the study of electromagnetism and electrochemistry. He discovered cathode rays in 1838, which triggered the study of blackbody radiation—the dawn of quantum mechanics

Professor James Clerk Maxwell (1831–1879)
A Scottish scientist who formulated the classical theory of electromagnetic wave that integrated electricity, magnetism, and light as different forms of the same phenomenon. His works are also known as the *second great unification in physics* after Sir Isaac Newton

Professor Ludwig Boltzmann (1844–1906)
An Austrian physicist and philosopher whose greatest achievement was in the development of statistical mechanics and second law of thermodynamics. In 1877, he proposed energy states of a physical system that can be discrete

Professor Max Planck (1858–1947)
A German theoretical physicist whose discovery of energy quanta awarded the Nobel Prize for Physics in 1918. Planck's hypothesis that energy is radiated and absorbed in discrete *quanta* precisely matched the observed patterns of blackbody radiation

Professor Albert Einstein (1879–1955)
A German-born theoretical physicist who developed the theory of relativity. In 1905, Albert Einstein interpreted Planck's quantum hypothesis realistically and used it to explain the photoelectric effect awarded the Nobel Prize for Physics in 1921

Professor Erwin Schrödinger (1887–1961)
An Austrian physicist who developed a number of fundamental results in the field of quantum theory: the Schrödinger equation provides the calculation on wave function of a system how it changes dynamically in time awarded the Nobel Prize for Physics in 1933

Professor Werner Heisenberg (1901–1976)
A German theoretical physicist and one of the key pioneers of quantum mechanics. He published his work in 1925 in a breakthrough paper for the creation of quantum mechanics awarded the Nobel Prize for Physics in 1932

Professor Richard Feynman (1918–1988)
American theoretical physicist, best known for his work in the path integral formulation of quantum mechanics and quantum electrodynamics. For his contributions to the development of quantum electrodynamics awarded the Nobel Prize for Physics in 1965

(1) Maxwell electromagnetic wave equation (Becherrawy 2012), given by

$$\nabla \times E = -\frac{\partial B}{\partial t} \tag{2.1}$$

(2) Maxwell electromagnetic wave equation, given by

$$F = m \cdot \ddot{x} \tag{2.2}$$

Wave–particle duality is the central philosophy of QM which describes every matter (quantum particle) not only in terms of particle dynamics but can also be considered (more importantly observed) as waves, which is governed by Schrödinger equation.

Time-dependent Schrödinger equation, given by (Schrödinger 1926),

$$i\hbar \frac{\partial}{\partial t} \psi(x, t) = \hat{H} \psi(x, t) \tag{2.3}$$

Such important property is clearly demonstrated by three important findings and experiments in the history of quantum mechanics, and they are as follows:

(1) Planck's blackbody radiation (Planck 1901) and the discovery of Planck's law:

$$E = h\upsilon \tag{2.4}$$

where E is Energy, h $= 6.55x10^{-34}$ *J-s (Planck's constant), υ-frequency of blackbody radiation.*

Figure 2.2 shows the intensity of light emitted from a blackbody at any given wavelength. Each curve represents behavior at a different body temperature. Planck was the first to explain the shape of these curves. His works on blackbody radiation and the discovery of Planck constant provided the foundation of quantum mechanics.

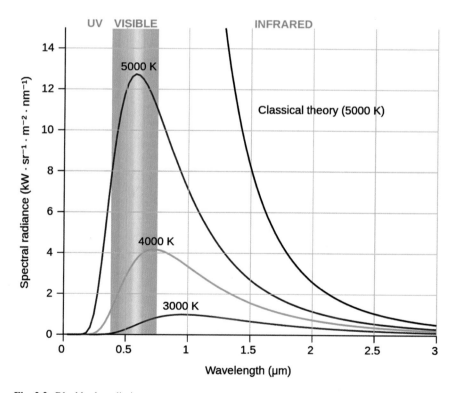

Fig. 2.2 Blackbody radiation

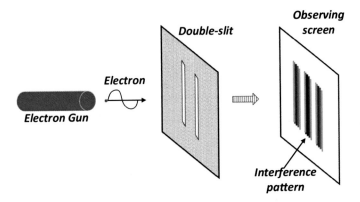

Fig. 2.3 Double-slit experiment

(2) Double-Slit Experiment

Double-slit experiment of photon particle is an important experiment in quantum mechanics that reveals the wave–particle duality of quantum particles (Fig. 2.3). Moreover, it displays the fundamentally probabilistic nature of quantum-mechanical phenomena. The experiment was first performed with light by Professor Thomas Young (1802). In 1927, The Davisson–Germer experiment demonstrated that electrons were shown the same behavior, which was later extended to atoms and molecules (Davisson and Germer 1928).

(3) Einstein's Photoelectric Experiment

Photoelectric effect is the emission of electrons when light (or laser beam) falls on a metal plate in a vacuum tube, which can form a complete with the generation of electricity. The electrons emitted in this manner can be called photoelectrons. In 1905, Professor Albert Einstein published a paper (Einstein 1905) proposed that light energy is carried in discrete quantized packets to explain experimental data from the photoelectric effect (Fig. 2.4). This model contributed to the development of quantum mechanics and awarded him the Noble Prize in 1921 for "The discovery of the photoelectric effect."

Quantum finance mentioned in Chap. 1 is a common belief in financial community. In fact, the two major branches of modern technical analysis techniques—chart pattern analysis versus technical indicators/oscillators analysis—are typical examples of how we percept and interpret worldwide financial markets by their wave (market patterns)–particle (market indicators) duality. The difference is: in quantum finance, we have concrete mathematical and computer algorithms to model and study the wave–particle duality of financial markets.

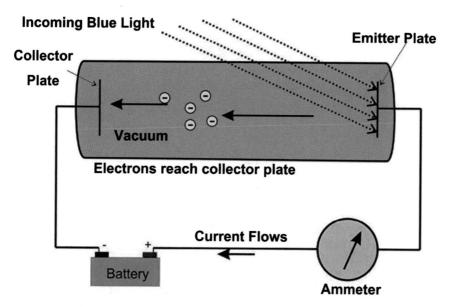

Fig. 2.4 Photoelectric experiment

2.1.3 Heisenberg's Uncertainty Principle

The uncertainty principle (also known as Heisenberg's uncertainty principle) is a unique and the most important property in quantum mechanics (Plotnitsky 2010).

In the principle of uncertainty, Professor Werner Heisenberg (1901–1976) described the fact that there is a natural limit of precision for the measurement (observation) of certain pairs of physical properties on any matters, which known as complementary variables (or what we called *quantum pairs*) such as position (x) and its momentum (p), given by

$$\Delta p \cdot \Delta x = \frac{\hbar}{2} \tag{2.5}$$

where $\hbar$ is the reduced Planck's constant $= \hbar/2\pi$.

As explained in his famous quote:

The uncertainty principle refers to the degree of indeterminateness in the possible present knowledge of the simultaneous values of various quantities with which the quantum theory deals; it does not restrict, for example, the exactness of a position measurement alone or a velocity measurement alone.

Werner Heisenberg (1901–1976)

In fact, uncertainty principle is an inherent property of matter due to its own wave property. It must be emphasized that measurement does not mean only a process in which a physicist observer takes part but rather any interaction between classical and quantum objects regardless of any observer.

This principle has profound influence in modern physics (especially experimental physics) such as mesoscale *numerical weather prediction* (NWP) in computational meteorology, superconductivity, and quantum optics.

In quantum finance, principle of uncertainty provides a way to describe the natural limit of precision for the observation of intrinsic pair in financial market, which will be revealed in Chap. 4.

2.1.4 Quantization of Energy

In quantum physics, quantization is the process of transition from a classical understanding of physical phenomena to a brand new interpretation in using quantum mechanics. Quantization converts classical fields into operators acting on quantum states of the field theory.

Rutherford–Bohr model (or *Bohr model* in short), presented by Professors Niels Bohr and Ernest Rutherford in 1913, proposed the quantization of energy levels in atoms (Bohr 1913) with three unique properties:

1. Discrete orbits (energy levels)—like planets in the solar system, electron is able to revolve in certain stable orbits around the nucleus which contrary to what classical electromagnetism suggests. These stable orbits are called *stationary orbits* and are attained at certain discrete distances from the nucleus. The electron cannot have any other orbit in between the discrete ones.
2. These discrete orbits are attained at distances for which the angular momentum of the revolving electron is an integral multiple of the reduced Planck's constant given by.
3. Electrons can only gain and lose energy by jumping from one allowed orbit to another, with the absorption or emission of discrete amount of electromagnetic radiation given by.

Figure 2.5 shows the Bohr model that the planetary-like model of the hydrogen atom ($Z = 1$), where the electron confined to an atomic shell encircles a small, positively charged atomic nucleus; and where an electron jumps between orbits is accompanied by an emitted or absorbed amount of electromagnetic energy (hf). The orbits in which the electron may travel are shown as gray circles, where n is the principal quantum number. The $3 \rightarrow 2$ transition depicted here produces the first line of the Balmer series, and for hydrogen ($Z = 1$) results in a photon of wavelength 656 nm (red light).

Fig. 2.5 The Bohr model

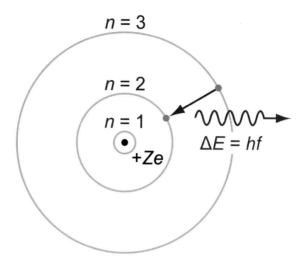

2.1.5 Quantum Entanglement

Quantum entanglement (Schrödinger 1935) is a quantum phenomenon which occurs when pairs or groups of quantum particles generate, interact, or share spatial proximity in ways such that the quantum state of each particle cannot be described independently of the state of the other(s), even when the particles are separated by a very large distance—described by Einstein as *spooky action at a distance* (Einstein et al. 1935). Quantum entanglement has been demonstrated experimentally on photons, neutrinos, electrons, and even on small diamonds.

Figure 2.6 shows a laser beam fired through a certain type of crystal that can cause individual photons to be split into pairs of entangled photons.

The utilization of entanglement in communication and computation is a very active research area. In fact, quantum entanglement has many applications in quantum information theory (Sen 2017; Laloë 2012):

1. Quantum computing (Yanofsky and Mannucci 2008) is computed using quantum-mechanical phenomena, such as superposition and entanglement. Such a computer is completely different from binary digital electronic computers based on transistors and capacitors. Whereas common digital computing requires that the data be encoded into binary digits (bits), each of which is always in one of the two definite states (0 or 1), quantum computation uses quantum bits or qubits, which can be within the superpositions of states.
   In January 2019, IBM launched IBM Q System One, its first integrated quantum computing system for commercial use.
2. Quantum cryptography (Assche 2006) is an application of quantum entanglement in real-time encryption and message communication between two remote parties which share the same quantum state.

Fig. 2.6 Quantum
entanglement

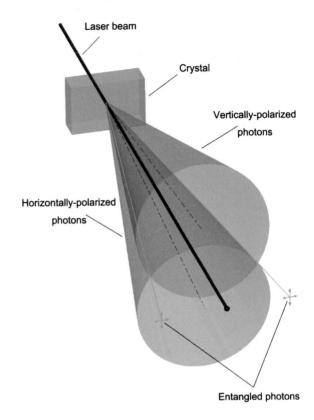

3. Quantum teleportation is a process by which quantum information (e.g., the exact
   state of an atom or photon) can be transmitted (exactly, in principle) from one
   location to another, with the help of classical communication and previously
   shared quantum entanglement between the transmitting and receiving locations.
4. Quantum finance is accompanied by the intrinsic property of quantum entanglement. Operational researches were carried out in the previous 15 years on how
   various financial markets, major forex markets, or major commodity markets
   such as gold and crude oil markets are affected when quantum entanglement
   occurred, and hence the prediction of major market reversal event.

   Important applications of spin-off from quantum theory include

- Semiconductors and microprocessors;
- Laser technology;
- Quantum computing;
- Quantum cryptography;
- Quantum finance;
- Quantum optics;
- Superconductivity (high-speed transportation); and

- Medical and images' research such as magnetic resonance imaging (MRI) and DNA editing.

2.2 From Quantum Mechanics to Quantum Field Theory

Quantum field theory is a theoretical framework that combines three important theoretical physics theories (Padmanabhan 2016).

They are

1. Quantum mechanics—provides basic mathematical, physical and philosophical framework;
2. Classical field theory—provides mathematical, physical model, and framework on quantum fields' composition and dynamics;
3. Special relativity—provides mathematical, physical model and framework regarding interchange between quantum particles and their intrinsic energies, the creation and annihilation mechanisms of subatomic particles, and their mathematical interpretations.
4. Einstein's famous mass–energy formulation:

$$E = mc^2 \tag{2.6}$$

Note: This formulation not only explains the intrinsic energy of a matter with mass m or its mass–energy equivalent property, but more importantly it provides a way to describe and interpret how fundamental (quantum) particles can be created or annihilated to/from pure energy. The interpretation is not only the central concept of special and general relativity but also is an important component in quantum field theory to explain the existence and dynamics of various subatomic particles.

As there are a great deal of topics and phenomena in quantum field theory, this book will focus on key topics which are essential for constructing the mathematical and physical framework on quantum finance.

They are

- Classical field versus quantum field,
- Feynman path integral, and
- Quantum oscillators.

2.3 Classical Field Versus Quantum Field

Before we learn more in-depth understanding of what is quantum field theory, let's have a look on what is the major difference between a *classical field versus quantum field*.

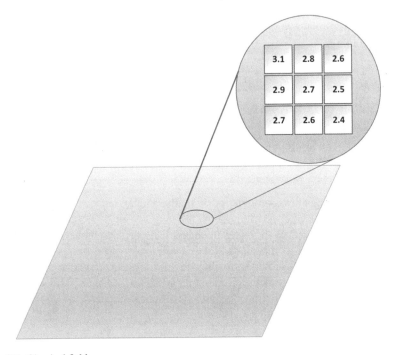

Fig. 2.7 Classical field

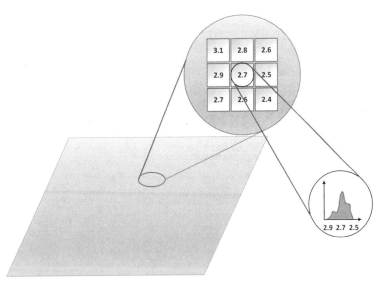

Fig. 2.8 Quantum field

A classical field (*field* in short) can be thought of as the assignment of a physical quantity at each point of space and time.

In the physical world, there are two major classical fields (nonrelativistic):

1. Newton's gravitational field (G field)

$$g(r) = -\frac{GM}{r^2} \qquad (2.7)$$

where G = gravitational constant, M = mass of the massive body (e.g., Earth) exerts on a comparably small object with mass m, and r = distance.

2. Maxwell's electromagnetic field (EM field)

$$E = -\frac{1}{4\pi\epsilon_0}\frac{q}{r^2} \qquad (2.8)$$

where q = charge of the test charged object, ϵ_0 = permittivity of free space, and r = distance.

In quantum field theory, particles are described by quantum fields that satisfy the Schrödinger equation, which is given by (nonrelativistic):

$$i\hbar\frac{\partial}{\partial t}\Psi(r,t) = \left[-\frac{\hbar^2}{2\mu}\nabla^2 + V(r,t)\right]\Psi(r,t) \qquad (2.9)$$

Figure 2.7 shows the 2D illustration of a classical field, in which there are discrete numbers at every point of the space, whereas in a quantum field, the field value at single point in the space is not an exact number but rather some sort of distribution— *probability density function (pdf)* or *wave function* in terms of quantum mechanics in Fig. 2.8.

2.4 Feynman's Path Integral

2.4.1 A Thought Experiment

Developed by Professor Richard Feynman in 1948 (Feynman 1948; Feynman and Hibbs 1965), the path integral formulation of quantum finance theory is concerned with the direct computation of scattering amplitude of a certain interaction process, rather than the establishment of operators and state spaces. To understand path integral, let's begin with a *thought experiment* on standard double-slit thought experiment (DSTE).

Figure 2.9 shows the configuration of a standard DSTE, photon particle emits from source (S) (at T = 0) either go through opening O_1 or O_2 and strike the projection screen at final point F (at T = t).

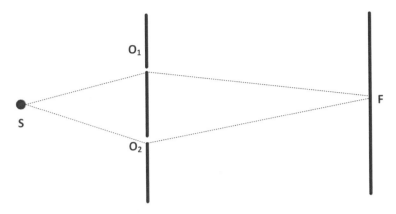

Fig. 2.9 Double-slit thought experiment (single layer)

By using the superposition principle in QM, the amplitude of detection at final point F is the sum of the amplitude for the paths S-O_1-F and S-O_2-F.

How about if we drill another opening O_3?

By common sense, it should be simply done by the superposition of one more path amplitude S-O_3-F to the evaluation function, as shown in Fig. 2.10.

The resulted amplitude (ψ_F) measured at F will be

$$\psi_F = \sum_i^n \psi_F(S - O_i - F) \tag{2.10}$$

Let's move on to the next level.

How about if we add one more layer and start with three openings?

Figure 2.11 shows the multi-slit thought experiment with double layer.

Generally, the path integration (path integral) equation multi-slit thought experiment with double layers will look like

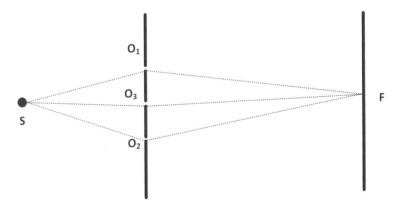

Fig. 2.10 Triple-slit thought experiment (single layer)

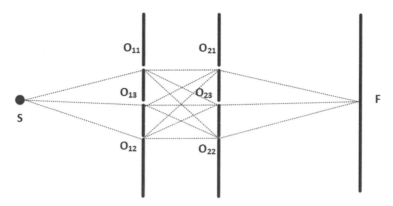

Fig. 2.11 Multi-slit thought experiment (double layers)

$$\psi_F = \sum_j^n \sum_i^n \psi_F\left(S - O_{1i} - O_{2j} - F\right) \qquad (2.11)$$

In fact, it is what Feynman did in 1948 during his famous Feynman's lectures on path integral and quantum mechanics.

Now then, what would happen if we add infinite number of layers; at the same time drill infinite number of openings onto it?

Figure 2.12 shows all layers between S and F will simply *disappear* into empty space. Why do they behave like this?

The answer is: photons emit from point source S and arrive at detector F through empty space. The wave function detected at F is the sum of *infinite paths* beginning from source S to detector F—known as *path integral formulation* proposed by Feynman in 1948.

$$\psi_F = \sum_i \psi_{All\ Path}(S - F) \qquad (2.12)$$

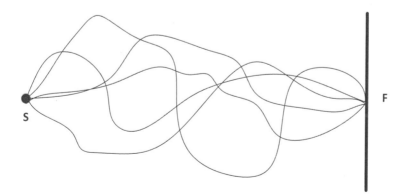

Fig. 2.12 Infinite-slit thought experiment (infinite layers)—path integral

2.4.2 Feynman's Path Integral Formulation

In quantum field theory (Padmanabhan 2016), the wave function value detected the path from point source q_S to a final point q_F in time T which is governed by the unitary operator e^{-iHT} where H is the Hamiltonian.

Using Dirac's *bra* and *ket* notation, we have

$$\langle q_F | e^{-iHT} | q_S \rangle \tag{2.13}$$

Using Feynman's path integral formulation (Feynman and Hibbs 1965), we divided the whole T into N segments of width $\delta t = T/N$. We have

$$\langle q_F | e^{-iHT} | q_S \rangle = \langle q_F | e^{-iH\delta t} e^{-iH\delta t} e^{-iH\delta t} \ldots | q_S \rangle \tag{2.14}$$

Then, we can break it down into N intervals:

$$\langle q_F | e^{-iHT} | q_S \rangle = \left[\prod_{k=1}^{N-1} \int dq_k \right] \langle q_F | e^{-iHT} | q_{N-1} \rangle \ldots \langle q_1 | e^{-iHT} | q_S \rangle \tag{2.15}$$

After relative calculation, we obtain the famous path integral formulation of QFT:

$$\langle q_F | e^{-iHT} | q_S \rangle = \int Dq(t) e^{i \int_0^T dt \left[\frac{1}{2} m\dot{q}^2 - V(q) \right]} \tag{2.16}$$

$$\langle q_F | e^{-iHT} | q_S \rangle = \int Dq(t) e^{i \int_0^T dt\, L(\dot{q}, q)} \tag{2.17}$$

where $L(\dot{q}, q) = \frac{1}{2} m\dot{q}^2 - V(q)$ and

$$\int Dq(t) = \lim_{n \to \infty} \left(-\frac{im}{2\pi \delta t} \right)^{N/2} \left[\prod_{k=1}^{N-1} \int dq_k \right] \tag{2.18}$$

2.4.3 Path Integral Formulation in Quantum Finance

In quantum finance, path integral formulation (Baaquie 2018) plays an important role in the modeling of the most fundamental financial instrument—worldwide interest rate (r) modeling.

Image the following scenario:

If we consider worldwide interest rate, say, US interest rate r as the propagation of quantum particle Qr.

The change of interest rate Qr during a period of time T from state Qr_S to Qr_F can be formulated by Feynman path integral formulation:

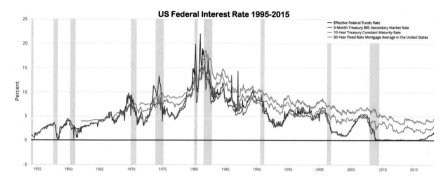

Fig. 2.13 US federal interest rate 1995–2015

$$Qr_F \left| e^{-iHT} \right| Qr_S = \int DQr(t) e^{i \int\limits_0^T dtL \left(\dot{Q}r, Qr \right)} \tag{2.19}$$

In fact, it is one of the major applications of QFT in finance for the past 15 years.

Besides path integral formulation, are there any other similar financial instrument(s) can be used?

Figure 2.13 shows the US federal interest rate from 1955 to 2015 in four different timeframes. In Chap. 3, we will explore how to model forward interest rates in different timeframes with quantum finance by using Feynman's path integral technique.

2.5 Quantum Oscillators

2.5.1 Quantum Harmonic Oscillator—The Basics

Quantum harmonic oscillator (QHO) is the quantum mechanics analog of a classical harmonic oscillator (Blaise and Henri-Rousseau 2011).

Simple harmonic oscillator:

$$F = m\ddot{x} = -kx \tag{2.20}$$

where k is the force constant.

Damped harmonic oscillator:

$$F = F_{ext} - kx - c\dot{x} \tag{2.21}$$

where c is the damping coefficient and $\zeta = c \left/ 2\sqrt{mk} \right.$ is the damping ratio

$\omega_0 = \sqrt{k \left/ m \right.}$ is the "Undamped angular frequency of the oscillator."

(Recall conditions for critically damped/overdamped/underdamped oscillations)

QHO is one of the most important model systems in QM and QFT.
In one dimension QHO, the Hamiltonian $\hat{H}$ is given by

$$\hat{H} = \frac{\hat{p}^2}{2m} + \frac{1}{2}k\hat{x}^2 = \frac{\hat{p}^2}{2m} + \frac{1}{2}m\omega^2\hat{x}^2 \qquad (2.22)$$

where m = mass of the quantum particle; k is the force constant; $\hat{x}$ is the position
operator; $\hat{p} = -i\hbar\partial/\partial x$ is the momentum operator; and ω is the angular frequency.
The solution of 1D-QHO is given by

$$\psi_n(x) = \frac{1}{\sqrt{2^n n!}} \cdot \left[\frac{m\omega}{\pi\hbar}\right]^{1/4} \cdot e^{-i\omega t} \cdot H_n\left(\sqrt{\frac{m\omega}{\hbar}}x\right) \qquad (2.23)$$

n = 0, 1, 2, ...
where n = 0, 1, 2, ...; ψ_n is the wave function and n is the quantum number.
The corresponding discrete energy levels are given by

$$E_n = \hbar\omega\left(n + \frac{1}{2}\right) = (2n+1)\frac{\hbar}{2}\omega, \quad n = 0, 1, 2, \ldots \qquad (2.24)$$

Note that E_0 corresponds to the ground state and all n > 0 are discrete excited
states.

Figure 2.14 shows the different dynamics in classical versus quantum harmonic oscillators. Figure 2.14A and B illustrate the classical harmonic oscillators;
Fig. 2.14C–H illustrate the quantum harmonic oscillators—solutions according to
Schrödinger equation, where the horizontal axis is particle position, and the vertical
axis is the real part (blue) or imaginary part (red) of the wave function.

2.5.2 N-Dimensional Quantum Harmonic Oscillators

Using the same logic of 1D QHO, the N-dimensional QHO is exactly analogous to
N independent one-dimensional harmonic oscillators with identical mass and spring
constant.

In this case, the quantities $x_1, \ldots, x_N$ would refer to the positions of each of the N
quantum particles.

The Hamiltonian is given by

$$H = \sum_{i=1}^{N}\left(\frac{p_i^2}{2m} + \frac{1}{2}m\omega^2 x_i^2\right) \qquad (2.25)$$

For a particular set of quantum numbers (n), the energy eigenfunctions for the N-dimensional oscillator are expressed in terms of the one-dimensional eigenfunctions
as

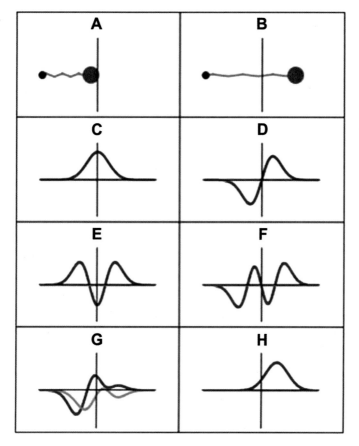

Fig. 2.14 Illustrations of classical versus quantum harmonic oscillators

$$\langle x|\psi_n \rangle = \prod_{i=1}^{N} \langle x_i|\psi_n \rangle \tag{2.26}$$

The energy levels of the systems are given by

$$E_n = \hbar\omega \left[(n_1 + n_2 + \cdots + n_N) + \frac{N}{2} \right] \tag{2.27}$$

where $n_i = 0, 1, 2 \ldots$ are the energy level in dimension i.

Figure 2.15 shows a 2D trampoline model of quantum harmonic oscillators. As shown, the quantum anharmonic oscillator model (in 2D perspective) can be illustrated as trampoline in which every quantum oscillator is connected with four neighboring oscillators in the 2D plane, resulted in the anharmonic vibration (from its stationary level) as vertical movement which analog to the amplitude of the wave function shown in the above QHO equations.

Fig. 2.15 2D trampoline
model of quantum harmonic
oscillators

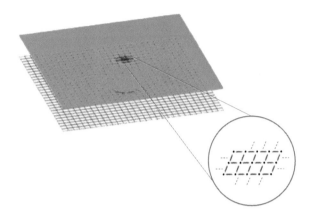

2.5.3   Quantum Anharmonic Oscillators

Even though quantum harmonic oscillator (QHO) can explain many quantum phenomena and wave–particle dualities for simple quantum particles, however, for real-
world applications, many quantum particles and quantum phenomena involve the
anharmonic oscillations (Gilyarov and Slutsker Slutsker 2010).

Figure 2.16 shows an HCl molecule as an anharmonic oscillator (Padilla and Pérez
2012) vibrating at energy level E_3. D_0 is the dissociation energy, r_0 is the bond length,
and U is the potential energy. Energy is expressed in wavenumbers. The hydrogen
chloride molecule is attached to the coordinate system to show bond length changes
on the curve.

The Hamiltonian of a typical quantum anharmonic oscillator (QAHO) is given by

$$H = \frac{p^2}{2m} + \frac{1}{2}m\omega^2 x^2 + \lambda x^k \tag{2.28}$$

where λ is the damping coefficient and k is the order of QAHO.

Commonly found QAHOs include

Cubic QAHO with k = 3:

$$H = \frac{p^2}{2m} + \frac{1}{2}m\omega^2 x^2 + \lambda x^3 \tag{2.29}$$

Quartic QAHO with k = 4:

$$H = \frac{p^2}{2m} + \frac{1}{2}m\omega^2 x^2 + \lambda x^4 \tag{2.30}$$

Sextic QAHO with k = 6:

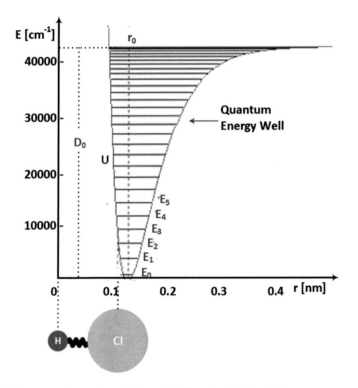

Fig. 2.16 Quantum anharmonic oscillations of HCI molecules' illustrations

$$H = \frac{p^2}{2m} + \frac{1}{2}m\omega^2 x^2 + \lambda x^6 \tag{2.31}$$

λx^{2m}-class QAHO—a special class of QAHO which provides a genius way of solving these high-order AHO equations.

$$H = \frac{p^2}{2m_q} + \frac{1}{2}m_q\omega^2 x^2 + \lambda x^{2m} \tag{2.32}$$

In quantum finance theory, we have deduced that the quantum dynamics of price return (r) for any financial market is in fact a kind of quartic quantum anharmonic oscillations, which is also a special case of λx^{2m}-class QAHO with m = 2. Detail mathematical derivation along with numerical analysis to solve important quantum finance equation will be studied in Chaps. 4 and 5.

2.6   Conclusion

In this chapter, we introduced a general overview of four major concepts and models in quantum theory: (1) quantum mechanics, (2) quantum field theory, (3) Feynman's path integral, and (4) quantum anharmonic oscillators.

These four concepts with models in quantum theory form critical mass to establish the basic theoretical and mathematical models in quantum finance theory in next chapter, namely, (1) Feynman's path integral model of quantum finance (also known as first generation of quantum finance); and (2) quantum anharmonic model of quantum finance (also known as second generation of quantum finance).

To avoid complex mathematical derivations, this chapter focused on the basic concepts of quantum theory with its related models which are critical to understand the mathematical and quantum model in quantum finance. For readers who are interested to explore in-depth mathematical models and quantum theory, please refer to the reference section at the end of this chapter.

Problems

2.1   Before quantum mechanics, what are the two main streams of theories govern the laws of physics and how quantum mechanics are difference with these theories?

2.2   We always say 1920s are the key years for the birth of quantum mechanics. Who are the two key figures in the course of the development of the basic concepts and theories in quantum mechanics and discuss their major contributions?

2.3   What is the physical importance of Schrödinger equation in terms of the interpretation of wave–particle duality phenomena?

2.4   What are the three most important experiments in the 1920s for the interpretation and illustration of wave–particle duality in quantum mechanics? Discuss and explain their findings and importance. Can they be interpreted in financial markets? How?

2.5   Discuss and explain briefly Heisenberg's uncertainty principle in quantum mechanics. How can such phenomena relate to modern finance?

2.6   What is the basic difference between a quantum field and classical field? How can quantum field be interpreted in terms of financial markets?

2.7   Discuss and explain quantization phenomena in quantum theory. How can it have been related to quantum finance? What are the ground-state and various excited energy levels in quantum finance?

2.8   Discuss and explain quantum entanglement in quantum theory. How can it be interpreted in quantum finance?

2.9   What is quantum computing? Discuss and explain briefly the basic concept and how it is related to quantum finance.

References

Assche, G. (2006) *Quantum Cryptography and Secret-Key Distillation*. Cambridge University Press; 1 edition.

Baaquie, B. (2018) *Quantum Field Theory for Economics and Finance*. Cambridge University Press. 1st edition.

Becherrawy, T. (2012) *Electromagnetism: Maxwell Equations, Wave Propagation and Emission*. Wiley-ISTE; 1 edition, 2012.

Blaise, P. and Henri-Rousseau O. (2011) *Quantum Oscillators*. Wiley.

Bohr, N. (1913) On the Constitution of Atoms and Molecules, Part I" (PDF). Philosophical Magazine. 26 (151): 1–24.

Davisson, C. J. and Germer, L. H. (1928) Reflection of Electrons by a Crystal of Nickel. *Proceedings of the National Academy of Sciences of the United States of America*. 14 (4): 317–322.

Einstein, A. (1905) Über einen die Erzeugung und Verwandlung des Lichtes betreffenden heuristischen Gesichtspunkt [On a Heuristic Viewpoint Concerning the Production and Transformation of Light]. *Annalen der Physik* (Berlin) (in German). Hoboken, NJ (published 10 March 2006). 322 (6): 132–148.

Einstein, A., Podolsky, B., Rosen, N. (1935) Can Quantum-Mechanical Description of Physical Reality Be Considered Complete? Physical Review 47 (10): 777–780.

Feynman, R. P. (1948) Space-Time Approach to Non-Relativistic Quantum Mechanics. *Reviews of Modern Physics*. 20 (2): 367–387.

Feynman, R. P.; Hibbs, A. R. (1965) *Quantum Mechanics and Path Integrals*. New York: McGraw-Hill.

Gilyarov V. L. and Slutsker, A. I. (2010) Energy features of a loaded quantum anharmonic oscillator. *Physics of the Solid State* 52(3): 585–590.

Heisenberg, W. (1925) Über quantentheoretische Umdeutung kinematischer und mechanischer Beziehungen. *Zeitschrift für Physik*. 33 (1): 879–893.

Laloë, F. (2012) Applications of quantum entanglement. In *Do We Really Understand Quantum Mechanics?* Cambridge University Press, 150–167.

PoorLeno (2019) Hydrogen Density Plots. https://commons.wikimedia.org/w/index.php?title=File: Hydrogen_Density_Plots.png&oldid=343305978. Accessed 21 Aug 2019.

Padmanabhan, T. (2016) *Quantum Field Theory: The Why, What and How (Graduate Texts in Physics)*. Springer.

Padilla A. and Pérez, J. (2012) A Simulation Study of the Fundamental Vibrational Shifts of HCl Diluted in Ar, Kr, and Xe: Anharmonic Corrections Effects. *International Journal of Spectroscopy*, 2012: 1–7.

Planck, M. (1901) Ueber das Gesetz der Energieverteilung im Normalspectrum [On the law of the distribution of energy in the normal spectrum]. *Annalen der Physik*. 4 (3): 553–563.

Plotnitsky, A. (2010) *Epistemology and Probability: Bohr, Heisenberg, Schrödinger, and the Nature of Quantum-Theoretical Thinking (Fundamental Theories of Physics)*. Springer.

Sen, A. (1926) Quantum Entanglement and its Applications. *Current Science* 112(7): 1361, 2017.

Schrödinger, E. An Undulatory Theory of the Mechanics of Atoms and Molecules. *Physical Review* 28: 1049–1070.

Schrödinger, E. (1935) Discussion of probability relations between separated systems. *Mathematical Proceedings of the Cambridge Philosophical Society*. 31 (4): 555–563.

Yanofsky, N. S. and Mannucci, M. A. (2008) *Quantum Computing for Computer Scientists*. Cambridge University Press; 1 edition.

Young, T. (1802) The Bakerian Lecture: On the Theory of Light and Colours. *Philosophical Transactions of the Royal Society of London*. 92: 12–48.

Chapter 3
An Overview of Quantum Finance Models

> It is a curious historical fact that modern quantum mechanics
> began with two quite different mathematical formulations: the
> differential equation of Schrödinger and the matrix algebra of
> Heisenberg. The two apparently dissimilar approaches were
> proved to be mathematically equivalent.
> Richard P. Feynman (1918–1988)

What is quantum finance?
What are the major quantum finance models?
Quantum finance – which way to go?

In this famous quotation on quantum mechanics, Professor Feynman pointed out the vital feature of quantum mechanics mathematical derivation, the two totally different approaches to formulate and model the quantum world which end up being mathematically equivalent.

Did it sound familiar?

As we learnt from quantum mechanics' basic concept, whether we observe the quantum world as particle dynamics or wave dynamics, it only affects the way we model the quantum world, and the mathematical tool along with the model we are using. The reality of the quantum world doesn't change, and from the mathematical perspective, they are mathematical equivalent.

As mentioned in Chap. 1, quantum finance is a newly developed interdisciplinary subject with the integration of

1. Quantum theory—the provision of quantum mechanics and quantum field theory as the quantum finance dynamics with field analysis; and
2. Computational finance—the provision of computational finance models as the system framework and the application basis.

To meet the demand from the exponential growth of program trading and intelligent financial forecast service, in 2012, the author had integrated contemporary quantum finance theory with AI technology to implement quantum finance forecast system for the real-time forecast of worldwide commodity markets.

In this chapter, we will first review a brief history of quantum finance.

© Springer Nature Singapore Pte Ltd. 2020
R. S. T. Lee, *Quantum Finance*,
https://doi.org/10.1007/978-981-32-9796-8_3

Second, we will study the two main branches of quantum finance models—Feynman's path integral model and the quantum anharmonic oscillator model.

Lastly, we will explore the future of quantum finance, and how it relates to the development of intelligent finance systems.

3.1 A Brief History of Quantum Finance

Quantum finance is a newly developed interdisciplinary subject introduced in 1990s by applying quantum mechanics and quantum field theory to theoretical economics—so-called *econophysics*.

Nevertheless, *econophysics style* of R&D was established much earlier.

In 1900, Professor Louis Jean-Baptiste Alphonse Bachelier (1870–1946), a French mathematician in his Ph.D. thesis Théorie de la speculation (translated as The Theory of Speculation) published by Annales Scientifiques de l'École Normale Supérieure, set the foundation of a mathematical model with Brownian motion in valuing stock options (Bachelier 1900).

It was historically the first paper to use advanced mathematics in the study of finance. He is also considered as the forefather of mathematical finance and also a pioneer in the study of stochastic processes.

His thesis contained three different versions of the first mathematical theory of Brownian motion. In modern terminology, Brownian motion was characterized as

1. A process with independent homogeneous increments whose paths are continuous;
2. The continuous-time process, which is the limit of symmetric random walks; and
3. The Markov process whose forward Kolmogorov equation is the heat equation.

Owing to the above reasons, most mainstream *econophysicists* consider finance as an application of *Brownian motion*—the fundamental phenomenon of statistical physics for the modeling of financial market.

The first published work on Econophysics—*An Introduction to Econophysics - Correlations and Complexity in Finance*—was written by Professors Rosario Mantegna and Eugene Stanley in 1999 (Mantegna and Stanley 1999). This pioneering text explored the use of statistical physics concepts such as stochastic dynamics, short- and long-range correlations, self-similarity, and scaling concepts of financial systems' description. These were the dynamic new specialty of econophysics.

For the last two decades, various methods and theories were proposed for stock price/returns analysis, interest rate modeling, option pricing, and portfolio analysis.

3.2 Major Works and R&D in Quantum Finance

Although statistical physics is the mainstream theory of Econophysics, active R&D with the adoption of quantum mechanics, quantum field theory with related concepts, and frameworks such as Feynman's path integral model and quantum oscillator model to model financial markets can be found which include the following:

1. B. Baaquie in his book *Quantum Finance* (Baaquie 2004) reviewed the application of Feynman's path integral theory for option pricing and interest rate modeling. Professor Baaqui is also the first scholar who consolidated a complete concept and theory of quantum finance using quantum field theory (Baaquie 2007, 2013, 2018);
2. Other research works on path integral including the sensitivity analysis using path-independent quantum finance model by Kim et al. (2011);
3. Quantum anharmonic oscillator modeling on finance analysis included Gao and Chen (2017) work on quantum anharmonic oscillator model for the stock market; Ye and Huang (2008) work on nonclassical oscillator model for persistent fluctuations in stock markets; Meng et al. (2015) work on quantum spatial-periodic harmonic model for daily price-limited stock markets;
4. Quantum wave function for stock market analysis by Ataullah et al. (2009);
5. Quantum statistical approach to simplified stock markets by Bagarello (2009);
6. A finite-dimensional quantum model for the stock market by Cotfas (2013);
7. Nakayama (2009) works on gravity dual for Reggeon field theory and nonlinear quantum finance;
8. Piotrowski and Sładkowski (2005) studied the quantum diffusion model of prices and profits;
9. Probability wave approach on security transaction volume–price behavior analysis by Shi (2006);
10. Bohmian quantum potential approach on stock market credibility analysis by Nasiri et al. (2018);
11. Schaden (2002) applied quantum theory to model secondary financial markets;
12. Zhang and Huang (2010) defined wave functions and operators of the stock market to establish the Schrödinger equation for stock price.

3.3 Two Pillars of Quantum Finance Models

In respect of quantum finance R&D of the past decades, there are basically two main quantum models with significant impacts and research results in financial modeling and applications. They are

1. Feynman's path integral approach (Zee 2011; Swanson 2014) and
2. Quantum anharmonic oscillator approach (Bloch 1997).

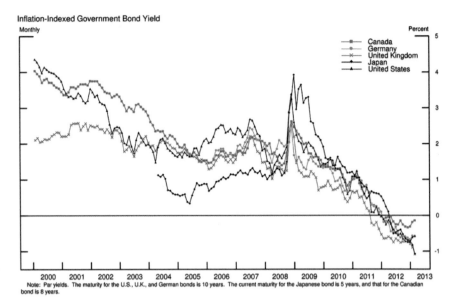

Fig. 3.1 Feynman's path integral approach in quantum finance for government bond and interest rate modeling

In the next two sections, we will review the basic concept and framework of these two approaches in quantum finance. Figures 3.1 and 3.2 show the typical applications of these two mathematical models on the modeling of forward interest rate and quantum price levels in secondary financial markets, respectively.

3.4   Feynman's Path Integral Approach in Quantum Finance

3.4.1   Forward Interest Rate in a Nutshell

Before we study the Feynman's path integral approach in quantum finance for the modeling of forward interest rate, let's have an overview of forward interest rate in finance.

From computational finance perspective, forward interest rate $f(t, x)$ is the interest rate fixed at time $t \leq x$ for an instantaneous (overnight) loan taken at future time x.

So, at any instant t, $f(t, x)$ is a function of x (not t) that can be visualized as a 1D path in the future time domain (x) satisfy $(x \geq t)$, as shown in Fig. 3.3.

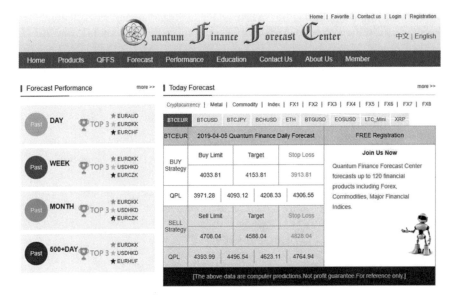

Fig. 3.2 Quantum anharmonic model in quantum finance for quantum price level modeling (QFFC 2019)

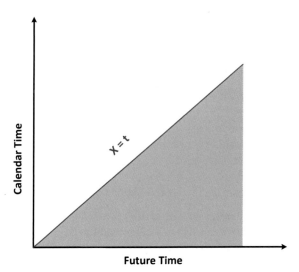

Fig. 3.3 Forward interest rate function domain (green area)

With the initial condition $f(t_S, x)$ is fixed by the financial market, each point (t, x) manifests itself in the *functional domain* (green area) and corresponds to one *forward interest rate* incidence $f(t, x)$.

3.4.2   Zero Coupon Bond (Treasury Bond)

Treasury bond is a risk-free financial instrument which has a single cash flow consisting of a fixed payoff of, say, $1 at some future time T; its price at time $t < T$ is denoted by $P(t, T)$, with $P(T, T) = 1$.

So, for a bond maturing at time T, its value $P(t, T)$ before maturity is given by discounting $P(T, T) = 1$ to the time t by the spot interest rate, given by

$$P(t, T) = E\left[e^{-\int_t^T dsr(s)}P(T, T)\right] = E\left[e^{-\int_t^T dsr(s)}\right] \qquad (3.1)$$

where

1. *E[] is the expectation value;*
2. *$r(s) = f(s, s)$ is the spot interest rate.*

Figure 3.4 shows the forward interest rate of Eurodollar futures from 1990 to 1996 with eight different maturities, ranging from 3 months up to 7 years.

3.4.3   Quantum Field Theory of Interest Rate Model

To model the interest rate in respect of quantum field theory using Feynman's path integral, first take a look at the forward interest rate of Eurodollar futures from 1990 to 1996 with eight different maturities, ranging from 3 months up to 7 years as shown in Fig. 3.5.

If we look carefully, we can find certain interesting features from this chart:

1. All these forward interest rate time series appeared to be highly chaotic but have a certain shared pattern. In Chap. 8—chaos and fractals in quantum finance—we will study such kind of chaotic phenomena—so-called *deterministic chaos*, a vital feature and phenomena in financial engineering.
2. All these curves are basically randomly evolved in time, but in a highly co-related manner, basically without any line crossings.

In other words, the interest rate function $f(t, x)$ can be assumed to be independent variable w.r.t. to all x and t values.

Let A(t, x) be a two-dimensional quantum field, and the quantum interest rate model is given by Baaquie (2004)

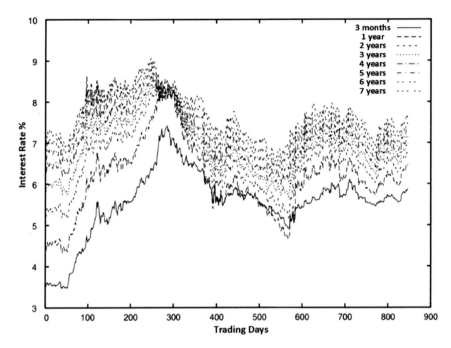

Fig. 3.4 Forward interest rate of Eurodollar Futures from 1990 to 1996

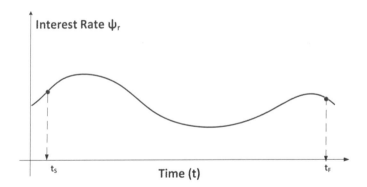

Fig. 3.5 Feynman's path integral model for interest rate

$$\frac{\partial f(t, x)}{\partial t} = \alpha(t, x) + \sigma(t, x)A(t, x) \tag{3.2}$$

3.4.4 Feynman's Path Integral Formulation of Forward Interest Rate

The Feynman's path integral for the forward interest rate is given by Baaqie (2013)

$$E(F[A]) = \frac{1}{Z} \int DA e^{S[A]} F[A] \tag{3.3}$$

where $\int DA = \prod_{(t,x)\in B} \int_{-\infty}^{+\infty} dA(t,x)$ and $Z = \int DA e^{S[A]}$

The time-dependent Hamiltonian is given by

$$H(t) = \frac{-1}{2} \int_t^\infty dx dx' M(x, x'; t) \frac{\delta^2}{\delta f(x)\delta f(x')} - \int_t^\infty dx \alpha(t, x) \frac{\delta}{\delta f(x)} \tag{3.4}$$

where $M(x, x'; t) = \sigma(t, x) D(x, x'; t) \sigma(t, x')$

Note:

1. S[A] is the stiff action function, given by Baaqie (2013)

$$S[A] = -\frac{1}{2} \int_{t_s}^{+\infty} dt \int_0^{+\infty} dz \left[A^2(t, z) + \frac{1}{\mu^2} \left(\frac{\partial A}{\partial z} \right)^2 (t, z) + \frac{1}{\lambda^4} \left(\frac{\partial^2 A}{\partial z^2} \right)^2 (t, z) \right] \tag{3.5}$$

2. D is the domain of (t, x) for the path integration;
3. The initial value of the forward interest rate $f(t_s, x)$ is an important boundary condition for Feynman's path integral calculation.
4. As shown in Eq. 3.3, to calculate the expectation value $E(F[A])$, one has to perform integration of F[A] over all possible functions $A(t, x)$ of all possible (t, x) in the domain space D.
5. In other words, to evaluate the forward interest rate using Feynman's path integral method, one has to perform many integrations infinitely, one integration of $\int dA(t, x)$ for each state variable (t, x), which is natural as the core concept of Feynman's path integral is the integration of (superposition in quantum theory) all possible paths that lead from S to F, as indicated in Chap. 2.

3.5 Quantum Anharmonic Oscillator Approach in Quantum Finance

3.5.1 Motivation

One of the major characteristics of quantum theory is the wave–particle duality which can be described by the Schrödinger equation.

Time-dependent Schrödinger equation:

$$i\hbar\frac{\partial}{\partial t}\Psi(x, t) = \left[-\frac{\hbar^2}{2m}\frac{\partial^2}{\partial^2 x} + V(x, t)\right]\Psi(x, t) \tag{3.6}$$

Time-independent Schrödinger equation:

$$\left[-\frac{\hbar^2}{2m}\frac{\partial^2}{\partial^2 x} + V(x)\right]\Psi(x) = E\Psi(x) \tag{3.7}$$

Instead of solving quantum dynamics' problem by integrating all (in fact is infinity) the possible paths from starting point S to end point F, why not focus on solving the Schrödinger equation directly using numerical method (numerical approximation)?

Imagine: if we can solve this equation numerically (i.e., instead of analytically), say, into an n-order polynomial with differential operators, together with the integration of finite difference method (FDM). Technically speaking, we can implement such algorithm into a computer program for actual implementation (not simulation).

Actually, such notion is more important to tackle real-world problems with numerous time series data such as weather and finance.

As we will see, anharmonic oscillation (AHO) provides an excellent analog of Schrödinger equation for the modeling of quantum dynamics (3.6) and quantum energy levels (3.7).

In fact, many atomic/subatomic particles' dynamics in the real world can be described in terms of quantum anharmonic oscillation (QAHO) as shown in Fig. 3.6.

It shows the quantum anharmonic oscillator model for HCI molecules and their corresponding quantum energy well.

3.5.2 QAHO Modeling—Prices Versus Returns

Different from Feynman's path integral method, quantum anharmonic oscillator (QAHO) method directly models the wave–particle dynamics—Schrödinger equation.

Owing to this reason, QAHO method tailored for the modeling of secondary financial markets such as the trading of stocks and futures among investors or OTC

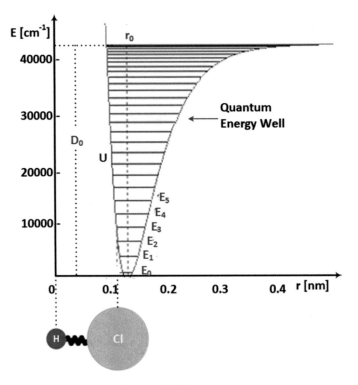

Fig. 3.6 Quantum anharmonic oscillations of HCI molecules

trading such as commodities, financial indices, and forex between market maker (MM) and investors.

In other words, the objective of QAHO modeling in quantum finance targets at quantum dynamics and quantum energy levels for market instruments involve in the secondary markets, forex market, for example (Fig. 3.7).

So, the natural and most fundamental question is: What market instrument(s) we should model? Prices, indices, price returns, price deviations…?

To answer this question, we should ask this vital question:

If financial market is really a quantum phenomenon, that is, the motions, fields, and dynamics of quantum particles, what are the financial particles we are talking about? Or what is the best analog of the displacement x (from the equilibrium state)?

Price or price returns or …?

Why?

3.5.3 Quantum Anharmonic Oscillations of Returns

The answer is … *price returns* (r or *return* in short).

Fig. 3.7 Worldwide financial products

If we want to make the best analog of quantum particle displacement from the equilibrium position, naturally we should choose price return (r) with two major reasons:

1. Return (r) in finance is an important financial instrument to reflect the relative change of financial behavior (say stock price) within a fixed period of time (say, daily returns), which directly analog to physical meaning of quantum theory for the definition of x which relates to displacement to the equilibrium position; and
2. Use of return (r) to model financial market in quantum finance is totally coherent classical and statistical finance theories which all use price return for financial market modeling and statistical analysis.

So how about Prices? Is there any significance of price in quantum finance?
The answer is: very important.

As analog to gravitational field, one can imagine if all financial instruments are quantum financial particles, what will be the quantum field?

The natural logic should be the *quantum financial energy field*, QFEF (or *quantum energy field* in short) created by itself (just like gravitation field created by *graviton*) and/or QFEF created by other financial particles (in other financial markets).

So, to analog the potential energy in gravity and electromagnetic field (EM field), quantum price energy levels (or quantum price levels, QPLs in short) are the quantum finance quantity to describe the quantized energy field acting onto the quantum financial particles.

3.5.4   A Quantum Anharmonic Oscillator Model of Financial Markets

Let's begin with QAHO modeling in quantum finance.

First, define *quantum financial particle* r—price return (e.g., stock).

For numerical computation, we define the fractional price change as

$$r(t, \Delta t) = \frac{p(t) - p(t - \Delta t)}{p(t - \Delta t)} \tag{3.8}$$

where $p(t)$ is the price (say stock price) at time t and Δt denote the time interval.

The analog motions and dynamics of (stock) price to a quantum particle (namely, *quantum financial particles*, QFPs), price return (r) will be a direct analog to the displacement (x) of a quantum particle from its equilibrium state, assuming they are performing anharmonic oscillation (as we will see shortly).

So, quantum dynamics of the quantum financial particle are given by Schrödinger equation:

$$i\hbar \frac{\partial}{\partial t} \psi(r, t) = H\psi(r, t) \tag{3.9}$$

H is the Hamiltonian operator given by

$$H = -\frac{\hbar^2}{2m} \frac{\partial^2}{\partial^2 r} + V(r, t) \tag{3.10}$$

3.5.5   Physical Meaning of Hamiltonian in Quantum Finance

The Hamiltonian (H) in Schrödinger equation

$$H = -\frac{\hbar^2}{2m} \frac{\partial^2}{\partial^2 r} + V(r, t) \tag{3.10}$$

consists of two parts, K.E. (first part) and P.E. (second part). K.E. corresponds to quantum motion of the quantum financial particle (QFP), and P.E. corresponds to potential energy (well) generated by the quantum energy field of QFP itself (or other QEF generated by other QFPs in multidimensional financial markets).

The reduced Planck's constant $\hbar$ can be considered as the uncertainty of irrational transaction and m represents the intrinsic properties of financial market such as market capital in a stock market.

Corresponding to the stationary market environment in classical computational finance theory (CFT), we can analog such quantum dynamics as time-independent Schrödinger equation, given by

$$H(r)\varphi(r) = E\varphi(r) \tag{3.11}$$

Or

$$\left[-\frac{\hbar^2}{2m}\frac{\partial^2}{\partial^2 r} + V(r)\right]\varphi(r) = E\varphi(r) \tag{3.12}$$

where $\Psi(r, t) = \varphi(r)e^{-iEt/\hbar}$; E is the eigenenergy of financial market; and $\varphi(r)$ is the corresponding eigenfunction.

Inherent from uncertainty principle in quantum theory, return (r) should be existed as wave function until we make the observation, which in fact will be affected by time frame of measurement that is again consistent with modern finance and fractal theories in particular.

3.5.6   Probability Density Function in Quantum Finance

In an efficient market and stationary market environment, quantum dynamics of the QFP r can be visualized as wave function with probability density function (pdf) given by

$$Q(r, t) = |\psi(r, t)|^2 = |\psi(r)|^2 \tag{3.13}$$

where Q(r, t) is the probability density function for a series of observations (measurements) of a financial market.

Figure 3.8 shows the statistics (i.e., *probability density function*) of XAUUSD (gold w.r.t. USD) daily returns in the past 2048 time series observations. We will study in detail quantum price level (QPL) evaluation in Chap. 5.

In an efficient market with stationary market environment, the probability density function should behave as normal distribution coherent classical statistical finance theory. The distribution itself should be characterized by its statistical indicators such as mean, variance, and kurtosis.

What is the physical meaning of distribution in terms of quantum theory?

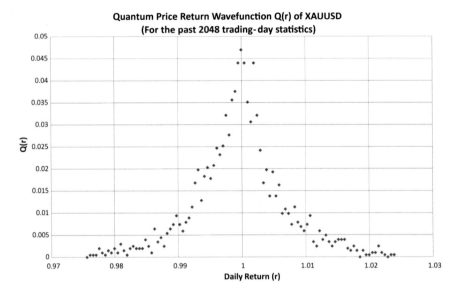

Fig. 3.8 Probability density function of XAUUSD

As recalled from the famous *double-slit experiment* of photon particles, the observed (detected) pattern in the backdrop screen is, in fact, the measurements (observations) of the quantum behavior of the wave-to-particle realization and resulted as probability density function (pdf). In other words, in quantum theory (quantum finance theory in particular), classical statistical (finance) theory is one of the *realizations* of the quantum (finance) theory, which can help us to examine the characteristics of the quantum (finance) behaviors.

3.5.7 Quantum Anharmonic Oscillator Formulation

In computational finance, there is an important financial instrument which is the *excess demand* (z), defined as follows.

At any time t, $z_+(t)$ and $z_-(t)$ denote the instantaneous demand and supply for the financial asset.

The excess demand (z) at any instance is given by

$$\Delta z = z_+ - z_- \tag{3.14}$$

Let $r(t)$ is the instantaneous returns, which is given by

$$\frac{dp}{dt} = r(t) = F(\Delta z) \tag{3.15}$$

For small Δz, F can be approximated by a scaling factor γ and become

$$r(t) = \frac{\Delta z}{\gamma} \tag{3.16}$$

where γ represents *market depth*, the excess demand z required to move quantum price p by one single quantum.

So we have

$$\frac{dr}{dt} = \frac{d^2 p}{dt^2} = \frac{1}{\gamma}\frac{d(\Delta z)}{dt} \tag{3.17}$$

After learning the financial dynamics of various parties in an efficient market, we will deduce that

$$\frac{dr}{dt} = \gamma\frac{d\,\Delta z}{dt} = -\gamma\delta r + \gamma\upsilon r^3 \tag{3.18}$$

where δ and υ are the damping terms and volatility factors of financial market.

The Brownian price return can be described by Langevin equation:

$$m_r\frac{d^2 r}{dt^2} = -\eta\frac{dr}{dt} - \frac{dV(r)}{dr} \tag{3.19}$$

where

m_r mass of the financial particle p;
η damping force factor;
$V(r)$ time-independent quantum potential.

For overdamping case where $\frac{d^2 r}{dt^2} = 0$, we have

$$-\frac{dV(r)}{dr} = \eta\frac{dr}{dt} = -\gamma\eta\delta r + \gamma\eta\upsilon r^3 \tag{3.20}$$

$$V(r) = \int \left(-\gamma\eta\delta r + \gamma\eta\upsilon r^3\right)dr = \frac{\gamma\eta\delta}{2}r^2 - \frac{\gamma\eta\upsilon}{4}r^4 \tag{3.21}$$

3.5.8 *Quantum Finance Schrödinger Equation—QFSE*

Combining Eqs. (3.17)–(3.21), we obtained the *Quantum Finance Schrödinger Equation* (aka QFSE) of a quantum finance particle as

$$\left[\frac{-\hbar}{2m}\frac{d^2}{dr^2} + \left(\frac{\gamma\eta\delta}{2}r^2 - \frac{\gamma\eta\upsilon}{4}r^4\right)\right]\varphi(r) = E\varphi(r) \qquad (3.22)$$

Certain main characteristics can be found in QFSE:

1. The time-independent Schrödinger equation of a quantum finance contains both K.E. and P.E. terms.
2. Different from classical quantum harmonic oscillator, quantum finance oscillator is an anharmonic quantum oscillator which consists of two high-order P.E. terms to represent (1) damping (trading restoration and market absorption) potential and (2) volatility (risk control) potential.
3. Although market is visualized (observed) as price, the quantum dynamics are constrained by price return $\left(r = \frac{dp}{dt}\right)$, which is consistent with classical financial theory.

3.5.9   Significance of QFSE in Quantum Finance

The significance of QFSE in quantum finance includes the following:

1. Once we formulate QFSE in terms of quantum anharmonic oscillation (QAHO), solving the quantum anharmonic oscillator will become clear. It is one of the popular mathematical and high-energy physics topics. The solution is also applied to modern physics technology which includes superconductivity, quantum theory on laser and optics, quantum computing, and naturally quantum finance.
2. This exquisite equation sets an important path to (actually) evaluate the "holy grail" of quantum finance—quantum finance price energy level (or quantum price level, QPL) for all financial instruments in the secondary financial markets.
3. In Chap. 4, we will study the complete quantum finance theory's mathematical modeling based on quantum anharmonic oscillation theory.
4. In Chap. 5, we will explore how to formulate, implement quantum finance theory to calculate quantum price levels (QPLs) for worldwide financial products, and more importantly, build a workable MT4 program for actual calculation in any PC machines.

3.6   Quantum Finance—Which Way to Go

3.6.1   Feynman Feynman's Path Integral Approach Versus Quantum Anharmonic Oscillator Approach

	Feynman's path integral	Quantum anharmonic oscillator
PROS	• Technically speaking, this method can be used to evaluate basically all financial instruments in primary financial market, such as forward interest rate and, option pricing • Strict implementation of Feynman's path integral technique on finance modeling and technically sensible	• Technically speaking, this method can be used to evaluate any financial instruments in secondary financial market including financial derivative, forex prices, and financial indices such as DJI and FTSE • With proper integration of numerical computational technique(s), QAHO can also be used to calculate the quantum energy levels—so-called quantum price levels (QPLs) for any financial markets • With the adoption of numerical computational techniques and finite difference method (FDM), QAHO can be easily implemented by computer/trading program, MQL/MT4, for example
CONS	• To calculate expectation value of wave function, integrations of all the possible paths are required • Technically possible, but mathematically too complex to calculate and implement for commercial use	• As QAHO is based on polynomial approximation of differential equation(s) of QM and QFT, calculation results are basically the best approximation of these quantum finance instrument values • Owing to the polynomial approximation nature and sensitivity to initial conditions (which is approximation in natural), error-prone for market forecast, especially when the time horizon extends

3.6.2 Quantum Finance—Which Way to Go

The comparison chart listed in Sect. 3.6.1 showed that both Feynman's path integral approach and quantum anharmonic oscillator approach have pros and cons. Further, both approach methods head to different area applications in finance engineering.

Owing to its intrinsic property of modeling the detail paths with their dynamics of every quantum financial particle, Feynman's path integral approach is well suited for modeling quantum dynamics on *primary financial markets* (*PFMs*) such as interest rates, forward interest rates of contracts and bonds, plus issue new securities in stock markets. It is the first generation of quantum finance modeling technique being studied and explored in the finance engineering realm, also known as *first-generation quantum finance*. However, the intrinsic property of its computational complexity entails Feynman's path integral method, especially when dealing with financial big data which consist of infinite numbers of possibility (i.e., paths), such methods are difficult to be applied to real-time worldwide financial engineering (Fig. 3.9).

Quantum anharmonic oscillator approach, on the other hand, models quantum financial world as anharmonic oscillations of quantum financial particles. The major breakthrough is the conversion of classical Schrödinger equation into simpler quantum anharmonic equations. Thus, the equations can be solved by simple numerical calculation using digital computational method to be revealed in Chaps. 4 and 5. Such major breakthrough leads us to apply quantum finance technology into real-world financial engineering, with the integration of contemporary AI tools to implement real-time financial prediction and quantum trading systems. This approach is the latest mathematical technique for quantum finance modeling, also known as *second generation of quantum finance*. However, since this model is tailored for solving QFSE and the evaluation of all related quantum financial energy level known as quantum price levels (QPLs) to be revealed in Chaps. 4 and 5, it is well suited for the application to all financial products in secondary financial markets (SFMs) such as worldwide forex markets, commodity markets, financial indices, and cryptocurrency market.

Like any other theory and technology, quantum finance—as a newly emerging theory—should and ought to adopt any possible and workable theory, methodology, and technology to make it more useful, practical, and comprehensible, not solely on research but also for public and financial community.

Especially for the exponential growth of program trading, there are tremendous needs for intelligent financial forecasting and trading tools, not solely for the institutional clients and fund houses but more importantly for independent traders and investors—to provide an open and fair environment for worldwide investment.

As one would expect, the implementation of quantum price levels calculation is a good start, but it is also just the beginning of the journey.

The major challenges we are facing are how to make use of it, together with other AI tools and technology, and to design and build intelligent financial forecast and smart trading programs, which will be studied in the following chapters.

Fig. 3.9 Quantum finance—which way to go? (Tuchong 2019)

3.7 Conclusion

Every discipline of technology has its own course and history. In this chapter, we discussed the brief history of quantum finance, from Professor L. Bachelier's *Theory of Speculation* in 1900 to the latest works on quantum anharmonic oscillator theory of quantum finance. We also studied the latest research and studies of quantum finance in the past 30 years.

Through these, we focused on the discussion of two major mathematical approaches and models of quantum finance—(1) Feynman's path integral approach and (2) quantum anharmonic oscillator approach. We reviewed their basic concepts, principal features, and financial and mathematical implications with their significances in the modeling of real-world financial markets.

The applicability of any new theory and technology must fulfill three challenges:

1. physically sound,
2. mathematically logical, and
3. computationally feasible.

Physically sound means that such model and theory must be aligned with the physical world of reality it represents, in our case, the quantum world of worldwide financial markets.

Mathematically logical means that all the mathematical models and frameworks should be aligned with contemporary theoretical, mathematical research and studies.

Computationally feasible means that the new model and theory must be practically feasible to be adopted as computer models, algorithms, and computer systems for real-time modeling of financial market.

In next chapter, we will study how quantum anharmonic oscillator of quantum finance can be physically and mathematically sound for the modeling of worldwide secondary financial markets. In Chap. 5, we will further explore how such technology is computationally feasible and effective to model and evaluate quantum price levels (QPLs), which provides a critical mass to implement real-time worldwide financial predication and quantum trading systems.

Problems

3.1   Why Richard Feynman in his famous quotation said the two apparently dissimilar approaches in quantum mechanics (1) differential equation formulation of Schrödinger and (2) matrix algebra formulation of Heisenberg are in fact to be mathematically equivalent?

3.2   What is computational finance? Discuss and explain what are the major differences between computational finance and classical finance.

3.3   State and explain the two major branches of quantum finance models and how they are related with quantum field theory and modern quantum mechanics.

3.4   What is econophysics? Discuss and explain how modern physics affect and contribute to the development of modern economic and finance theories. Give three modern economic/finance theories as example.

3.5   What is/are the major contribution(s) of L. Bachelier in modern finance?

3.6   Why mainstream econophysicists traditionally considered finance as an application of Brownian motion? And how such notion is important to the development of finance and economic theories in the past century?

3.7   State and discuss three applications and R&D works of quantum finance done in the past two decades.

3.8   What is Feynman's path integral approach in quantum finance? And how it can be applied to forward interest rate modeling?

3.9   What are the differences between primary versus secondary financial markets? Discuss and explain why Feynman's path integral approach is best to be applied in primary financial markets instead of secondary financial markets.

3.10   What is forward interest rate? Why is it important to financial markets?

3.11   Figure 3.5 shows the forward interest rates of Eurodollar futures from 1990 to 1996, with eight different maturities ranging from 3 months to 7 years. State and explain three characteristics found from Fig. 3.5 in terms of

(i)   the coherence of the eight chart patterns;

(ii) the ordering of these patterns in terms of the period of maturity; and

(iii) the crossings of chart patterns.

3.12 State and explain the formulation of forward interest rate modeling using Feynman's path integral in quantum finance.

3.13 Describe and explain the physical meaning and logic behind interest rate modeling using Feynman's path integral in quantum finance.

3.14 What are the major pros and cons for the modeling of forward interest rate using Feynman's path integral in quantum finance?

3.15 In addition to the modeling of forward interest rate, what are the other potential applications of Feynman's path integral in quantum finance?

3.16 What is quantum anharmonic oscillator approach in quantum finance? State and explain how it works.

3.17 For financial markets modeling, we usually use price returns (r) instead of price (p) directly, why? Give two examples of financial markets to support your explanation.

3.18 What is the major difference between quantum anharmonic oscillator approach versus Feynman's path integral approach in quantum finance?

3.19 Discuss and explain why quantum anharmonic oscillator approach is tailored for the modeling of secondary financial markets instead of primary financial markets.

3.20 Discuss and explain physical and mathematical meanings of price (p), price returns (r), and quantum price levels (QPLs) in quantum finance using anharmonic oscillator model, using Dow Jones Index (DJI) as example to support your explanation.

3.21 What is the physical meaning of Hamiltonian (H) in Schrödinger equation of quantum finance? What are K.E. and P.E. parts? And how they are related in financial market's dynamics in quantum finance?

3.22 What is the physical meaning and importance of eigenenergy values of the time-independent Schrödinger equation in quantum finance?

3.23 What is the physical meaning and importance of *probability density function* (and its values) in quantum finance? And how can we observe/ measure it in real-world financial markets?

3.24 How can we interpret classical statistical finance theory using quantum finance theory? And how it is related to wave–particle duality phenomena in quantum theory?

3.25 What is *market depth* in modern finance? What is its importance? What is the mathematical interpretation of *market depth* in quantum finance?

3.26 State and explain the formulation and physical meaning of each mathematical term of Langevin equation in quantum finance.

3.27 State and explain the main characteristics of quantum finance Schrödinger equation (QFSE).

3.28 What are the significances of quantum finance Schrödinger equation (QFSE)? Give two examples of how to apply QFSE for the modeling of financial markets.

3.29   Discuss and explain the pros and cons for the modeling of quantum finance using (1) Feynman's path integral approach and (2) quantum anharmonic oscillator approach.

3.30   For the modeling of forex/cryptocurrency market, such as short-term financial prediction, which approach is more referable? Why?

References

Ataullah, A., Davidson I. and Tippett, M. (2009) A wave function for stock market returns. *Physica A: Statistical Mechanics and its Applications.* 388(4): 455–461.

Baaquie, B. E. (2004) *Quantum Finance.* Cambridge University Press.

Baaquie, B (2007) Quantum Finance: Feynman's Path Integrals and Hamiltonians for Options and Interest Rates. Cambridge University Press, 1st edition.

Baaquie, B. E. (2013) Financial modeling and quantum mathematics. *Computers and Mathematics with Applications.* 65(10): 1665–1673.

Baaquie, B. (2018) Quantum Field Theory for Economics and Finance. Cambridge University Press. 1st edition.

Bachelier, L. (1900) Théorie de la speculation. *Annales Scientifiques de l'École Normale Supérieure,* 3(17): 21–86.

Bagarello, F. (2009) A quantum statistical approach to simplified stock markets. *Physica A: Statistical Mechanics and its Applications.* 388(20): 4397–4406.

Bloch, S. C. (1997) *Introduction to Classical and Quantum Harmonic Oscillators.* Wiley, 1st edition.

Cotfas, L. (2013) A finite-dimensional quantum model for the stock market. *Physica A: Statistical Mechanics and its Applications.* 392(2): 371–380.

Gao, T. and Chen, Y. (2017) A quantum anharmonic oscillator model for the stock market. *Physica A: Statistical Mechanics and its Applications.* 468: 307–314.

Kim, M. J., Kim, S. Y. Hwang, D. I., Lee, S. Y. (2011) The sensitivity analysis of propagator for path independent quantum finance model. P*hysica A: Statistical Mechanics and its Applications.* 390(5): 847–863.

Mantegna, R. M. and Stanley, H. E. (1999) Introduction to Econophysics: Correlations and Complexity in Finance. Cambridge University Press, 1st edition.

Meng, X., Zhang, J. and Guo, H. (2015) Quantum spatial-periodic harmonic model for daily price-limited stock markets. *Physica A: Statistical Mechanics and its Applications.* 438: 154–160.

Nakayama, Y. (2009) Gravity dual for Reggeon field theory and nonlinear quantum finance. *International Journal of Modern Physics A.* 24(32): 6197–6222.

Nasiri, S., Bektas, E. and Jafari, G. R. (2018) The impact of trading volume on the stock market credibility: Bohmian quantum potential approach. *Physica A: Statistical Mechanics and its Applications.* 512: 1104–1112.

Piotrowski, E. W. and Sładkowski, J. (2005) Quantum diffusion of prices and profits. *Physica A: Statistical Mechanics and its Applications,* 345(1–2): 185–195.

QFFC (2019) Official site of Quantum Finance Forecast Center. http://qffc.org. Accessed 22 Aug 2019.

Schaden, M. (2002) Quantum finance. *Physica A: Statistical Mechanics and its Applications.* 316(1–4): 511–538.

Shi, L. (2006) Does security transaction volume-price behavior resemble a probability wave? *Physica A: Statistical Mechanics and its Applications.* 366: 419–436.

Swanson, M. S. (2014) *Feynman's Path Integrals and Quantum Processes (Dover Books on Physics).* Dover Publications.

Tuchong (2019) Quantum Finance – which way to go? http://stock.tuchong.com/image?imageId= 383418791542128669. Accessed 22 Aug 2019.

Ye, C. and Huang, J. P. (2008) Non-classical oscillator model for persistent fluctuations in stock markets. *Physica A: Statistical Mechanics and its Applications*. 387(5): 1255–1263.

Zee, A. (2011) *Quantum Field Theory in a Nutshell*. Princeton University Press, 2nd edition.

Zhang, C. and Huang, L. (2010) A quantum model for the stock market. *Physica A: Statistical Mechanics and its Applications*. 389(24): 5769–5775.

Chapter 4
Quantum Finance Theory

The mathematical framework of quantum theory has passed
countless successful tests and is now universally accepted as a
consistent and accurate description of all atomic phenomena.
Erwin Schrödinger (1887–1961)

If financial markets really exhibit quantum properties
How can we model it??
What do we want to achieve?

Since Professor Erwin Schrödinger proposed his influential Schrödinger equation in 1925 (Schrödinger 1926)—the core of quantum mechanics and quantum field theory—the studies and research of this "spooky" quantum world were thrived. The advance of computing technology nurtured the study of this unique subatomic world with the possibility of solving these highly complex quantum equations using contemporary computer technology is no more a reverie.

If we say 1920s was quantum theory's first golden age, the launch of IBM System ONE—first commercial-used quantum computer road show in Las Vegas at CES2019 early this year coined quantum theory's second golden age—the age of quantum computing.

Why now?

As there is a saying, things wouldn't happen simply by chance. The only thing is whether we know their cause-and-effect—the *causality*.

If we say quantum computer is the *machine* that drives this second golden age of quantum theory, the quantum dynamics model we will study in this chapter—the quantum anharmonic oscillator (QAOH) model—can be considered as the *soul* of this giant *machine*.

Based on concepts and theories of quantum mechanics and quantum field theory studied in previous chapters, in this chapter, we will introduce the core concept of this book—quantum finance theory. First, we will study basic conceptual model of quantum finance theory—the quantum anharmonic oscillator model and introduce quantum price level (QPL) concept for the modeling of quantum finance energy levels. After that, we will revisit Schrödinger equation and combine with the author's

© Springer Nature Singapore Pte Ltd. 2020
R. S. T. Lee, *Quantum Finance*,
https://doi.org/10.1007/978-981-32-9796-8_4

latest work on quantum dynamics modeling of various parties in a secondary financial market (SFM), along with a step-by-step mathematical derivation of quantum finance Schrödinger equation (QFSE).

4.1   Quantum Finance Theory

4.1.1   The Concept

In quantum finance, we model the dynamics of financial instruments (such as currencies, financial indices, cryptocurrencies) of worldwide financial markets as quantum financial particles (QFPs) with wave–particle duality characteristics.

The motions and dynamics significance of these quantum financial particles are subject to their intrinsic quantum energy fields, the so-called quantum price fields (QPFs) and appeared to us as quantum price levels (QPLs) in financial markets. They are similar to quantum particles that are affected by the superposition of their own energy levels and the energy field generated by other neighboring quantum particle(s). Figure 4.1 illustrates the quantum financial energy fields created by three types of financial particles: US dollars, Euros, and British pounds.

From technical finance perspective, these quantum price levels correspond to the support and resistance (S & R) levels as we know of.

In other words, one of the major objectives of quantum finance theory is to establish an effective and logical quantum finance model and help us to locate all these

Fig. 4.1 Illustration of quantum financial field for worldwide financial products

QPLs of worldwide financial markets using quantum mechanics and quantum field theories. Such quantum finance model must be logically sound and should be a coherent body of classical finance concepts and models.

If dynamics and motions of all fundamental particles found in the physical world are governed by quantum mechanics and quantum field theory as mentioned in Chap. 3, it is reasonable and rational to say that the worldwide financial markets—more precisely—are the price/index movements of all different financial instruments in every single time step, which are believed as results of the *collective conscious, behavior*, and *decision-making* of all participants in the markets that can also be modeled by quantum theory without exception.

The only problem is: How can we model them in terms of quantum mechanics and quantum field theory?

More importantly, how to implement these mathematical models in a logical, sensible, and effective manner so that they can apply and implement them in any digital computers?

4.1.2 The Notion of Quantum Price

There is one pivotal analog of quantum finance theory and classical quantum theory is the notion of the fundamental particles in quantum finance—quantum price.

From the finance perspective, we are all familiar with that the *current price* (*spot price*) of any financial product not only represents instantaneous *price value* but also the *price level* (a kind of *energy level* in finance) that resides at every particular moment as shown in Fig. 4.2.

Fig. 4.2 Stock market price movement (Tuchong 2019a)

In other words, in classical finance, *price* itself is not only a scalar value, but in some way; it also represents the *intrinsic energy* it possesses (analog to the potential energy in classical mechanics). In terms of classical finance, the meaning of *price* for any financial market has three different interpretations:

1. It represents the instantaneous price value of that particular financial product at a particular time;
2. It represents the *time* of the particular financial product which reaches that particular price value. That's why when we say the price of a particular financial product (e.g., stock), we also speak about the time;
3. It represents that intrinsic energy of that financial product at a particular time, which similar to the interpretation for the potential energy of objects and matters in the physical world.

As analog to quantum theory, it is similar to the particle existed and influenced by its own *quantum potential well* (Harrison and Valavanis 2016). Figure 4.3 shows a typical example of the quantum potential energy well generated by a quantum particle (e.g., an atom).

The truth is: the analog of financial particles in quantum finance theory using quantum theory is even more sensible and understandable in the finance community than the notion of quantum particles in its quantum field. The only concern is: How can we make use of classical finance concepts and beliefs to establish quantum dynamics of quantum finance particles?

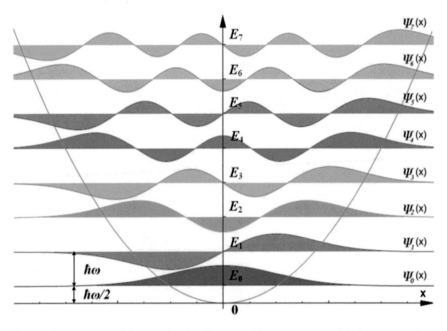

Fig. 4.3 Quantum potential energy level well generated by a quantum particle (e.g., an atom)

4.1.3 Revisit of Schrödinger Equation

It was recalled that in quantum theory, any quantum particle shown in Fig. 4.4 at position x, the time-dependent Schrödinger equation (Zee 2011) is given by

$$i\hbar \frac{\partial}{\partial t}\psi(x, t) = \hat{H}\psi(x, t) \tag{4.1}$$

where $\psi(x, t)$ is the wave function, $\hat{H}$ is the Hamiltonian operator, and $\hbar$ is the reduced Planck constant, whereas the corresponding time-independent counterpart is given by

$$\hat{H}\,\varphi(x) = E\varphi(x) \tag{4.2}$$

where

$$\psi(x, t) = \varphi(x)\,e^{-iEt/\hbar}$$

Note that the time-independent Schrödinger equation is a standard eigenfunction where E is the eigenenergy with discrete energy levels, and the Hamiltonian operator $\hat{H}$ is given by

$$\hat{H} = \frac{-\hbar}{2m}\frac{\partial^2}{\partial x^2} + V(x) \tag{4.3}$$

where $\hat{H}$ comprises kinetic energy, K.E. (first term) + potential energy, P.E. (second term).

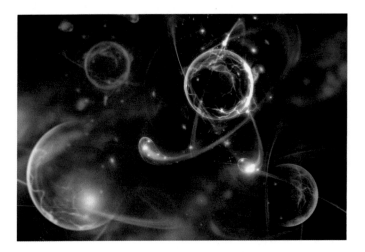

Fig. 4.4 Illustration of quantum particle in the subatomic world (Tuchong 2019b)

4.2   Quantum Finance Model

4.2.1   The Mathematical Model

Let r be the *price return* of a particular *quantum financial particle (QFP)* at time t
(say USD/CAD or simply US Index).

We can rewrite Schrödinger equation in (4.1) as

$$i\hbar \frac{\partial}{\partial t} \psi(r, t) = \hat{H} \psi(r, t) \tag{4.4}$$

and the corresponding $\hat{H}$ is given by

$$\hat{H} = \frac{-\hbar}{2m} \frac{\partial^2}{\partial r^2} + V(r, t) \tag{4.5}$$

where

1. $\hat{H}$ comprises K.E. (kinetic energy, the first term) and P.E. (potential energy, the second term);
2. $\hbar$ is the Planck constant representing the uncertainty of financial behavior;
3. m is the mass representing the intrinsic potential of financial market, such as the market capitalization of a particular financial product in financial market.

As mentioned in Chap. 3, the usage of *price returns (r)* instead of *price (p)* itself is threefold:

1. Use *price returns (r)* instead of price (p) on mathematical and statistical modeling is a kind of de facto standard in classical financial and statistical finance theory;
2. It is more sensible and easier to implement, especially if we want to compare the performance of two (or more) financial products. By using price return (r) instead of price (p), even when the two financial products have totally different orders of magnitude, e.g., Dow Jones Index versus CADUSD, one is at order 1×10^4 and the other at order of 1×10^1;
3. In terms of quantum theory, the physical meaning of x in time-dependent Schrödinger equation (4.1) is the displacement of quantum particle relative to its equilibrium position, and not the distance of travel in classical mechanics as in Newton's laws of motion. Therefore, it is obvious and more sensible to use price return (r) instead of price (p) as physical interpretation of r (say daily return) can be considered as the relative price change of that particular financial product at present compared with its closing price of the previous trading day.

4.2.2 Physical Meaning of Wave Function ψ

Wave function (ψ) is the most important component in the entire quantum theory. It is the realization of wave–particle duality of quantum particles, which can be considered as the bridge to link up between particle and wave realms in the subatomic world as shown in Fig. 4.5.

What is the wave function (ψ) in quantum finance?

Can we observe it, or measure it?

As recalled that in the "double-slit experiment" (Aharonov et al. 2017), what we saw in the backdrop screen is the realization of wave function from quantum particle projections.

However, if we measure it "closely", all wave-like phenomena will disappear. Can we also recall that?

So, what shall we do?

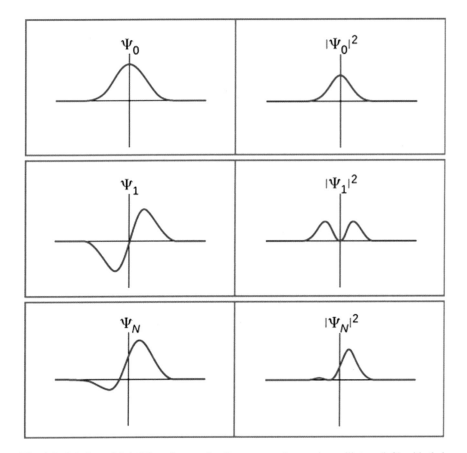

Fig. 4.5 Solution of Schrödinger's equation for quantum harmonic oscillators (left) with their amplitudes (right)

In quantum theory, we measure the *pdf* (probability density function, ρ) of observations instead, which is given by

$$\rho(x, t) = |\psi(x, t)|^2 = |\varphi(x)|^2 \tag{4.6}$$

The wave function in quantum finance can be observed by using similar method.

For example, for a time series of 1000 trading days, we measure the closing price return of these 1000 time steps and plot the probability density function (*pdf*) of price return r versus *pdf* of occurrence ϵ [0, 1] in order to analog ψ.

That is,

$$\rho(r, t) = |\psi(r, t)|^2 = |\varphi(r)|^2 \tag{4.7}$$

4.3   Financial Dynamics

4.3.1   Key Players in Secondary Financial Markets

Once we have the financial model, the next step is to explore *the dynamics* which means all motions and activities occur inside the model. That is: what are the *dynamics* in a typical secondary financial market such as forex, commodity, or cryptocurrency?

In other words, what are the major participants in a financial market? What are their behaviors?

For example, in forex market—the biggest OTC (over the counter) market in the worldwide finance (Hull 2016), what are the key participants?

Figure 4.6 shows a framework in a typical secondary financial market (SFM) such as worldwide forex markets (Strumeyer 2017).

4.3.2   Market Maker (MM)

Market maker (MM), also known as liquidity provider, is a company or an individual that quotes both a buy and a sell price in a financial instrument or commodity held in inventory, hoping to make a profit on the bid-offer spread, or turn (Čekauskas et al. 2012; Zhu et al. 2009).

The U.S. Securities and Exchange Commission defines market maker as firms that stand ready to buy and sell stock on a regular and continuous basis at a publicly quoted price.

Most foreign exchange trading firms are market maker and so are many banks.

Fig. 4.6 Key Players in a typical secondary financial market (SFM)

The most common type of market maker is brokerage houses that provide purchase and sale solutions for investors, in an effort to keep financial markets liquid.

Market maker can also be an individual intermediary, but due to the size of securities needed to facilitate the volume of purchases and sales, the vast majority of market makers work on behalf of large institutions.

They sell to and also buy from clients, compensated by means of price differentials for services in providing liquidity, reducing transaction costs, and facilitating trade.

They earn profit by the difference between the price at which they are willing to buy a stock and the price that the firm is willing to sell it, also known as the market *spread* or *bid-ask spread*.

Market maker also provides liquidity to their own firm's clients, for which they earn a commission.

In terms of investment dynamics, the main function is to maintain healthy market liquidity or facilitate the efficient absorption of buy/sell orders (Fig. 4.7).

4.3.3 Arbitrageurs (ARs)

Arbitrageurs are traders that take advantage of a price difference between two or more markets.

They look for an imbalance in the pricing of a security in two different markets and buy it at the cheap price in one market to immediately sell it at the higher price in the other market, making a profit on the difference (Hong et al. 2012; Matsushima 2013).

This kind of operation is called arbitrage.

Fig. 4.7 The Market Maker (MM)

Arbitrageurs are typically very experienced investors since arbitrage opportunities are difficult to find and require relatively fast trading. Arbitrageurs also play an important role in the operation of capital markets, as their efforts in exploiting price inefficiencies keep prices more accurate than they otherwise would be.

However, in an *efficient market* nowadays with open information and high-speed trading, there are basically no rooms for arbitragers trading (Malkiel 2003).

Example of Arbitrage Trading

Suppose that the exchange rates of EUR/USD in Frankfurt quotes at $1.25, while in New York it quotes at $1.3 at the same time.

Hence, the euro is undervalued in Frankfurt and overvalued in New York. An arbitrageur would exploit this imbalance to exchange dollars for euros in Frankfurt and immediately exchange the euros for dollars in New York.

The profit for each dollar in this double exchange is $0.8 - 0.769 = 0.031$ or a 3% on the amount invested (Fig. 4.8).

4.3.4   Speculators (SPs)

Speculators take risk on purpose by betting on future movements of the security's price (Brunetti et al. 2016; Yi 2017).

For example, a speculator buys a stock under the conviction that its price will rise without any logical and rational reasons most of the time.

Many speculators pay little attention to the fundamental value of a security and instead purely focus on price movements.

In finance, speculation is also the practice of engaging in risky financial transactions in an attempt to profit from short-term fluctuations in market value of a tradable financial instrument—rather than attempting to profit from the underlying financial attributes embodied in the instrument such as capital gains, dividends, or interest.

Fig. 4.8 The Arbitrageurs (ARs)

Speculators are important to market because they bring liquidity besides assume market risk and, conversely, they can also have a negative impact on markets when their trading actions result in a speculative bubble that drives up an asset's price to unsustainable levels. If a speculator believes that a particular asset is going to increase in value, they may choose to purchase as much of the asset as possible. This activity, based on the perceived increase in demand, drives up the price of the particular asset. If this activity is seen across the market as a positive sign, it may cause other traders to purchase the asset as well, further elevating the price. This can result in a speculative bubble, where the speculator activity has driven the price of an asset above its true value.

In terms of investment behavior, speculators differ from *common investors* in the sense that they don't have any risk control mindset. In other words, there is no *damping factor* against *market volatility* in their investment strategies (Fig. 4.9).

4.3.5 Hedgers (HGs)

Hedgers trade so to reduce or eliminate the risk of taking a position on security.

The main goal is to protect the portfolio from losing value at the expense of lowering the possible benefits.

This attitude (or trading strategy) is called *hedging*.

A hedging strategy usually involves taking contrarian positions in two or more securities. That's why many people also call hedgers as contrarians, but actually they are not always 100% acting against the market trends (Radalj 2006; Röthig 2011).

Speculators and hedgers are different terms that describe traders and investors. Speculation involves trying to make a profit from a security's price change, whereas

Fig. 4.9 Speculators (SPs)

hedging attempts to reduce the amount of risk, or volatility, associated with a security's price change (Yung and Liu 2009; Lin et al. 2009).

Hedging involves taking an offset position in a derivative in order to balance any gains and losses of the underlying asset. Hedging attempts to eliminate the volatility associated with an asset price by taking offset positions contrary to what the investor currently has. The main purpose of speculation, on the other hand, is to profit from betting on the direction in which an asset will be moving.

For example, taking a long position on a stock and a short position on another stock with inverse price behavior or the portfolio risk control strategy we have just mentioned (Fig. 4.10).

4.3.6   Investors (IVs)

An *investor* is a person who allocates capital with the expectation of a future financial return.

Types of investments include equity, debt securities, real estate, currency, commodity, token, derivatives such as put and call options, futures, forwards, etc.

That is, someone who provides a business with capital and someone who buys a stock are both investors.

There are two types of investors, individual (independent) investors and institutional investors.

Individual investor is a nonprofessional investor who buys and sells securities, mutual funds, or exchange-traded funds (ETFs) through traditional or online brokerage firms or savings accounts. Retail investors invest much smaller amounts than large institutional investors, such as mutual funds, pensions and university endowments,

Fig. 4.10 Hedgers (HGs)
(Tuchong 2019c)

and trade less frequently. But wealthier retail investors can now access alternative investment classes like private equity and hedge funds.

Institutional investors (Davis and Steil 2004; Gabaix et al. 2006) are major players in the financial market. They are the pension funds, mutual funds, and also some private equity investors. Institutional investors account for about three-quarters of the volume of trades on the New York Stock Exchange. They move large blocks of shares and have tremendous influence on stock market's movements.

In terms of investment dynamics, investors normally act as *Trend Followers* together with certain degree of sense of risk control (i.e., certain degree of *damping factor* against *market volatility* in their investment strategies) (Fig. 4.11).

4.4 Financial Dynamics and the Notion of Excess Demand

Once we have the key player in financial market, how can we model their financial dynamics?

The notion of *excess demand* (z).

In classical finance and microeconomics, *excess demand* is a function expressing excess demand for a product—the excess of quantity demanded over quantity supplied—in terms of the product's price and possibly other determinants. In a mathematical perspective, it is the product's demand function minus its supply function.

Fig. 4.11 Investors (IVs)

In a pure exchange economy, the excess demand is the sum of all agents' demands minus the sum of all agents' initial endowments (Tian 2016; Momi 2010).

The price of the product is said to be the equilibrium price if it is such that the value of the excess demand function is zero: which means, when the market is in equilibrium, the quantity supplied equals the quantity demanded. In this case, it is said that the market clears. If the price is higher than the equilibrium price, excess demand will normally be negative, meaning that there is a surplus (positive excess supply) of the product, and not all of it being offered to the marketplace is being sold. If the price is lower than the equilibrium price, excess demand will normally be positive, meaning that there is a shortage of demand in the financial market.

Mathematical Derivations

At any time t, $z_+(t)$ and $z_-(t)$ denote the instantaneous demand and supply for the financial asset.

The excess demand (z) at any instance is given by

$$\Delta z = z_+ - z_- \tag{4.8}$$

Let r(t) is the instantaneous returns, which is given by

$$\frac{dp}{dt} = r(t) = F(\Delta z) \tag{4.9}$$

For small Δz, F can be approximated by a scaling factor γ and become

$$r(t) = \frac{\Delta z}{\gamma} \tag{4.10}$$

where γ can be used to represent the *market depth*, the excess demand z required to move the quantum price p by one single quantum. Note that, when γ is high, it means that the market has a higher *absorbability* to excess demand z against price changes.

So, we have

$$\frac{dr}{dt} = \frac{d^2p}{dt^2} = \frac{1}{\gamma}\frac{d(\Delta z)}{dt} \tag{4.11}$$

4.5 Quantum Dynamics in Financial Markets

According to the investment behaviors of all these five key participants in financial markets, their corresponding quantum dynamics can be interpreted as follows.

4.5.1 Market Makers (MMs)

The quantum dynamics for market makers is given by

$$\left.\frac{dz_+}{dt}\right|_{MM} = -\alpha_+ z_+ \text{ and} \tag{4.12a}$$

$$\left.\frac{dz_-}{dt}\right|_{MM} = -\alpha_- z_- \tag{4.12b}$$

Note:

1. Market makers (MMs) provide market facilitator services to absorb ALL outstanding excess order z_+ and z_-.
2. α_+ and α_- are the market absorbability factors.
3. In terms of quantum dynamics, basically it is a quantum harmonic oscillator (QHO) with $\dot{z}_\pm \propto z_\pm$.

Combining with Eq. (4.8), we have

$$\begin{aligned}
\left.\frac{d\Delta z}{dt}\right|_{MM} &= \left.\frac{d(z_+ - z_-)}{dt}\right|_{MM} \\
&= \left.\frac{dz_+}{dt}\right|_{MM} - \left.\frac{dz_-}{dt}\right|_{MM} \\
&= -\alpha_+ z_+ + \alpha_- z_-
\end{aligned} \tag{4.13}$$

For an *efficient market*, we can assume

$$\alpha_+ = \alpha_- = \alpha_{MM} \tag{4.14}$$

So, we have

$$\left. \frac{d\,\Delta z}{dt} \right|_{MM} = -\alpha_{MM}\,\Delta z \tag{4.15}$$

Using Eq. (4.10), we have

$$\left. \frac{d\,\Delta z}{dt} \right|_{MM} = -\gamma \alpha_{MM}\, r \tag{4.16}$$

4.5.2 Arbitrageurs (ARs)

As mentioned, in an *efficient market* nowadays with open information and high-speed trading, there are basically no rooms for arbitrageurs' trading.

Principle of No Arbitrage on Security Markets

There are no-arbitrage opportunities in Reality.

Extended No Arbitrage on security markets states:

1. Arbitrage is not possible.
2. There are no transaction costs, no taxes, and no restrictions on short selling.
3. It is possible to borrow and lend at equal risk-free interest rates.
4. All securities are perfectly divisible.

The principle of no-arbitrage conditions states that market participants cannot claim risk-free return/profit unless there are conditions that facilitate to those claims. The no-arbitrage rule maintains a key position in analyzing price and price movements in financial markets as well as being a central component of such theories and concepts as efficient market hypothesis.

So, assume *Principle of No Arbitrage* is in every quantum time step, there is no quantum dynamics for arbitrageurs.

In reality, even though there might have a chance of arbitrageurs for an instance of time, but within an efficient market and high-speed trading, such dynamics will soon be *died-down* and absorbed by the market makers (MMs).

In respect of quantum finance dynamics of the quantum financial particles (QFPs) and the high speed of electronic transactions in every moment of time, it is practically impossible nowadays for arbitrageurs to make profit by looking for the price difference between different financial markets of the same product. The quantum dynamics of the same product between different markets synchronize to their equilibrium states at every moment.

4.5.3 Speculators (SPs)

Contradictory, speculators always emerge in every financial market.
   Their quantum dynamics are given by

$$\frac{d\,\Delta z}{dt}\bigg|_{SP} = -\delta_{SP} r \tag{4.17}$$

Since speculators have no idea of risk control, their quantum dynamics only contain the harmonic oscillator term (the delta term, $\delta_{SP} r$), without any higher order volatility term.
   Note that, although speculators might happen to be trend follower ($+\delta_{SP}$), most of the time their risk-taking nature will drive them to irrational speculation of market reversals and act against the market ($-\delta_{SP}$).
   As shown in Eq. (4.17), the existence of the only first-order return r means that the entire quantum dynamics of speculators (SPs) is a simple quantum harmonic oscillator (QHO) without any anharmonic component.
   From the physical meaning perspective, the existence of the only first order of return (r) in Eq. (4.17) means that dynamics of the speculators only exist the *force constant* which is the same as the *force constant* for spring in Newton's mechanics.

4.5.4 Hedgers (HGs)

Hedgers (HGs) represent experienced and skillful traders (also known as *sophisticated traders*) that apply sophisticated hedging strategies across different products and markets. Further will be studied in Chap. 6.
   Although they do not always act against the trend, their skills usually demonstrated by reverse trading or prediction of market reversal and act before common investors.
   So, their quantum dynamics are given by

$$\frac{d\,\Delta z}{dt}\bigg|_{HG} = -\left(\delta_{HG} - \upsilon_{HG} r^2\right) r \tag{4.18}$$

Note that the quantum dynamics for a hedger has two terms: (1) quantum harmonic oscillatory term (delta term)—proportion to return (r) and (2) quantum anharmonic term (υ) stands for the market volatility, risk control factor proportional to r^2.

In terms of quantum dynamics, the existence of both first-order and third-order terms of return r cause it to become a typical *quantum anharmonic oscillator* (QAOH) (Bhargava et al. 1989; Gao and Chen 2017).

4.5.5   Investors (IVs)

As mentioned, *investors* represent common investors and rational investors with certain degree of risk control.

Their unusual strategies (1) follow the trend to gain profit and (2) minimize risk. So, their quantum dynamics are given by

$$\frac{d\,\Delta z}{dt}\bigg|_{IV} = \left(\delta_{IV} - \upsilon_{IV} r^2\right) r \tag{4.19}$$

Note that, similar to hedgers, the quantum dynamics of an investor has two terms: (1) quantum harmonic oscillatory term (delta term)—proportion to return (r) and (2) quantum anharmonic term (υ) stands for the market volatility, risk control factor proportional to r^2.

But different from hedgers, common investors are usually *trend followers* (TFs), so they are basically acting toward returns (r).

4.5.6   Overall Dynamics for All Different Parties

By combining quantum dynamics of all key participants in the financial market, the overall quantum dynamics in a typical financial market is given by

$$\frac{d\,\Delta z}{dt} = \frac{d\,\Delta z}{dt}\bigg|_{MM} + \frac{d\,\Delta z}{dt}\bigg|_{SP} + \frac{d\,\Delta z}{dt}\bigg|_{HG} + \frac{d\,\Delta z}{dt}\bigg|_{IV} \tag{4.20}$$

$$\frac{d\,\Delta z}{dt} = -\gamma \alpha_{MM} r - \delta_{SP} r - \left(\delta_{HG} - \upsilon_{HG} r^2\right) r + \left(\delta_{IV} - \upsilon_{IV} r^2\right) r \tag{4.21}$$

That is,

$$\frac{d\,\Delta z}{dt} = -\delta r + \upsilon r^3 \tag{4.22}$$

Combining Eq. (4.11), we have

$$\frac{dr}{dt} = \gamma \frac{d\,\Delta z}{dt} = -\gamma \delta r + \gamma \upsilon r^3 \tag{4.23}$$

where
$$\delta = \gamma \alpha_{MM} + \delta_{SP} + \delta_{HG} - \delta_{IV} \text{ (damping term) and}$$
$$\upsilon = \upsilon_{HG} - \upsilon_{IV} \text{ (volatility term)}$$
The Brownian price return can be described by Langevin equation:

$$m_r \frac{d^2 r}{dt^2} = -\eta \frac{dr}{dt} - \frac{dV(r)}{dr} \tag{4.24}$$

where

m_r mass of the financial particle p;
η damping force factor;
$V(r)$ time-independent quantum potential.

For the consistency of Eqs. (4.23) and (4.24), i.e., overdamping case where $\frac{d^2 r}{dt^2} = 0$, we have

$$-\frac{dV(r)}{dr} = \eta \frac{dr}{dt} = -\gamma \eta \delta r + \gamma \eta \upsilon r^3 \tag{4.25}$$

$$V(r) = \int \left(-\gamma \eta \delta r + \gamma \eta \upsilon r^3 \right) dr = \frac{\gamma \eta \delta}{2} r^2 - \frac{\gamma \eta \upsilon}{4} r^4 \tag{4.26}$$

So, the time-independent Schrödinger Eqs. (4.4)–(4.5) of a quantum finance particle can be written as follows:

Quantum Finance Schrödinger Equation, QFSE

$$\left[\frac{-\hbar \, d^2}{2m \, dr^2} + \left(\frac{\gamma \eta \delta}{2} r^2 - \frac{\gamma \eta \upsilon}{4} r^4 \right) \right] \varphi(r) = E \varphi(r) \tag{4.27}$$

Note:

1. The time-independent Schrödinger equation of a quantum finance contains both KE and PE terms.
2. Different from classical quantum harmonic oscillator, the quantum finance oscillator is a quantum anharmonic oscillator which consists of two high-order PE terms that represent (1) damping (trading restoration and market absorption) potential and (2) volatility (risk control) potential.
3. Although the market is visualized (observed) as price, the quantum dynamics are controlled by price return ($r = \frac{dp}{dt}$), which is consistent with classical financial theory.

4.6 Conclusion

In this chapter, we introduced the main theme of this book: quantum finance theory using quantum anharmonic oscillator (QAOH) model. We began with quantum financial particles (QFPs) concept and their intrinsic quantum energy fields, the so-called quantum price field (QPF). We also studied the notion of quantum price in quantum finance and its relationship with quantum energy level. After that, we examined the Schrödinger equation and explored the physical meaning of wave function (ψ).

We then studied the core of this chapter—the five key players in secondary financial markets (SFMs)—market makers, arbitrageurs, speculators, hedges, and investors. First, we analyzed their roles, characteristics, and behaviors in the financial market. After that, we introduced classical financial dynamics and the notion of *excess demand*.

Based on the definition of excess demand, we derived quantum dynamics of these five key parties and deduced the overall quantum dynamic equation of their combined behaviors in secondary financial market. By combining with Eq. (4.11) and the assumption of overdamping case, we derived the quantum finance Schrödinger equation (QFSE) in form of an order-4 quantum anharmonic oscillator system.

An important issue of our step-by-step derivation of the QFSE is that we began the entire mathematical modeling quantum dynamics of secondary financial market based on classical *belief* and *understanding* in terms of classical finance and microeconomic theory. By doing so, we can ensure that the QFSE derived can be consistent with basic financial concept and theory, despite we are now using completely new perspective and tool to model the dynamics for secondary financial markets.

In the next chapter, we will explore how to use QFSE, in conjunction with numerical computational technique and *finite difference method* (FDM) to derive and evaluate quantum finance energy level (QFEL), together with quantum price levels (QPLs) for any financial products in secondary financial markets.

Problems

4.1 In the past, quantum theory including quantum mechanics and quantum field theory are queried by many mainstream scientists and believed to be some sort of "*spooky*" science. However, nowadays many scientists and general community accept the general idea of quantum theory and the existence of subatomic world. Why?

4.2 What is quantum computer? What is the major difference between quantum computer and the contemporary computers we are using?

4.3 What is the analog between quantum particles in quantum theory and the dynamics of the *financial particles* in financial markets? Give a live example such as forex market for illustration.

4.4 As a basic concept in finance, *"prices"* in a financial market, such as Dow Jones index (DJI), have three physical and mathematical meanings and interpretations, what are they? How can these three characteristics of prices relate to quantum finance theory?

4.5 What are the major similarities and differences between *prices* in financial markets versus quantum particles in quantum theory?

4.6 *Quantization* phenomena are easily found in financial markets and commonly accepted in technical analysis. Why? Give three examples of technical analytical tools and explain how they work in the realm of quantum finance.

4.7 State the time-dependent Schrödinger equation and its Hamiltonian operator.

(i) Identify kinetic energy (KE) and potential energy (PE) terms in the formulation;

(ii) State and explain how these two terms are interpreted in quantum finance.

4.8 In computational finance and quantum finance, we usually use price returns (r) instead of prices (p) for the mathematical formulation and financial analysis. Why? What is/are the advantage(s) of using *returns* instead of *prices* for mathematical formulation? Use two financial products as examples for illustration.

4.9 What are the physical meanings of wave function (ψ) in (1) quantum theory of quantum particles; (2) quantum finance of quantum financial particles?

4.10 How can we observe and measure wave function (ψ) values (and distribution) in terms of (1) quantum theory of quantum particles such as photons; (2) quantum finance theory of quantum financial particles such as Dow Jones index?

4.11 The below figure shows the solutions of Schrödinger's equation for quantum harmonic oscillators (left) and their amplitudes (right).

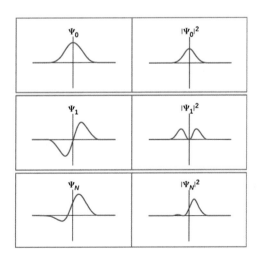

(i)   Write a MATLAB program to solve the Schrödinger's equation for quantum harmonic oscillators and plot these figures;

(ii) Describe their physical meanings in terms of quantum finance theory of financial particles such as Dow Jones index (DJI);

(iii) Can we simply model quantum finance particles such as DJI using quantum harmonic oscillators in reality? Why?

4.12   What is a secondary financial market (SFM)? What are the major differences between primary financial market (PFM) and SFM? In each case, please give two examples for illustration.

4.13   What is a market maker (MM)? Discuss and explain the role(s) and importance of market maker in financial markets.

4.14   What are arbitrageurs? What are the major differences between arbitrageurs and common investors in terms of (1) trading/investment strategy; (2) risk control? Give two examples of financial markets for illustration.

4.15   What are speculators? What are the major differences between speculators and common investors in terms of (1) trading/investment strategy; (2) risk control? Why speculators are important in financial markets? Give two examples of financial markets for illustration.

4.16   What are hedgers? What are the major differences between hedgers and common investors in terms of (1) trading/investment strategy; (2) risk control? Why hedgers are important in financial markets? Give two examples of financial markets for illustration.

4.17   Can an investor be both a hedger and speculator at the same time? Why or why not? Give two examples of financial markets for illustration.

4.18   What are *common investors* (or *investors* in short)? What are the characteristics of investors in terms of (1) trading/investment strategy; (2) risk control? Why common investors are vital in financial markets? Give two examples of financial markets for illustration.

4.19   Discuss and explain excess demand (z) in terms of (1) physical meaning in financial markets; (2) mathematical formulation. How can it relate to quantum finance?

4.20   State and explain the quantum finance formulations of the five key participants in a typical financial market.

4.21   State and explain the *principle of no arbitrage*. Explain why it is important in the formulation of the quantum dynamics of arbitrageurs in quantum finance.

4.22   What is an *efficient market*? Explain why it is important in the formulation of the quantum dynamics of market maker in quantum finance.

4.23   What is quantum anharmonic oscillator (QAHO)? What is the major difference between quantum harmonic oscillator (QHO) and its anharmonic counterpart QAHO in terms of (1) quantum dynamics and (2) energy levels?

4.24   For the formulation of real-world financial market, should we choose quantum harmonic oscillator (QHO) or quantum anharmonic oscillator (QAHO) for system modeling and mathematical formulation? Why?

4.25 Combine the quantum finance mathematical formulation of all five key participants in a typical financial market and derive the quantum finance Schrödinger equation.

4.26 State the time-independent quantum finance Schrödinger equation (QFSE).

   (i) Describe and explain the physical meaning of each term in QFSE;
   (ii) What is the physical meaning of (1) Planck constant, (2) mass, (3) damping factor η, gamma γ, and (4) delta δ terms in QFSE?

4.27 Discuss and explain the importance of quantum finance Schrödinger equation (QFSE) in terms of (1) modeling of secondary financial markets and (2) real-time financial prediction. Give two examples of secondary financial markets as explanation.

References

Aharonov, Y. et al. (2017) Finally making sense of the double-slit experiment. *Proceedings of the National Academy of Sciences of the United States of America* 114(25): 6480–6485.

Bhargava, V.T.A. et al. (1989) Anharmonic oscillators in higher dimension: Accurate energy eigenvalues and matrix elements. *Pramana* 32(2): 107–115.

Brunetti, C. et al. (2016) Speculators, Prices, and Market Volatility. *Journal of Financial and Quantitative Analysis* 51(5): 1545–1574.

Čekauskas, K. et al. (2012) The Effects of Market Makers and Stock Analysts in Emerging Markets: Market Makers and Stock Analysts. *International Review of Finance* 12(3): 305–327.

Davis, E. P. and Steil, B. (2004) *Institutional Investors (The MIT Press)*. MIT Press.

Gabaix, X. et al. (2006) Institutional Investors and Stock Market Volatility. *The Quarterly Journal of Economics* 121(2): 461–504.

Gao, T. and Chen, Y. (2017) A quantum anharmonic oscillator model for the stock market. *Physica A: Statistical Mechanics and its Applications.* 468: 307–314.

Harrison, P. & Valavanis, A. (2016) Quantum Wells, Wires and Dots: Theoretical and Computational Physics of Semiconductor Nanostructures, John Wiley & Sons.

Hong, H. et al. (2012) Do arbitrageurs amplify economic shocks? *Journal of Financial Economics* 103(3): 454–470.

Hull, J. C. (2016) *Fundamentals of Futures and Options Markets*. Pearson, 9th edition.

Lin, B. et al. (2009) Who bets against hedgers and how much they trade? A theory and empirical tests. *Applied Economics* 41(27): 3491–3497.

Malkiel, B.G. (2003) The Efficient Market Hypothesis and Its Critics. *The Journal of Economic Perspectives* 17(1): 59–82.

Matsushima, H. (2013) Behavioral aspects of arbitrageurs in timing games of bubbles and crashes. *Journal of Economic Theory* 148(2): 858–870.

Momi, T. (2010) Excess demand function around critical prices in incomplete markets. *Journal of Mathematical Economics* 46(3): 293–302.

Radalj, K.F. (2006) Hedgers, speculators and forward markets: Evidence from currency markets. *Environmental Modelling and Software* 21(9): 1381–1386.

Röthig, A. (2011) On speculators and hedgers in currency futures markets: who leads whom? *International Journal of Finance & Economics* 16(1): 63–69.

Schrödinger, E. (1926) An Undulatory Theory of the Mechanics of Atoms and Molecules. *Physical Review* 28: 1049–1070.

Strumeyer, G., Swammy, S. (2017) *The Capital Markets: Evolution of the Financial Ecosystem (Wiley Finance)*. Wiley.

Tian, G. (2016) On the existence of price equilibrium in economies with excess demand functions. *Economic Theory Bulletin* 4(1): 5–16.

Tuchong (2019a) Stock Market Price Movement. http://stock.tuchong.com/image?imageId=455757697366360685. Accessed 21 Aug 2019.

Tuchong (2019b) Illustration of quantum particle in the subatomic world. http://stock.tuchong.com/image?imageId=488506530928001038. Accessed 21 Aug 2019.

Tuchong (2019c) Hedgers. http://stock.tuchong.com/image?imageId=75952820146313693. Accessed 21 Aug 2019.

Yi, M. (2017) Speculator-triggered crisis and interventions. *Journal of Macroeconomics* 52: 135–146.

Yung, K. & Liu, Y. (2009) Implications of futures trading volume: Hedgers versus speculators. *Journal of Asset Management* 10(5):318–337.

Zee, A. (2011) *Quantum Field Theory in a Nutshell*. Princeton University Press, 2nd edition.

Zhu, M. et al. (2009) Does the market maker stabilize the market? *Physica A: Statistical Mechanics and its Applications* 388(15): 3164–3180.

Chapter 5
Quantum Price Levels—Basic Theory and Numerical Computation Technique

> It is true that in quantum theory we cannot rely on strict causality. But by repeating the experiments many times, we can finally derive from the observation's statistical distributions, and by repeating such series of experiments, we can arrive at objective statements concerning these distributions.
> Werner Heisenberg (1901–1976)

What are quantum price levels?
Do they really exist?
How can we model them?
How can we make use of them?

Professor Werner Heisenberg in this famous quotation stated an important fact: To prove or disprove any new concept or theory, simple concepts and ideas are just not enough. What we need are some *hard evidences* from repeating the experiments many times to explore the general observation's statistical distributions. Based on these statistical distributions, we discover some critical patterns. From that, we can conclude with some objective statements and findings to prove or disprove our proposed concept or theory.

It is consistent with German philosopher Immanuel Kant's remarkable quotation regarding the relationship between *theory* and *experience*:

> Experience without theory is blind, but theory without experience is mere intellectual play.
> Immanuel Kant (1724–1804)

In his famous quotation, Immanuel Kant pointed out that the important relationship between *concepts* and *ideas* versus *experiences* and *experimental observations* is that although concepts, ideas, and theories are the foundation and critical mass for new knowledge, without solid objective experimental observations and experiences, they are merely intellectual fantasies.

The merit of science against pseudoscience is that scientific theories and concepts can be proved, deduced, and observed in the form of mathematical proven, deduction, and tested objectively through experiments. Thus, scientific attitude is even more critical in quantum theory because basic concept and theory of subatomic world and

© Springer Nature Singapore Pte Ltd. 2020
R. S. T. Lee, *Quantum Finance*,
https://doi.org/10.1007/978-981-32-9796-8_5

its quantum phenomena are very different from our general belief of this physical world.

The truth is: the mathematical framework of quantum theory has passed countless successful tests with experiments and is now accepted universally as a uniform and accurate description of all quantum phenomena in the subatomic world.

In this book, we follow the same footsteps.

In Chaps. 1–3, we introduced the world of quantum finance and studied its basic concept and theory. Based on the quantum anharmonic oscillator (QAOH) model learnt in Chap. 4, we examined its quantum dynamics and mathematical formulation in a typical secondary financial market.

In this chapter, we discuss how such mathematical formulation can be applied in terms of objective experimental results with distribution to evaluate the quantum finance energy levels (QFELs) and quantum price levels (QPLs).

First, we will begin with the investigation of QPL as an analog to quantum energy levels (QELs) in classical quantum world.

Next, we will explore Schrödinger equation and the physical meaning of wave function in this equation. From that, we will study the experiments (time series trading results) of financial products for over 2000 trading days and obtain the observation's statistical distributions. By repeating such series of experiments onto various financial products, we can arrive at objective statements concerning these distributions. After that, we will combine these observation results with finite difference method (FDM) to derive the value of λ in the quantum finance Schrödinger equation (QFSE).

Finally, we will employ the latest research on λ^{2m} QAOH model, together with *Cardano's depressed cubic equation solver* to solve quantum finance Schrödinger equation (QFSE) and calculate all quantum finance energy levels (QFELs); and hence the corresponding QPL using numerical computation method.

5.1   Quantum Price Levels (QPLs)

5.1.1   The Concept

What are quantum price levels?

As a direct analog to energy levels in an atom shown in Fig. 5.1, quantum finance energy levels (QFELs) can be considered as invisible energy levels that exist in every financial market; and quantum price levels (QPLs) can be interpreted as the realization of these financial energy levels shown in every secondary financial market.

They are similar to quantum energy levels in an atom; these quantum price levels exist intrinsic in nature as shown in Fig. 5.2.

In other words, they coexist with the continuance of financial particles in every financial market.

From the finance perspective such as forex market, which means when market opens every time, the quantum financial particles of various currency pairs (e.g.,

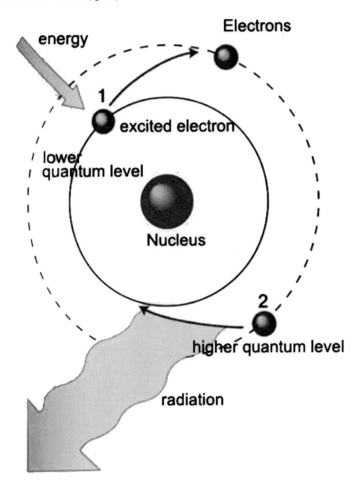

Fig. 5.1 Quantum energy levels in an atom

CADUSD, AUDEUR, JPYUSD) will automatically exist and generate their quantum energy field instantaneously, visualized as quantum price levels (QPLs) shown in every financial market and start their quantum finance anharmonic oscillations and motions.

More importantly, these QPLs exist in discrete states with level 0 as the *ground states* (E_0) (during market open) and all excited states E_n in discrete energy levels.

5.1.2 The Quantum Dynamics

In terms of quantum dynamics, these energy states are eigenenergy levels described in Schrödinger equation of quantum financial particles—the Quantum Finance

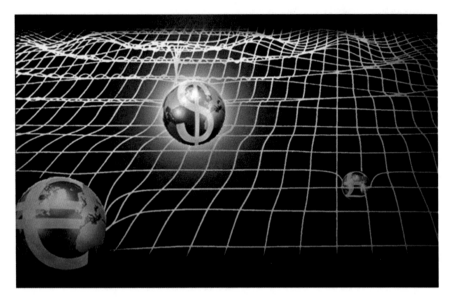

Fig. 5.2 Quantum price levels in quantum financial energy field

Schrödinger Equation (QFSE) derived in Chap. 4, which is given by the following order-4 quantum anharmonic oscillator (5.1):

$$\left[\frac{-\hbar}{2m}\frac{d^2}{dr^2} + \left(\frac{\gamma\eta\delta}{2}r^2 - \frac{\gamma\eta\upsilon}{4}r^4\right)\right]\varphi(r) = E\varphi(r) \tag{5.1}$$

Like an electron in an atom, without external force in action, the electron will remain in its stable ground-state *orbit*.

Similarly, without external financial stimulus such as financial events, major financial news, significant worldwide events, the release financial index figures such as PMI, CPI, PPI, etc., these financial particles will remain in their stable and equilibrium states (so-called *ground states*) and *vibrates* in the form of *random-like quantum anharmonic oscillations* as described by QFSE.

However, when there is a significant *financial stimulus* (either positive or negative events and stimulus), these financial particles will *jump* to another higher or lower energy levels (i.e., QPLs), not continuous, but in the form of discrete jumps.

After that, *market makers* (MMs) will *absorb* the stimulus and financial particles will be back to equilibrium states but reside at a new QPL.

5.1.3   QPLs Versus S & R Levels

Do such *quantum price levels* (QPLs) really exist?

Fig. 5.3 Support and resistance levels in technical analysis

The truth is … we don't know.

But if we ask any experienced financial analysts and traders, they will all tell us that financial markets are neither random motions nor continuous movements. Discrete and *quantized price levels* do exist in all financial markets, ranging from stock, forex, worldwide commodity, and even the latest cryptocurrency markets. That's why we have the concept of *support and resistance levels* (S & R levels) for over centuries in the financial realm.

In fact, the technical analysis developed since the eighteenth century is based on this fundamental concept and idea (Murphy 1999).

The difference is that, with the exponential growth of program trading, especially the popularity of high-frequency program trading (HFPT), the financial markets become so complex and chaotic that we don't believe one can observe (locate) all these *hidden energy levels* manually (Fig. 5.3).

5.2   Schrödinger Equation Revisit

As recalled in quantum dynamics (Zee 2011; Schmitz 2019), for a particle at position x, the time-dependent Schrödinger equation is given by

$$i\hbar \frac{\partial}{\partial t}\psi(x, t) = \widehat{H}\psi(x, t) \tag{5.2}$$

where $\psi(x, t)$ is the wave function, $\widehat{H}$ is the Hamiltonian operator, and $\hbar$ is the reduced Planck constant.

Note that the position x in Schrödinger equation is the displacement of quantum particle from its equilibrium states, rather than the displacement stated in the classical

Newtonian dynamics, whereas the corresponding time-independent counterpart is given by

$$\widehat{H}\varphi(x) = E\varphi(x) \tag{5.3}$$

where $\psi(x, t) = \varphi(x)e^{-iEt/\hbar}$.

Note that the time-independent Schrödinger equation is a standard eigenfunction where E is the eigenenergy with discrete energy levels and the Hamiltonian operator $\widehat{H}$ is given by

$$\widehat{H} = \frac{-\hbar}{2m}\frac{\partial^2}{\partial x^2} + V(x) \tag{5.4}$$

where $\widehat{H}$ composes of kinetic energy, K.E. (kinetic energy, the first term) + potential energy, P.E. (potential energy, the second term).

5.3 Quantum Finance Model

Let r be the price return of a particular *quantum financial particle* (*QFP*) at time t (say USD/CAD or simply US Index).

In typical secondary financial markets such as forex or stock market with over 2000-trading day historical records. Normally, we can use the daily returns to represent r in QFSE. In that case, the daily return r will be given by

$$r = \frac{Today\,Closing\,Price}{Yesterday\,Closing\,Price} \tag{5.5}$$

Of course, in case of real-time high-frequency program trading (HFPT), such return price r can use the hourly or 5-min returns as the basis of timeframe.

So, we can rewrite Schrödinger equation in (5.2) as

$$i\hbar\frac{\partial}{\partial t}\psi(r, t) = \widehat{H}\psi(r, t) \tag{5.6}$$

and the corresponding Hamiltonian operator $\widehat{H}$ is given by

$$\widehat{H} = \frac{-\hbar}{2m}\frac{\partial^2}{\partial r^2} + V(r, t) \tag{5.7}$$

Note:

• The Hamiltonian operator $\widehat{H}$ composes of K.E. and P.E.;

- $\hbar$ is the Planck constant representing the uncertainty of investors' behavior in the financial market;
- m is the mass representing the intrinsic potential of financial market (such as the overall market capitalization of secondary financial market).

5.4 Physical Meaning of Wave Function Ψ

In quantum mechanics and quantum field theory, wave function (ψ) is the most important component in the complete mathematical model, as it is the *realization* of wave–particle duality of quantum particles in this unique subatomic world of reality.

What is the corresponding wave function in quantum finance?

Can we observe it, or measure it?

Remember in the *double-slit experiment* of quantum particle, what we have seen in the backdrop screen is the *realization* of the wave function of quantum particle projections.

However, if we measured it *closely* at every single path of motion of these quantum particles, all the wave-like phenomena will disappear. Remember that?

In other words, if finance particles really possess such *quantum duality property*, technically speaking, we cannot observe or measure these discrete quantum energy levels (quantum price levels, QPLs, in our case) *intentionally and directly*. What we can observe is the *random-like* price fluctuations in market movement chart.

So, what should we do?

The answer is to

1. follow Professor Heisenberg's suggestion in his famous quote cited at the beginning of this chapter; and
2. do the experiment many times and observe wave functions' distribution.

When we studied classical quantum mechanics in Chap. 2, remember we talked about evaluating the quantum wave function of a quantum particle; all we can do is by measuring *pdf* (probability distributed function, ρ) of the observations instead, which is given by

$$\rho(x, t) = |\psi(x, t)|^2 = |\varphi(x)|^2 \tag{5.8}$$

In quantum finance, wave function can be observed and evaluated by using similar method.

For example, for a time series of 2048 trading days of Gold versus US Dollar (XAUUSD), we measure the daily closing price returns (r) of these 2048 time step and plot the pdf of r versus pdf of occurrence ϵ [0, 1] to analog the quantum finance wave function ψ of XAUUSD.

That is,

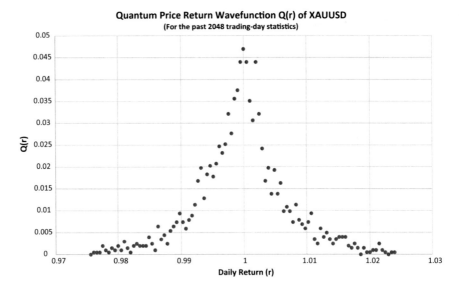

Fig. 5.4 Quantum price return wave function Q(r) of XAUUSD for the past 2048-trading day (As of January 6, 2019 market information)

$$\rho(r, t) = |\psi(r, t)|^2 = |\varphi(r)|^2 \tag{5.9}$$

Figure 5.4 shows the *quantum price return wave function* Q(r) of XAUUSD for the past 2048 trading days (as of January 6, 2019 market information).

It is calculated by evaluating the distribution function of *daily price returns* (r) and plot against the *total number of occurrences*, given by

$$r = \frac{no.\, of\, occurences\, of\, event\, r}{total\, no.\, of\, events\, E} \tag{5.10}$$

where E is the total number of events, 2046 events in our case (with the exclusion of the boundary cases).

Several features can be found:

1. The wave function falls into a standard normal distribution.
2. Maximum (also the mean) occurrence of r is close to 1, which is also the ground state (without any stimulus). Such finding is quite surprising in the financial perspective. As one may think of the returns with the most frequency events should be either greater than or less than 1, it won't be 1 which means *no change* from yesterday's price. But the truth is, after calculating all the 129 worldwide financial products, almost all of them have similar patterns, in which the maximum returns occur when r is close to (or equal to) 1.
3. Basically, the wave function is symmetric with the *central axis* which corresponds to the mean (which is also the maximum) wave function value that can be reflected

by the fact that quantum finance Schrödinger equation (QFSE) is symmetric with respect to the ground state. In other words, once we evaluate the *excited* states of quantum finance energy levels ($E_1, E_2, E_3, \ldots$) which corresponds to *positive QPL* (*the QPLs above the ground-state QPL when market open*), all *negative QPL* (*the QPL below the ground-state QPL*) can be deducted automatically, as we will see in the upcoming numerical calculations.

5.5 Solving Quantum Finance Schrödinger Equation

Quantum finance Schrödinger equation (QFSE) is given by

$$\left[\frac{-\hbar}{2m}\frac{d^2}{dr^2} + \left(\frac{\gamma\eta\delta}{2}r^2 - \frac{\gamma\eta\upsilon}{4}r^4\right)\right]\varphi(r) = E\varphi(r) \tag{5.11}$$

Once we have the method to evaluate $\varphi(r)$ (i.e., Q(r) in the last section), the center question is to find all corresponding quantum price levels (QPLs).

That is, all the eigenenergy values in QFSE.

Numerous physicists and mathematicians in the past 50 years had devised many methods and techniques to solve this important equation, such as the Hill determinant, Bargmann representation, coupled-cluster method, and variation–perturbation expansion (Grosse and Martin 2005; Muller-Kirsten 2012; Popelier 2011; Bouard and Hausenblas 2019; Rampho 2017; Kisil 2012; Witwit 1996; Rohwedder 2013; Carbonnière 2010).

However, most of them are either technically or mathematically complex in numerical computations.

In 2007, Dasgupta et al. in their paper *Simple systematics in the energy eigenvalues of quantum anharmonic oscillators* published in the *Journal of Physics A: Mathematical and Theoretical* provided a genius numerical method to solve a class of Schrödinger equations known as "λx^{2m} *quantum anharmonic oscillators.*"

In the next section, let's take a look at how we can adopt this numerical calculation method to solve QFSE within seconds by simple PC computers.

We will be surprised by how genius and simple numerical methods can be by changing the reference horizon.

5.6 Numerical Computation of Quantum Anharmonic Oscillators

A typical λx^{2m} quantum anharmonic oscillators (aka AHO) is given by (Dasgupta et al. 2007):

$$H^{(m)}(\lambda)\psi = -\frac{d^2\psi}{dx^2} + \left(x^2 + \lambda x^{2m}\right)\psi = E\psi \tag{5.12}$$

in which the excited energy levels can be closely approximated by the following polynomials:

$$\left(\frac{E^{(m,n)}}{2n+1}\right)^{(m+1)} - \left(\frac{E^{(m,n)}}{2n+1}\right)^{(m-1)} = \left(K_0^{(m,n)}\right)^{(m+1)}\lambda \tag{5.13}$$

where $E^{(m,n)}$ is the nth excited state energy of λx^{2m} AHO and $K_0^{(m,n)}$ are constants.

If we look QFSE closely, it is in fact a typical *quartic anharmonic oscillator* in Chap. 2 (Sect. 2.5.3) with a quartic term in P.E. dynamics.

So, we can convert QFSE (5.11) into a λx^{2m} AHO:

$$\frac{d^2\varphi_r}{dr^2} + \left(r^2 + \lambda r^{2m}\right)\varphi_r = E\varphi_r \tag{5.14}$$

Put m = 2, we have

$$\frac{d^2\varphi_r}{dr^2} + \left(r^2 + \lambda r^4\right)\varphi_r = E\varphi_r \tag{5.15}$$

Note: We normalize quadratic term r^2 with quartic term r^4 and combine it to coefficient λ. Note that this λ is different from the previous one in the original QFSE. Put in order to honor with Dasgupta's work, the author intends to keep it in here (and also in the book). Besides, K.E. is also normalized with coefficient set to 1, which is a usual practice in numerical derivation of Schrödinger equation as one may find out that we can discard K.E. during the course of evaluation of different energy levels.

Once we have QFSE in the form of (5.15), we can make use of numerical solution of quantum energy levels in (5.13) by setting m = 2 and further simplify the equation into

$$\left(\frac{E(n)}{2n+1}\right)^3 - \left(\frac{E(n)}{2n+1}\right) = (K_0(n))^3\lambda \text{ or}$$

$$\left(\frac{E(n)}{2n+1}\right)^3 - \left(\frac{E(n)}{2n+1}\right) - (K_0(n))^3\lambda = 0 \tag{5.16}$$

where

$$K_0(n) = \left[\frac{1.1924 + 33.2383n + 56.2169n^2}{1 + 43.6196n}\right]^{1/3} \tag{5.17}$$

Note:

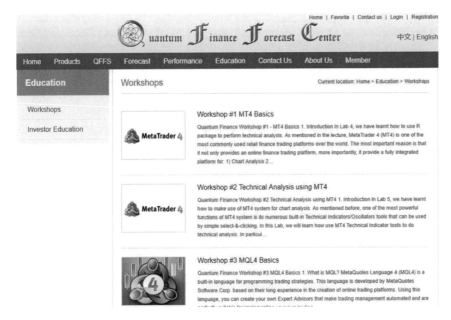

Fig. 5.5 Education corner of QFFC.org (QFFC 2019)

The numerical Eqs. (5.16) and (5.17) can be easily calculated by using MATLAB (Chapra 2017; Marghitu and Dupac 2012) and any other computational packages, and of course, using MQL (Young 2015) and integrated with MT (metatrader) platforms—the biggest international online trading and program development platform for real-world financial applications. Details can be found from the *education corner* of *quantum finance forecast center* official site, QFFC.org.

In summary, once we know coefficient λ, all the quantum energy levels (or quantum price levels) can be found (Fig. 5.5).

The question is: How can we find λ?

5.7 Evaluate Quantum Price Levels Using Numerical Computation Technique

In order to find λ, we have to revisit QFSE (5.15), given by

$$\frac{d^2\varphi_r}{dr^2} + \left(r^2 + \lambda r^4\right)\varphi_r = E\varphi_r \qquad (5.15)$$

As mentioned, this quantum finance Schrödinger equation (QFSE) has four characteristics:

1. QFSE is a complex quantum anharmonic oscillator with the combination of harmonic term $(E\varphi_r)$ + anharmonic term (i.e., $(r^2 + \lambda r^4)\varphi_r$)—also known as *perturbation term*.
2. It consists of two parts: K.E. part (the first term, kinetic energy) and P.E. part (the second term, potential energy).
3. The quantum finance wave function φ_r. If we observe (measure) it with sufficient samples (say over 2000 samples, e.g., 2000 trading days), it will become a typical normal distribution with the maximum value appears at the mean, which is also the stable (ground) state with K.E. $= 0$.
4. The quantum finance wave function is symmetric with the ground state as the central axis when the samples are sufficiently large.

Using computer to solve this complex equation, we can apply numerical technique with finite difference method (FDM).

5.8 Finite Difference Method (FDM)

With the advance of computer technology, finite difference method (aka FDM) provides an intuitive practical method to solve complex differential and high-order equations by numerical approximation and implemented by computer algorithms (Dimov et al. 2004; Duffy 2006; Tavella and Randdall 2000).

Although the technique is using numerical approximation, when the number of samplings/observations is sufficiently large, the finite step (Δx) is sufficiently small (as compared with sampling/observation horizon x). Such method can give a very close approximation to the "actual" analytical solution.

In fact, FDM is commonly used in many large-scale real-time problems such as worldwide weather prediction (also known as *numerical weather prediction*), earthquake modeling, traffic control systems, and naturally worldwide financial analysis.

The finite difference method is a threefold process:

1. For any function $y = Q(x)$, subdivide the x-axis into N subdivisions, i.e., x_1, x_2, x_3, ... x_N.
2. Set the value of y at x_i as y_i.
3. The finite difference formulations for all differential operators will be

$$dx = \Delta x \tag{5.18a}$$

$$\frac{dy}{dx}(x_i) = \frac{y_{i+1} - y_i}{\Delta x} \tag{5.18b}$$

$$\frac{d^2 y}{dx^2}(x_i) = \frac{y_{i+1} + y_{i-1} - 2y_i}{\Delta x^2} \tag{5.18c}$$

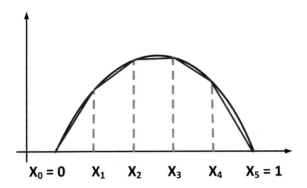

Fig. 5.6 Basic concept of FDM

Let's take a look at how we can apply FDM to solve our QFSE. Figure 5.6 illustrates the basic idea of FDM.

5.9 Finite Difference Method (FDM) to Evaluate QF Wave Function

Figure 5.7 illustrates the quantum price return wave function Q(r) (φ_r in our QFSE) mentioned in Sect. 5.4, that is, the wave function distribution statistic of XAUUSD by plotting the pdf of daily returns in the past 2048 trading days.

But the difference is that this time we display this pdf function by dividing the x-axis (r) into 100 equal divisions, with each width Δx given by

$$\Delta x = \frac{3\sigma}{50} \tag{5.19}$$

where σ is the standard deviation of r for the past 2048 trading days (totally, we have 2046 r sample observations by excluding the boundary records).

This figure also shows the regression curve of the wave function for illustration purpose.

As mentioned, when the number of observations/sampling is sufficiently large, all four characteristics of QFSE mentioned in Sect. 5.7 hold.

Besides, certain important findings can be concluded from Fig. 5.7:

1. φ_{Max} at r $\cong$ 1 (ground state, denotes as r_0).
2. $\varphi_r = \varphi(r_0)$ is symmetric with $r \cong 1$ as symmetry axis, especially when r-segment close to the symmetry axis.
3. So, we can take the first left and right r-segment for calculation, denoted as r_{-1} and r_{+1}, respectively.

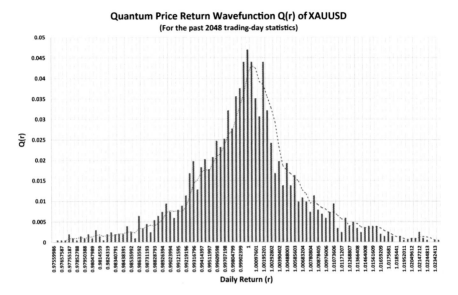

Fig. 5.7 Quantum price return wave function Q(r) of XAUUSD

Figure 5.8 shows three major r-segments in the QF wave function of r; they are $r_0, r_{-1}, and \ r_{+1}$.

It also corresponds to ground state $\varphi(r_0)$ and first +ve and −ve r states, $\varphi(r_{-1})$ and $\varphi(r_{+1})$, respectively.

According to QFSE (5.15), the wave function is symmetric with r_0 as the symmetry axis and $\varphi(r_0)$ is the max wave function values.

Since we have all 2046 r observations and also their distribution, i.e., $\varphi(r\text{'s})$, technically we have sufficient information to evaluate λ for every financial product.

To evaluate λ using finite difference method, we have two options:

1. Using FDM to evaluate the boundary continuously of r_0 and r_{+1} segments (i.e., The rightmost approximation of $\varphi(r_0) \cong$ leftmost approximate of $\varphi(r_{+1})$).
2. Using FDM and symmetric property of φ_r, we make use of the symmetric property of r_{-1}-segment and r_{+1}-segment to "cancel out" the K.E. term, and then using FDM to evaluate λ using the symmetric property of their P.E. component.

   Which option we could choose? Why?

5.10 Numerical Evaluation of $\lambda|_{XAUUSD}$ Using FDM

For illustration purpose, we will continue to use XAUUSD as an example to demonstrate how to evaluate λ using FDM.

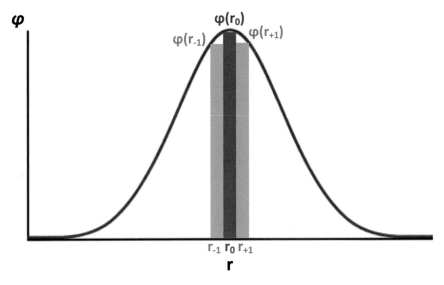

Fig. 5.8 Illustration of FDM calculation in quantum finance

For the 2048 trading days of XAUDUSD (as of January 16, 2019), we have the following statistics information in Table 5.1.

As recalled from the QFSE, we have

$$\frac{d^2\varphi_r}{dr^2} + \left(r^2 + \lambda r^4\right)\varphi_r = E\varphi_r \tag{5.15}$$

Since QFSE (5.15) is symmetric with respect to the central axis r_0, when we consider quantum dynamics for r_{+1} and r_{-1} segments, their K.E. terms can be canceled out, so we have

$$\left(r_{+1}^2 + \lambda r_{+1}^4\right)\varphi_{r_{+1}} = \left(r_{-1}^2 + \lambda r_{-1}^4\right)\varphi_{r_{-1}} \text{ or}$$

Table 5.1 Statistic results of the quantum price return wave function Q(r) of XAUUSD (as of January 16, 2019)

Product: XAUUSD		
No. of r = 2046	$r_0 = 0.999604$	$\varphi(r_0) = 0.047785$
$\Delta r = 0.000793$	$r_{+1} = 1.000396$	$\varphi(r_{+1}) = 0.038825$
$Max(\varphi) = 0.047785$	$r_{-1} = 0.998811$	$\varphi(r_{-1}) = 0.039821$
$Max(\varphi)_{No} = 50$	$\mu = 0.999821$	$\sigma = 0.013213$

Data source Forex.com MT4 system

$$\lambda = \left| \frac{r_{-1}^2 \varphi_{r-1} - r_{+1}^2 \varphi_{r+1}}{r_{+1}^4 \varphi_{r+1} - r_{-1}^4 \varphi_{r-1}} \right| \tag{5.20}$$

For XAUUSD, after calculation, we have $\lambda = 1.16813758$.

Table 5.2 shows λ values for all 120 forex products using MQL program (One can visit QFFC.org *education corner* for detailed implementation procedure).

5.11 Numerical Computation of Quantum Energy Levels E_n

Once we have λ, we can use (5.16)–(5.17) to evaluate all energy levels E_n.

$$\left(\frac{E(n)}{2n+1} \right)^3 - \left(\frac{E(n)}{2n+1} \right) - (K_0(n))^3 \lambda = 0 \tag{5.21}$$

$$K_0(n) = \left[\frac{1.1924 + 33.2383n + 56.2169n^2}{1 + 43.6196n} \right]^{1/3} \tag{5.22}$$

Note that (5.16) is a typical cubic polynomial which can be easily solved by MATLAB using "root" command.

For XAUUSD, by using $\lambda = 1.16813758$, we can write a simple MATLAB program (in the next section) to calculate all first 21 energy levels. Table 5.3 shows the experimental results for the calculation of the first 21 quantum finance energy levels (QFELs) of XAUUSD.

5.12 Numerical Computation of Quantum Finance Energy Levels (MATLAB Version)

In this MATLAB program shown in Fig. 5.9, we use λ value 1.16813758 to calculate the first 21 K and quantum finance energy levels (QFELs) of XAUUSD.

To solve the cubic equation, MATLAB command *roots( )* is used to evaluate the solution set. Since (5.21) is a typical cubic equation, normally it will have solution set of three values, with one real number and two complex solutions.

So, we set the real number solution as QFEL value.

Table 5.2 λ values for ALL 120 forex products using MQL program

CODE	λ values	CODE	λ values	CODE	λ values
XAGUSD	1.16813758	US2000	1.01691648	GBPDKK	0.50015095
CORN	0.98147439	AUDCAD	0.99800233	GBPHKD	0.49946476
US30	1.00927814	AUDCHF	1.00650666	GBPJPY	1.02721719
AUDUSD	1.01090471	AUDCNH	0.99788161	GBPMXN	1.00743969
EURCHF	0.9922947	AUDJPY	1.01297607	GBPNOK	0.97866528
GBPCAD	0.98033867	AUDNOK	1.01297576	GBPNZD	1.01766392
NZDJPY	0.99409385	AUDNZD	0.99883417	GBPPLN	0.98982647
USDCNH	1.00129406	AUDPLN	0.99972703	GBPSEK	1.00541074
XAUAUD	0.97310053	AUDSGD	0.99145652	GBPSGD	0.99543469
XAUCHF	1.28307613	CADCHF	1.05615292	GBPUSD	0.99737283
XAUEUR	1.03339416	CADJPY	0.97655725	GBPZAR	0.99672306
XAUGBP	1.10858157	CADNOK	0.9981341	HKDJPY	1.01256568
XAUJPY	1.20798503	CADPLN	1.02762915	NOKDKK	1.00723481
XAUUSD	0.87114449	CHFHUF	0.99232627	NOKJPY	1.00878002
COPPER	0.98546677	CHFJPY	0.94512371	NOKSEK	0.99266368
PALLAD	0.97495035	CHFNOK	1.00053241	NZDCAD	1.0105619
PLAT	0.93898709	CHFPLN	1.00582673	NZDCHF	0.97178881
UK_OIL	1.01219635	CNHJPY	1.00000253	NZDUSD	1.00708451
US_OIL	1.07644811	EURAUD	1.00721165	SGDHKD	1.0028679
US_NATG	1.76511177	EURCAD	0.96576866	SGDJPY	0.95994849
HTG_OIL	0.90630263	EURCNH	1.01233192	TRYJPY	0.5018959
COTTON	1.02930805	EURCZK	0.99233097	USDCAD	1.00299693
SOYBEAN	0.50226883	EURDKK	0.99994162	USDCHF	0.96609929
SUGAR	0.99525331	EURGBP	0.99797319	USDCZK	0.99456678
WHEAT	0.99615377	EURHKD	1.00358691	USDDKK	0.99719426
IT40	1.01850019	EURHUF	0.98265533	USDHKD	1.00178794
AUS200	0.99426146	EURJPY	0.91939902	USDHUF	1.01153898
CHINAA50	0.9806911	EURMXN	1.02025986	USDILS	1.0047121
ESP35	0.93834053	EURNOK	0.99508525	USDJPY	0.50079764
ESTX50	1.00351004	EURNZD	0.50156959	USDMXN	0.99266275
FRA40	1.00704187	EURPLN	1.06863464	USDNOK	0.9984592
GER30	1.03777101	EURRON	0.99952845	USDPLN	1.01260473
HK50	0.99188819	EURRUB	0.99533066	USDRON	1.00335247
JPN225	0.9884408	EURSEK	1.03002348	USDRUB	0.98921247
N25	0.98915404	EURSGD	1.00701412	USDSEK	1.0196364
NAS100	0.99279678	EURTRY	1.01094015	USDSGD	1.00527642

(continued)

Table 5.2 (continued)

CODE	λ values	CODE	λ values	CODE	λ values
SIGI	1.0158226	EURUSD	1.01141223	USDTHB	0.9990309
SPX500	1.00436699	EURZAR	1.04497648	USDTRY	1.02358136
SWISS20	1.00564252	GBPAUD	1.0092642	USDZAR	0.9828679
UK100	0.98794556	GBPCHF	0.99417175	ZARJPY	1.08799327

Source Experimental results using MQL program, 2019

Table 5.3 K values and QFEL values of the first 21 quantum finance energy levels

Product: XAUUSD ($\lambda = 1.16813758$)

Energy level	K	QFEL
0	1.060410426	1.409932766
1	1.266594551	4.744287679
2	1.491211949	8.908118719
3	1.663522514	13.59094957
4	1.806129863	18.6925368
5	1.929228428	24.15086474
6	2.038364753	29.92294434
7	2.136927359	35.97686567
8	2.227155031	42.28781818
9	2.310613024	48.8358504
10	2.388443595	55.60450183
11	2.4615088	62.57991521
12	2.530477086	69.75023292
13	2.595878459	77.10517084
14	2.658141083	84.635708
15	2.717616385	92.33385484
16	2.77459678	100.1924762
17	2.829328496	108.2051536
18	2.882021043	116.3660765
19	2.932854345	124.6699548
20	2.981984198	133.1119479

Source Experimental results using MATLAB development tool, 2019

```
% *****************************************************************************************
%
%  Date: 5.1.2019
%  Subject: Calculation of Quantum Price Level using QFSE
%  Author:  Dr. Raymond LEE
%  Version; 1.0
% *****************************************************************************************
% Define parameters
  Maxj = 10000;       % Max no. of iteration (approx of infinity)
  MaxN = 21;          % Max Quantum Price Energy Levels
  l   = 1.16813758;   % Lambda value for the QFSE of XAUUSD
  oldE = 0.0;         % Aux E variable
%
% Define and initialize K and Energy Level arrays
%
  E  = zeros(3,MaxN);  % Energy levels array
  K  = zeros(1,MaxN);  % K parameter array

% Calculate ALL K2 parameters
  for n=1:1:MaxN
   K(n) = ((1.1924+33.232383*(n-1)+56.22169*(n-1)*(n-1))/(1+43.6106*(n-1)))^(1/3);
  end

% Calculate ALL Energy Levels
  for n=1:1:MaxN
   a = (1/(2*(n-1)+1))^3;
   b = 0;
   c = -1/(2*(n-1)+1);
   d = -1*l*(K(n)^3);
   E(:,n) = roots([a,b,c,d]);
  end
```

Fig. 5.9 MATLAB program for the numerical computation of QFEL for XAUUSD

5.13 Depressed Cubic Equation Solver—Gerolamo Cardano

Actually, the MATLAB version is just for illustration.

In real-time situation, we will implement quantum energy level calculation using MQL.

Why?

The answer is simple, so that we can fully automate the time series financial data extraction process with statistic calculation of quantum finance wave function and calculate all quantum finance energy levels in a single MT program.

The only problem is: Without the fantastic cubic solver like *roots( )* in MATLAB, how can we solve the cubic (5.21) in MQL?

The solution is … Cardano's method.

Gerolamo Cardano (1501–1576) was a close friend of Leonardo da Vinci, an Italian mathematician, physician, biologist, physicist, chemist, astrologer, astronomer, philosopher, writer, and gambler.

He was one of the most influential mathematicians of the Renaissance and was one of the key figures in the foundation of probability as well as the earliest introducer of binomial coefficients and binomial theorem in the western world.

Cardano was the first mathematician to make systematic use of negative numbers.

One of his most influential discoveries is the solution to one particular case of the cubic equation, so-called "depressed cubic equation" (cubic equation without quadratic term):

$$ax^3 + bx + c = 0 \tag{5.23}$$

5.14   Cardano's Method for Calculating QFEL

Published in his book *Ars Magna* in 1545, Cardano devised a genius way to solve the *depressed cubic equation* as follows:

Given a depressed cubic equation:

$$t^3 + pt + q = 0 \tag{5.24}$$

Cardano introduced variables u and v such that

$$u + v = t \tag{5.25}$$

With additional condition,

$$3uv + p = 0 \tag{5.26}$$

By simple substitution of (5.19) into (5.18a, 5.18b, 5.18c), we have

$$u^3 + v^3 + (3uv + p)(u + v) + q = 0 \tag{5.27}$$

Using (5.26), we have

$$u^3 + v^3 = -q \text{ and} \tag{5.28a}$$

$$u^3 v^3 = -\frac{p^3}{27} \tag{5.28b}$$

The tricky point is that the combination of (5.28a and 5.28b) will lead to the following quadratic equation where u^3 *and* v^3 are the roots of the equation:

$$z^2 + qz - \frac{p^3}{27} = 0 \tag{5.29}$$

By solving (5.29), we can find u^3 *and* v^3:

$$u^3 = -\frac{q}{2} + \sqrt{\frac{q^2}{4} + \frac{p^3}{27}}, \quad v^3 = -\frac{q}{2} - \sqrt{\frac{q^2}{4} + \frac{p^3}{27}} \tag{5.30}$$

Finally, we have

$$t = \sqrt[3]{-\frac{q}{2} + \sqrt{\frac{q^2}{4} + \frac{p^3}{27}}} + \sqrt[3]{-\frac{q}{2} - \sqrt{\frac{q^2}{4} + \frac{p^3}{27}}} \tag{5.31}$$

At the time of Cardano, there is no concept of complex number. As we now know, the solution of this cubic equation will normally has three roots, one real and two symmetric complex roots. In our case, we take the first real root as solution, which can be done easily by any programming platform, MQL naturally.

5.15 Numerical Computation of Quantum Energy Levels (MQL Version)

By substituting (5.21) with Cardano's formula,

$$\left(\frac{E(n)}{2n+1}\right)^3 - \left(\frac{E(n)}{2n+1}\right) - (K_0(n))^3 \lambda = 0 \tag{5.21}$$

Finally, we have

$$E(n) = \sqrt[3]{-\frac{q}{2} + \sqrt{\frac{q^2}{4} + \frac{p^3}{27}}} + \sqrt[3]{-\frac{q}{2} - \sqrt{\frac{q^2}{4} + \frac{p^3}{27}}} \tag{5.31}$$

where

$$p = -(2n+1)^2 \tag{5.32a}$$

$$q = -\lambda(2n+1)^3[K_0(n)]^3 \text{ and} \tag{5.32b}$$

$$K_0(n) = \left[\frac{1.1924 + 33.2383n + 56.2169n^2}{1 + 43.6196n}\right]^{1/3} \tag{5.32c}$$

Figure 5.10 shows the program section in MQL for the calculation of first 21 energy levels of a financial product (with given λ determined in the previous section).

Once all QF energy levels are calculated, the determination of the quantum price levels will be a straightforward scaling problem.

```
//********************************************************************************************************
// 6. Using Quantum Finance Schrödinger EQT to find the first 21 Quantum Energy Levels
//
//   By solving the Quartic Anharmonic Oscillator as cubic polynomial equation of the form
//      a*x^3 + b*x^2 + c*x + d = 0
//
//   Using (Dasqupta et. al. 2007) QAHO solving equation:
//   (E(n)/(2n+1))^3 - (E(n)/(2n+1)) - K(n)^3*L = 0
//
//   and  Solving the above Depressed Cubic Eqt using Cardano's Method
//********************************************************************************************************

for (eL=0;eL<21;eL++)
{
  p = -1 * pow((2*eL+1),2);
  q = -1 * L * pow((2*eL+1),3) * pow(K[eL],3);

  // Apply Cardano's Method to find the real root of the depressed cubic equation
  u = MathPow((-0.5*q + MathSqrt(((q*q/4.0) + (p*p*p/27.0)))),p3);
  v = MathPow((-0.5*q - MathSqrt(((q*q/4.0) + (p*p*p/27.0)))),p3);

  // Store the QPE
  QPE[eL] = u + v;

  // Printout the QF Energy Levels
  Print("TP",nTP+1," ",TP_Code[nTP]," Energy Level",eL," QPE = ",QPE[eL]);
  FileWrite(QPLog_FileHandle," Energy Level",eL," QPE = ",QPE[eL]);
}
```

Fig. 5.10 Program section in MQL for the calculation of first 21 energy levels of a financial product

Figure 5.11 shows the overall algorithm for the determination of first 21 quantum finance energy levels (QFELs) and hence the quantum price levels for 120 financial products using MQL.

For details, please visit the programming workshop at the *Education Corner* of QFFC.org.

5.16   Numerical Algorithm to Calculate QPL Using MQL

For each financial product, do the following:

- Read the daily time series and extract (Date, O, H, L, C, V)
- Calculate dally price return r(t)
- Calculate quantum price return wave function Q(r) (size 100)
- Evaluate λ value for the wave function Q(r) using FDM and Eq. (5.21) and evaluate other related parameters:

  – sigma (std dev of Q)

Fig. 5.11 Flowchart for the determination of the first 21 QFELs and QPLs

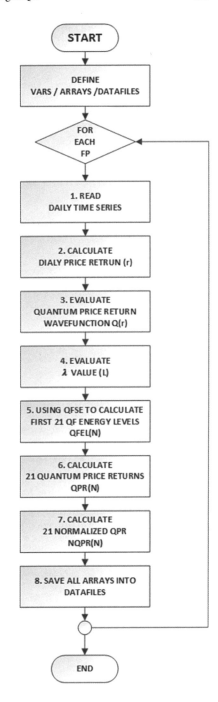

  – maxQPR (max quantum price return—for normalization)

- Once λ is found, using Quantum Finance Schrodinger Equation (numerical solution) by solving the depressed cubic equation using Cardano's method (5.32a–5.32c) to calculate first 21 quantum finance energy levels, QFEL(n), n = [1 ... 20]
- Calculate quantum price return, QPR(n)

$$p = -(2n + 1)^2 \tag{5.33}$$

$$QPR(n) = \frac{QFEL(n)}{QFEL(0)} \tag{5.34}$$

where n = [1 ... 20]
- Calculate normalized QPR(n)

$$NQPR(n) = 1 + 0.21 * sigma * QPR(n) \tag{5.35}$$

where n = [1 ... 20]
- Save two level of data files:

  – For each financial product, save the QPL table contains QPE, QPR, and NQPR for the first 21 energy levels
  – For all financial product, create a QPL summary table containing NQPR for all FP, which will be used for financial prediction using recurrent neural networks.

5.17 QPLs for XAUUSD

Table 5.4 shows QPE, QPR, and NQPR for the first 21 energy levels of XAUUSD by using the 2048 daily time series data from Forex.com.

According to quantum finance theory and the symmetric property of QFSE, at the beginning of each trading day, the first 21 QPL$_+$ is calculated by

$$QPL_0 = P_{Open} * NQPR(0) \tag{5.30a}$$

$$QPL_{+n} = P_{Open} * NQPR(n), \quad n = [1 \ ... \ 20] \tag{5.30b}$$

$$QPL_{-n} = P_{Open}/NQPR(n), \quad n = [1 \ ... \ 20] \tag{5.30c}$$

In real application, every day at 08:00 HKT/UTC0, QFFC will calculate the forecast H/L for worldwide 129 financial products, together with daily eight closest QPLs for each FP, upload onto QFFC official site for public access.

Table 5.4 QPE, QPR, and NQPR for the first 21 energy levels of XAUUSD by using the 2048 daily time series data from Forex.com

Product: XAUUSD ($\lambda = 1.16813758$)			
Energy level	QPE	QPR	NQPR
0	1.40993277	1	1.00277473
1	4.7443013	3.36491314	1.00933673
2	8.90806181	6.3180756	1.01753097
3	13.590797	9.63932275	1.02674654
4	18.69227098	13.25756193	1.03678619
5	24.15047183	17.12881096	1.04752787
6	29.9224128	21.22258132	1.05888698
7	35.97618549	25.51624187	1.07080074
8	42.28698048	29.99219642	1.08322032
9	48.83484717	34.63629495	1.09610645
10	55.60332572	39.43686325	1.10942674
11	62.57855942	44.38407341	1.12315392
12	69.74869114	49.4695157	1.13726466
13	77.10343711	54.68589633	1.15173872
14	84.63377674	60.0268174	1.16655835
15	92.33172074	65.48661249	1.18170782
16	100.1901342	71.06022114	1.19717309
17	108.2025989	76.74309121	1.21294154
18	116.3633045	82.53110164	1.22900172
19	124.666961	88.42050054	1.24534322
20	133.108728	94.40785487	1.26195653

Source Experimental results using MQL for implementation, 2019

5.18 Conclusion

In this chapter, we studied the basic theory of quantum price level (QPL).

More crucially, this chapter shows how to evaluate the quantum finance energy level (and hence the QPL) by solving quantum finance Schrödinger equation (QFSE) using simple numerical computation methods.

Prior to this book, quantum finance and path integral analysis seem to be a mathematically complex and far-reaching techniques for data scientists, quants, and financial analysts to comprehend; let's alone with the actual calculation/ program design of QPL.

But now, quantum finance and hence QPL can be easily and practically applied to real-world financial situation by using common PC systems.

Any quants and analysts can now use their own PC/MAC systems to program and calculate any QPL, and QPL they want to integrate into their own trading systems.

In fact, quantum price level (QPL) evaluation is only the first step of quantum finance.

More important is how to make use of it, together with other AI tools to design and implement real-time financial prediction and intelligent trading systems.

Problems

5.1 Why experiments and experimental results are key factors in quantum theory? And how can we do that in quantum finance? Give two examples in financial markets in your explanation.

5.2 What is *pseudoscience*? What is the major difference between *science* and *pseudoscience*? Give two examples of pseudoscience examples in finance and explain why they are not scientific methods.

5.3 What are the three major similarities and differences between quantum energy levels (QELs) in quantum theory versus quantum price level (QPL) in quantum finance?

5.4 What are *support and resistance levels* (S & R) in technical analysis? Why they are important in modern finance and financial trading?

5.5 Discuss and explain the similarities and differences between quantum price levels (QPLs) and *support and resistance levels* (S & R). Give two examples in financial markets in your explanation.

5.6 What is the physical meaning of wave function (ψ) in quantum finance? And how wave function can be interpreted by double-slit experiment?

5.7 What is a λx^{2m} quantum anharmonic oscillator? What is the major difference between a typical quantum anharmonic oscillator and a λx^{2m} quantum anharmonic oscillator? What is the mathematical significance of λx^{2m} quantum anharmonic oscillator?

5.8 State and explain the physical meaning of each term in the mathematical formulation of excited energy levels in a λx^{2m} quantum anharmonic oscillator.

5.9 State and explain the mathematical derivation of the quantum finance Schrödinger equation (QFSE) with the integration of λx^{2m} quantum anharmonic oscillator formulation. Explain how it can be used to evaluate quantum price levels (QPLs) in quantum finance.

5.10 How can we evaluate the wave function (ψ) of any financial product in quantum finance? The following figure shows the wave function distribution of Gold (XAUUSD) using the past 2048-trading day time series.

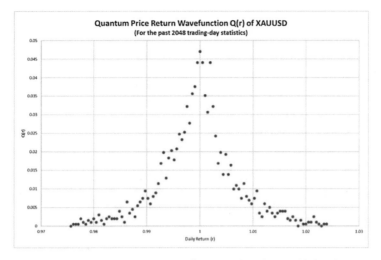

(i)   Using MQL (tutorials can be found in QFFC.org official site) or R to plot wave function distribution of (1) CORN; (2) Dow Jones Index (DJI) and crude oil (US Oil);

(ii) Compare these three wave function distribution charts with XAUUSD, what are the major similarities and difference between these four distribution charts? Why?

5.11 Given the following experiment results for the wave function distribution of Silver (XAGUSD) in 2046-trading days:

No. of r = 2046	$r_0 = 0.999604$	$\varphi(r_0) = 0.047785$
$\Delta r = 0.000793$	$r_{+1} = 1.000396$	$\varphi(r_{+1}) = 0.038825$
$Max(\varphi) = 0.047785$	$r_{-1} = 0.998811$	$\varphi(r_{-1}) = 0.039821$
$Max(\varphi)_{No} = 50$	$\mu = 0.999821$	$\sigma = 0.013213$

Calculate λ term for XAGUSD.

5.12 Write the flowchart to show the steps and processes to calculate the 20 closest quantum price levels (QPLs) of XAGUSD using MT4 system with 2000 past trading day time series.

5.13 Based on the λ term calculated for XAGUSD in Question 5.11, and the numerical formulation of QAHO given by

$$\left(\frac{E(n)}{2n+1}\right)^3 - \left(\frac{E(n)}{2n+1}\right) - (K_0(n))^3 \lambda = 0$$

where $K_0(n) = \left[\frac{1.1924+33.2383n+56.2169n^2}{1+43.6196n}\right]^{1/3}$

And today opening price for XAGUSD is 14.31, calculate the 21 closest *quantum price levels (QPLs)* for XAGUSD today.

5.14 Below given the quantum finance Schrödinger equation in QAHO format.

$$\frac{d^2\varphi_r}{dr^2} + \left(r^2 + \lambda r^4\right)\varphi_r = E\varphi_r$$

(i)   State and explain the physical meanings of each term in terms of quantum finance theory;

(ii) What is importance of the anharmonic term in the QFSE?

(iii) Discuss and explain the major characteristics of QFSE in QAHO format;

(iv) What is the importance of QFSE in QAHO format in terms of (1) modeling of quantum financial markets; (2) computational and implementation perspective for the evaluation of QPL?

5.15 What is *finite difference method* (FDM)? How it works?

5.16 State and explain the basic assumptions and formulations of finite difference method.

5.17 Discuss and explain how to apply finite difference method (FDM) for the evaluation of quantum finance wave function.

5.18 Based on the wave function distributions evaluated in 5.10 for CORN, DJI, and US Oil:

(i)   Choose the best *dr* to evaluate their λ values.

(ii) How to choose the best *dr* interval? Why? and What is the importance?

5.19 What is a *depressed cubic equation*? Discuss and explain Cardano's technique for solving *depressed cubic equation*.

5.20 By solving the *depressed cubic equation* to evaluate QFEL, normally three roots will be evaluated with two of them are imaginary roots.

(i)   What is the physical meaning of these three roots of QFEL in terms of quantum theory and quantum finance?

(ii) Is it possible to have three roots are all real numbers? Why? and what is the physical meaning in terms of quantum finance?

(iii) Which root we should take as the solution of QFEL? Why?

5.21 MQL programming assignment.
Study the MQL tutorial from QFFC.org official site. Write an MQL program to

(i)   Read the previous 2000-trading day time series of CORN, DJI, and US Oil.

(ii) Plot their wave function distribution charts.

(iii) Evaluate λ values of their wave function distribution.

(iv) Implement Cardano's technique to evaluate QFELs and QPRs of these products.

(v)   Using the opening price of these financial products, modify your MQL to calculate the first 41 quantum price levels (i.e., QPL_{-20}, QPL_{-19}, ..., QPL_0, ..., QPL_{+19}, QPL_{+20}).

(vi)   State and explain how you can test the accuracy and performance of using these QPLs to replace *support and resistance levels* (S & R) for financial analysis.

(vii) Modify your MQL to implement (vi) for the performance evaluation of these QPLs for CORN, DJI, and US Oil and compare their overall performance.

(viii) Perform the same procedure as (vii) to write an MQL program to calculate and evaluate the performance of QPLs for ALL forex products in your MT4 platform. Check for any patterns and characteristics you can find out and explain why.

References

Bouard, A. & Hausenblas, E. (2019) The nonlinear Schrödinger equation driven by jump processes. *Journal of Mathematical Analysis and Applications* 475 (1): 215–252.

Carbonnière, P. et al. (2010) The VCI-P code: an iterative variation–perturbation scheme for efficient computations of anharmonic vibrational levels and IR intensities of polyatomic molecules. *Theoretical Chemistry Accounts* 125(3): 543–554.

Chapra, S. C. (2017) Applied Numerical Methods with MATLAB for Engineers and Scientists. McGraw-Hill Education, 4th edition.

Dasgupta, A. et al. (2007) Simple systematics in the energy eigenvalues of quantum anharmonic oscillators. *Journal of Physics A: Mathematical and Theoretical* 40(4): 773–784.

Dimov, I. et al. (2014) *Finite difference methods, theory and applications*. Lecture Notes in Computer Science book series (LNCS, volume 9045), Springer.

Duffy, D. J. (2006) Finite Difference Methods in Financial Engineering: A Partial Differential Equation Approach. Wiley.

Grosse, H. and Martin, A. (2005) *Particle Physics and the Schrödinger Equation (Cambridge Monographs on Particle Physics, Nuclear Physics and Cosmology)*. Cambridge University Press.

Kisil, V. V. (2012) Hypercomplex Representations of the Heisenberg Group and Mechanic. *International Journal of Theoretical Physics* 51(3): 964–984.

Marghitu, D. B. and Dupac, M. (2012) Advanced Dynamics: Analytical and Numerical Calculations with MATLA. Springer.

Muller-Kirsten, H. J. W. (2012) *Introduction to Quantum Mechanics: Schrödinger Equation and Path Integral*. World Scientific.

Murphy, J. J. (1999) *Technical Analysis of the Financial Markets: A Comprehensive Guide to Trading Methods and Applications*. New York Institute of Finance.

Popelier, P. (2011) *Solving the schrodinger equation: has everything been tried?* World Scientific Pub Co Pte.

QFFC (2019) Official site of Quantum Finance Forecast Center. http://qffc.org. Accessed 21 Aug 2019.

Rampho, G. J. (2017) The Schrödinger equation on a Lagrange mesh. *Journal of Physics: Conference Series* 905: 12037.

Rohwedder, T. (2013) The continuous Coupled Cluster formulation for the electronic Schrödinger equation. *ESAIM: Mathematical Modelling and Numerical Analysis* 47(2): 421–447.

Schmitz, W. (2019) *Particles, Fields and Forces: A Conceptual Guide to Quantum Field Theory and the Standard Model (The Frontiers Collection)*. Springer.

Tavella, D. & Randall, C. (2000) *Pricing financial instruments: the finite difference method*. John Wiley & Sons, New York.

Witwit, M. R. M. (1996) Energy Levels for Nonsymmetric Double-Well Potentials in Several Dimensions: Hill Determinant Approach. *Journal of Computational Physics*, 123(2): 369–378.

Young, A. R. (2015) Expert Advisor Programming for MetaTrader 4: Creating automated trading systems in the MQL4 language. Edgehill Publishing.

Zee, A. (2011) *Quantum Field Theory in a Nutshell*. Princeton University Press, 2nd edition.

Chapter 6
Quantum Trading and Hedging Strategy

> *Self-discipline is what separates the winners and the losers.*
> Thomas Peterffy (born at 1944)

What are financial trading and hedging strategies?
What is quantum trading?
What is trading discipline?

Anyone working in program trading may be familiar with Mr. Thomas Peterffy. Mr. Peterffy was born in 1944, a Hungarian-born American billionaire businessman. He was the founder, chairman, CEO, and the largest shareholder of Interactive Brokers. He emigrated to United States and worked as an architectural draftsman and later became a computer programmer. In 1977, he purchased a seat at the American Stock Exchange and played a role in developing the first electronic trading platform for securities in US. He stated in his inspiring quotation a very important fact in financial investment—*self-discipline*. The truth is: if a trader has neither self-discipline nor self-control, one may incur vast losses in the long run.

As the author always mentioned in investment talks and seminars, humans are species of *habit* and *arrogance*. Subsequent profits were yielded from financial market, we inclined to be arrogant and persist in investing without any self-discipline and self-reflection, until one day it was lost in all respects. It is similar to an old saying in Wall Street: "*Failure is the common factor of successful investment. The only one who can truly win (profit) from the market as a long run is the one that has self-discipline and know when to cut loss and stop.*"

How can we maintain *self-discipline* in financial trading?

Although there are many ways to achieve this, the most effective method is scientific *trading techniques* and *strategies*. The logic behind is: If we perform trading or hedging by our own subjective judgements, we won't incessantly stop loss when the market goes to the opposite direction. As regards to behavior psychology, we always have the belief and see what we believe. For instance, if we believe the stock market is bearish and do a short sell by our own judgement (or feeling) without any scientific

© Springer Nature Singapore Pte Ltd. 2020
R. S. T. Lee, *Quantum Finance*,
https://doi.org/10.1007/978-981-32-9796-8_6

backup and evidence, even though the market goes to the opposite direction, we tend to always find evidence either from market charts or financial news to support our judgement that "*the market will go down very soon, just wait for a couple of hours*", until we have already lost a fair amount.

The merits of scientific trading strategy and hence program trading is that we won't let our subjective opinions affect our trading activities. If the market goes to the opposite direction, we simply follow the objective trading strategy to stop loss or trigger the hedging algorithm.

In this chapter, quantum finance forecast center (QFFC.org) provides a more preferable solution—daily/weekly financial forecast for worldwide 120+ financial products. Every day at 00:00UTC (08:00 HKT), QFFC will compute the daily high/low of worldwide 120+ financial product including worldwide forex, major commodity, international financial indices, major cryptocurrencies, and broadcast via QFFC official site QFFC.org. Worldwide investors and traders can base on these forecast results, which together with quantum trading and hedging strategies is revealed in this chapter to perform their quantum trading.

This chapter is organized as follows: First, we begin with basic concept of financial trading and hedging strategies, with an overview of latest R&D on AI-based trading and hedging strategies. Second, we introduce seven major classical trading the hedging strategies, which is the author's collection of over 20 years of experiences on financial analysis and trading in various financial markets. In fact, they are also the building blocks of quantum trading methodology. Third, we explore basic concepts and techniques on quantum trading—the main theme of this chapter. In the conclusion section, we study investment attitude and the importance of objectivity in trading and investment.

6.1   Financial Trading and Hedging Strategies in a Nutshell

In classical finance, trading strategy refers to the design and implementation of buying/selling algorithm (or composition of these actions) to gain profit in financial market.

Conventional trading strategies are based on fundamental or technical analysis (Murphy 1999) or both. They are usually verified by backward testing, where the process should follow scientific method, and by forward testing performed in a simulated trading environment.

Technical trading strategies are relied on technical indicators to generate trading signals. For example, a simple trading strategy may be a moving average (MA) crossover whereby a short-term moving average crosses above or below a long-term moving average (Kaufman 2013).

Hedging strategy (or *hedging* in short), on the other hand, refers to the design of composite trading algorithm to provide risk control and/or possible profit in adverse market movements.

Fig. 6.1 Intelligent trading and market analysis (Tuchong 2019)

Owing to the requirement of complex financial analysis and computations in 70s–80s, hedging was solely performed by major fund houses and financial institutions.

But thanks to the advance of computer technology, there is an improvement on personal computer and workstations along with the popularity of online trading platforms; nowadays, hedging becomes a possible and useful tool even for independent investors and traders.

In fact, modern finance includes hedging being one of the major strategies for the design and implementation of trading strategy, and it becomes a standard trading tool for the design of high-level trading algorithms and strategies (Leshik and Cralle 2011; Nelken 1999; Rhoads 2011) (Fig. 6.1).

6.2 Traditional Trading and Hedging Strategies

Traditional finance involved trading and hedging strategies in 70s–80s (i.e., before the age of online and program trading) on investment funds, bonds, and stocks which was mainly performed by major fund houses and financial institution in major financial markets.

Conventional trading and hedging strategies include the following:

1. Single-product trading/hedging strategy—long buy/short sell of stocks, or composition of these actions with their options counterparts.
2. Mutual product trading/hedging strategy—also known as *pair-trading*—is the long buy/short sell of one particular stock (e.g., Coca-Cola) with the counter-direction buy/sell or a mutual stock (highly correlated stock with different expected returns/variance returns) as safeguards and risk controls (e.g., Pepsi in this example).

Fig. 6.2 Program trading

Fig. 6.2 Program trading

3. Trading of a particular stock together with risk control by counteraction on co-related financial index, e.g., the trading of GM Inc. with hedging on DJIA.
4. Multi-market trading/hedging strategy (*portfolio trading*) become favored by major fund houses from 80s due to the widespread of *portfolio analysis*, portfolio trading which involves buy/sell composition of a batch of financial products (called *portfolios*) to maximize returns on one hand, and minimize risks—return variances—on the other hand.

Also, the surface of major investment funds, mutual funds products, and services began in 80s (or commonly called *wealth investment*) in financial community-enabled worldwide investment. The importance of financial values in portfolio investment resulted in a mass of R&D conducted in major universities and financial institutions in the past decades. Typical examples include R&D on chaos and fractals on finance, as well as quantum finance (Fig. 6.2).

6.3 Latest R&D on Trading and Hedging Strategies

Over the years, financial engineering ranging from financial signals, chart patterns study for the modeling of financial prediction and trading systems is one of the most stimulating topics for both academia and financial community. Not only because of its utmost importance in terms of financial and commercial values, but it vitally poses a real challenge to worldwide researchers and quantitative analysts (a.k.a. *quants*) throughout the world.

The flourishing of AI technology in the past 30 years such as various hybrid intelligent financial trading and hedging systems with the integration of various AI and related technology was proposed. Latest R&D on intelligent trading and hedging system include the following:

1. Artificial neural network (ANN) was proposed by Li et al. (2017) for the development of quantitative trading strategy.

2. AI clustering technique was proposed by Hsu et al. (2009) for the design of profit refiner on futures trading strategy.
3. Genetic algorithm (GA) technology was proposed by Allen and Karjalainen (1999) for the determination of technical trading rules and Shiau (2009) using multi-objective GA for the optimization of reservoir hedging rules.
4. Principle component analysis (PCA) technology was proposed by Byun et al. (2015) for multi-currencies trading in the forex market.
5. Fuzzy logic system (FLS) was proposed by Naranjo et al. (2015) for the development of fuzzy capital management and trading system; Vella and Ng (2015) used dynamic fuzzy money management approach for controlling the intraday risk-adjusted performance of AI trading algorithms.
6. Data mining system was proposed by Pauna (2019) for time price series for algorithmic trading systems.
7. Decision support system (DSS) was proposed by Schmidt et al. (2010) for experimental analysis of online trading algorithms.
8. Reinforcement-learning system (RLS) was proposed by Yang et al. (2018) for the design of investor sentiment reward-based trading systems.
9. Stochastic technique was proposed by Helseth (2016) for the design of linearized progressive hedging algorithm on secondary financial markets.
10. Hybrid AI system was proposed by Chang and Lee (2017) with the integration of Markov decision process (MDP) and GA to formulate trading strategies for stock markets; Evan et al. (2013) with the integration of ANN and GA developed algo-trading model for intraday foreign exchange speculation; Liu et al. (2017) with the integration of fuzzy logic rules and GA developed a method to quantify moving average strategy of crude oil future markets; Zhang and Maringer (2016) integrated recurrent reinforcement learning (RRL) with GA for equity trading.

6.4 Technical Analysis and Program Trading

Although both *fundamental* and *technical analyses* are the most popular and commonly used in financial analysis, the design of trading algorithm, technical analytical tools—indicators such as RSI, MA, and MACD are commonly used for program trading because they are

1. Easy to implement and program for program trading;
2. Fully integrated to time series financial data which are now openly provided by many online trading platforms (e.g., MT platform—the largest forex online trading platform in the world with over hundreds of financial service platforms/operators);
3. Easy to understand and learn, only basic financial and statistical knowledge is required to understand technical analysis;

Fig. 6.3 Program trading using technical indicators on MT4 trading platform

4. Easy to program and visualize on the trading platform. Unlike traditional stock market in 60s–70s, only financial analysts and brokers can access real-time figures (patterns) of financial markets. Nowadays, every investor and trader with basic computing (programming) knowledge can not only visualize the financial markets, but can also design and program their own trading and hedging algorithms into their investment portfolio. The additional knowledge they require are methods and techniques such as financial forecasts, AI-based trading algorithms, etc. which is also the aim of this book (Fig. 6.3).

6.5 Technical Analytical Tools and Indicators

Technical analysis is a trading discipline employed to evaluate investments and identify trading opportunities by analyzing statistical trends gathered from trading activity, such as price movement and volume.

Technical analysts (*TA*) believe that all knowledge regardless of type (fundamental, economic, political, psychological, or other) is already reflected and discounted in market prices.

Unlike fundamental analysts, they believe that it is meaningless to study company financial statements, earnings and dividend reports, industry developments, and other data in an attempt to determine the *intrinsic value* of a stock or other market instrument since it is common to have a wide divergence between intrinsic value and actual market price.

For example, it is not unusual for a company's stock to be trading at a price well above or below its book value per share.

Price movements are simply the reflection of changes in supply and demand without concern of their underlying forces. Technical analysts are only interested in what happens to the price according to supply and demand. Say, if demand is greater than supply, price will increase. On the other hand, if supply is greater than demand, price will decline. The study of market prices is all that is necessary.

The components of technical analysis are as follows:

- Major market patterns (e.g., major reversal patterns),
- Trend lines and channels,
- Supports and resistance (S & R) analysis, and
- Technical indicators and oscillators (e.g., MA, RSI, MACD, stochastics, Bollinger Bands).

Nowadays, the design of trading and hedging algorithms (and hence programs) mainly makes use of technical indicators, oscillators, supports, and resistance lines for easy implementation, since most of them have already built-in functions bundled with the program trading development platform, such as MQL trading indicator function library in MT platform.

6.6 Seven Major Financial Trading and Hedging Algorithms

In previous chapters, we have studied basic concepts and theory of quantum finance, especially on how to apply quantum finance theory to model and evaluate quantum price (energy) levels (QPLs) which is an important tool in technical analysis.

Before we move on to learn how to design and implement AI-based quantum financial forecast and trading system, let's study a practical financial investment topic—trading and hedging strategy.

First, we will review seven major trading and hedging techniques—which is the author's collection of over 20 years of trading experience in stock, commodity, and forex trading, along with the numerous technical trading courses conducted in China and Hong Kong.

Traders study various kinds of trading techniques and strategies throughout the years which are summarized into following seven types:

1. Trend trading strategy,
2. Breakout trading strategy,
3. Reversal trading strategy,
4. Channel trading strategy,
5. Averaging trading strategy,
6. Stop-loss trading strategy, and
7. Hedge trading strategy.

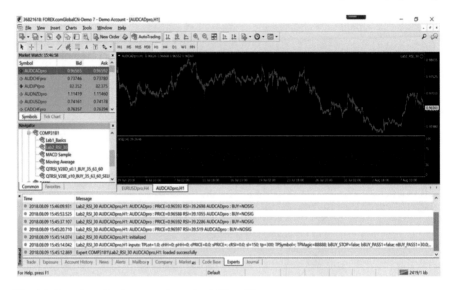

Fig. 6.4 Snapshot of program trading using RSI oscillator

Note that these seven trading strategies despite are independent in nature, and most of the time they can be combined to use in order to produce the best performance. Figure 6.4 shows the snapshot of program trading using RSI oscillator.

6.6.1   M1.1 Trend Trading Strategy—Key Patterns

Trend trading is also known as *forward trading*. The main concept is to trade following the market trend, if exists. Trend trading is a trading strategy that attempts to capture gains through the analysis of a security's momentum in a particular direction. Trend traders enter into a long position when a security is trending upward and/or enter a short position when a security is trending downward.

Trend trading strategy assumes that a financial instrument will continue to move along its current trend and often contain a take-profit or stop-loss provision if there are any signs of a reversal. It can be used by short-, intermediate-, or long-term trading. Regardless of their chosen timeframe, traders will remain in their position until they believe the trend has reversed, although reversals may occur at different times for each timeframe.

In technical analysis, we have studied all major patterns and trends identification techniques.

Trend trading strategy #1 using major key patterns include the following:

- Nine major reversal patterns such as head-and-shoulder patterns, round-top/bottom patterns, ascending/descending triangles, rectangles, etc.;

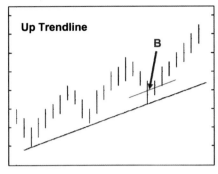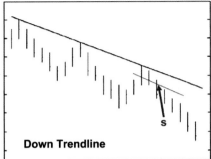

Fig. 6.5 M1.1 Trend trading strategy—key patterns

- Trend lines and channels;
- Consolidation patterns;
- Gaps; and
- Supports and resistances.

Once a trend is clearly identified, we have to execute the sell/buy command according to the trend direction.

Let's have a look at the up/down trend lines as shown in Fig. 6.5.

Note that the buy and sell should execute

- When the trend is well developed, i.e., at least hits two low/high.
- Execute at least 10% above (uptrend)/below (downtrend) price.

6.6.2 M1.2 Trend Trading Strategy—MA Crossing

In technical analysis, most of the time financial patterns are moving in trends, either bullish or bearish (or oscillation in some cases).

In other words, once we have identified the current pattern is bullish/bearish, we can trigger the buy/sell command according to the trend.

Typical example is *moving average* (MA) trend identification.

The decision of using MA crossing is simple:

1. If day-line down-crossing MA line is bearish, trigger sell signal;
2. If day-line up-crossing the MA line is bullish, trigger buy signal.

Again, one should place the selling signal at the price at least 10% below the down-cross point as a kind of buffer.

Same cases apply to other technical indicators and oscillators.

Figure 6.6 shows a typical example of the price (S&P 500) bar chart down-cross the 200-day MA line. It shows the fast signal line (i.e., the *day-line*) down-crossing

S&P 500 Index

Fig. 6.6 M1.2 Trend trading strategy—MA crossing

the slow 200-day *simple moving average* (SMA)—means *bearish*, trigger the sell signal.

As an exercise, based on what we have learnt here, try to figure out other bullish/bearish trends from any market chart and determine the buy/sell thresholds.

6.6.3 M1.3 Trend Trading Strategy—Signal Lines Crossing

Instead of single technical indicator up-/down-crossing, many TAs are using multiple technical signal crossing method, in which one short-term indicator acts as fast-moving signal and the other long-term indicator acts as slow-moving signal.

The decision of using signal crossing is simple:

1. If a fast signal line down-crossing a slow signal line—*bearish*, trigger sell signal and
2. If a fast signal line up-crossing a slow signal line—*bullish*, trigger buy signal.

Figure 6.7 shows two MA signal lines: 5-day MA (fast-moving) versus 20-day MA (slow-moving) lines. When 5-day fast-moving MA line down-crosses a 20-day slow-moving MA line from above, a *bearish* trend appears.

Fig. 6.7 M1.3 Trend trading strategy—signal lines crossing

Again, can place a sell command 10% below the down-cross point.

One may wonder: For MA signal lines crossing, what are the two MA lines being used?

Actually, there is no definitive answer. But many TAs prefer to use the Fibonacci number sequence (i.e., 1, 1, 2, 3, 5, 8, 13, 21, 34, 55, 89, 144…) to select the MA lines. For example, the signal crossing of MA-5 versus MA-21 (or MA-20 for simplicity) is used to simulate the comparison between weekly (short-term) trend with monthly (mid-term) trend.

6.6.4   M2.1 Breakout Trading Strategy—Key Patterns

Breakout trading strategy, similar to trend trading strategy is to trade by following the trend.

But different from trend trading strategy, it will only execute when the price has reached the *breakout* point.

In technical analysis, we have studied certain patterns such as *flag consolidation* pattern with well-defined *breakout* price as shown in Fig. 6.8.

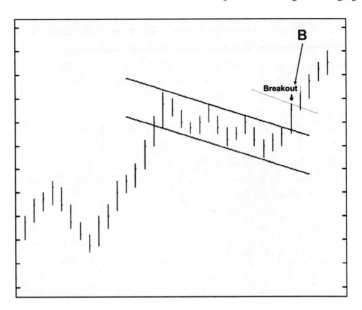

Fig. 6.8 M2.1 Breakout trading strategy—key patterns (e.g., flag consolidation)

It shows a typical flag consolidation pattern in technical analysis with well-defined channel lines. Once the pattern is confirmed, one can trigger a buy order while the market breaks from the breakout price.

Similar to trend trading strategy, the breakout price must be at least 10% away from the breakout price.

This strategy can also be used once we have clearly identified the support/resistance lines. In other words, when the price of an asset moves above a resistance area or moves below a support area. Breakouts indicate the potential for the price to start trending in breakout direction.

For example, a breakout to the upside from a chart pattern could indicate the price will start trending higher. Breakouts that occur on high volume (relative to normal volume) show greater conviction which means the price is more likely to trend in that direction.

6.6.5 *M2.2 Breakout Trading Strategy—S & R Lines*

Figure 6.9 shows two typical cases of breakouts from support and resistance lines.

Note that, once we identify the support/resistance lines, we can execute the buy/sell command when the price breakout from breakout thresholds of, say, 10% buffer.

In fact, breakout trading using supports and resistance lines is commonly exercised on stock and commodity trading using the day chart to identify *support* and *resistance* lines by applying standard technical analysis techniques. However, it may be difficult

Fig. 6.9 M2.2 Breakout
trading strategy—S & R
lines

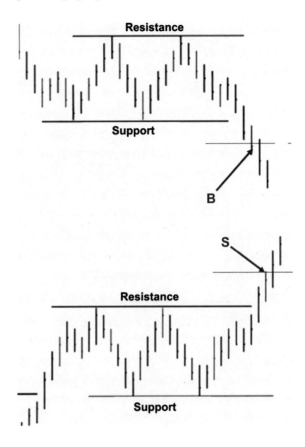

to identify at forex and cryptocurrency trading as they are more chaotic, especially
during US trading time zone, i.e., US time 09:00–16:00.

6.6.6 M3.1 Reversal Trading Strategy—Key Reversal Patterns

Reversal trading as its name is executed at least at one of these three scenarios:

1. Clearly identify the key reversal pattern(s);
2. Clearly identify effective support/resistance; and
3. Reversal occurs in over-buy/over-sell regions of technical indicator such as RSI.

Naturally, reversal trading strategy is a perfect tool once we have identified the
eight *key reversal patterns* by technical analysis, which include key reversals, head-
and-shoulders tops and bottoms, rounding tops and bottoms (saucers), ascending and
descending triangles, rectangles, double and triple tops and bottoms, diamond, rising
and falling wedges, V-shape formations (spikes).

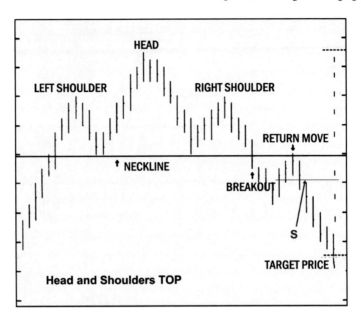

Fig. 6.10 M3.1 Reversal trading strategy—key reversal patterns

Figure 6.10 shows a typical head-and-shoulder top pattern.

Once the pattern is clearly identified, we can place a sell command 10% below the return move of the head-and-shoulder pattern.

As an exercise, try to figure out the position of sell/buy signal point for all the key reversal patterns using technical analysis.

6.6.7 M3.2 Reversal Trading Strategy—S & R Lines

Reversal trading strategy using support and resistance lines (S & R lines) is exactly the counterpart of breakout trading strategy #2—S & R lines in the sense that the S & R lines continue as an effective price level to support or resist from price breakout.

In many worldwide financial products trading such as forex and commodity trading, this situation usually occurs during Asia market period, i.e., 09:00 HKT–16:00 HKT in which the financial markets are oscillating within the daily S & R lines.

Figure 6.11 shows a typical reversal trading based on S & R Lines.

Once the S & R lines are identified, buy/sell command can be triggered after third reversal from the S & R. (Note: at least two times of reversals are required for S/R confirmation).

Again, a 10% threshold is required to dispatch signal.

Fig. 6.11 M3.2 Reversal trading strategy #2—S & R lines

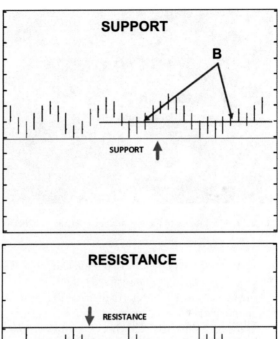

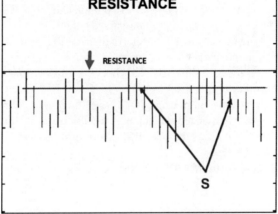

6.6.8 M3.3 Reversal Trading Strategy—Technical Indicator RSI

Relative strength index (RSI) is a price momentum indicator developed by Mr. J. Welles Wilder in 1978. Since its introduction of Wilder's New Concepts in Technical Trading Systems book published in 1978, it has been widely used by worldwide technical analysts, particularly those who are commodities and future oriented.

From the technical analysis perspective, relative strength index (RSI) is a very reliable technical indicator to indicate the strength of supply versus demand in secondary financial market. RSI compares bullish and bearish price momentum plotted against the graph of an asset's price.

Traditional interpretation and usage of RSI are that reading values of 70 or above indicate that a security is becoming overbought or overvalued and may be primed

Fig. 6.12 M3.3 Reversal trading strategy—technical indicator RSI

for a trend reversal or corrective pullback in price. RSI reading values of 30 or below indicate an oversold or undervalued condition.

The fact is: most TAs believe that 70–30 is sufficient to reflect the over-buy/over-sell in the market.

Figure 6.12 shows the daily bar chart of Dow Jones Index, DJI (top) versus 14-day RSI (bottom).

It is apparent in using 70–30 as selling/buying threshold is an acceptable choice.

For experienced traders, they usually trigger selling signal only when RSI rise to above 70, reach the max RSI, and return back to 70 RSI line. It is called *second-time-hit-top*.

Similar to RSI 30 case, it is called *second-time-touch-base*.

6.6.9 *M4.1 Channel Trading Strategy—Trend Channel Trading*

A *trend channel* (or *channel* in short) is a channel drawn on a security price series chart by graphing two trend lines drawn at resistance and support levels. Trading channels can be drawn using a variety of methodologies. Generally, traders believe that security prices will remain within a trading channel. Therefore, traders use trading channels to develop buy-and-sell trading signals.

In many instances, prices repeatedly move about the same distance away from a trend line before returning to the trend line. In these cases, a straight line can be drawn connecting the peaks of rallies in an uptrend or the bottoms of declines in a downtrend. That line is often parallel to the trend line and is called a return or channel line. Between the channel line and trend line creating a trend channel is a range within which prices move.

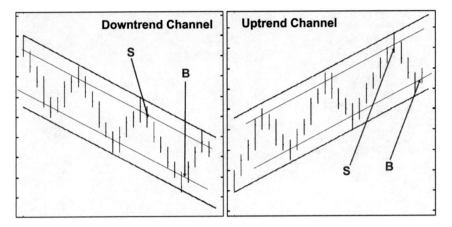

Fig. 6.13 M4.1 Channel trading strategy—channel patterns

Well-defined trend channels appear most frequently in charts of actively traded securities. Thinly traded securities offer little opportunity for trend channels to develop. Inexperienced TAs often use trend channels to determine good profit-taking levels. For example, in an uptrend, they will sell a stock when it reaches the upper level of its trend channel. Experienced TAs watch price movements within the two boundary lines of the trend channel looking for a warning signal that the trend direction is changing.

Channel trading strategy can be applied when one of the following occurs:

- Clearly identify the channel pattern and S/R lines.
- Clearly identify the price oscillator between S/R lines pair.
- Clearly identify the Bollinger Band.

Figure 6.13 shows typical cases for channel trading using uptrend and downtrend channels.

Again use at least 10% as threshold and trigger the signal when the price oscillates at least two times to confirm the effectiveness of the channel.

6.6.10 M4.2 Channel Trading Strategy—S & R Channel Trading

As mentioned in previous section, prices oscillating between supports and resistance lines (S & R lines) are commonly found during Asia trading period, i.e., HKT 09:00–16:00 in which most international forex prices "oscillating" between daily S-R lines, especially when there is no significant market news and financial figures release.

Figure 6.14 shows the day-trade chart for AUDCAD in MT platform

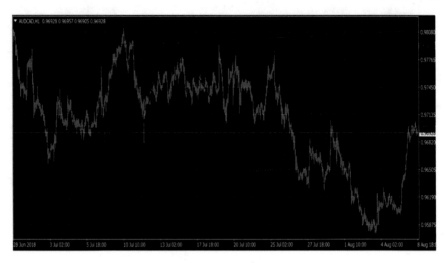

Fig. 6.14 M4.2 Channel trading strategy—S & R channel trading

of Forex.com.

We can clearly identify an effective S-R line pair to trigger two buy/sell signals.

However, at the third time of oscillation, a breakout occurs in which the previous support line now becomes the resistance (i.e., support → resistance). In this case, a sell signal can be triggered when the price arrived at the breakthrough 10% threshold of the breakout line.

The fact is: in traditional technical analysis, support and resistance lines are completely two different kinds of trading patterns. In other words, a support cannot turn into resistance lines (or vice versa). However, due to the popularity of program and algorithm trading, support and resistance lines mentioned in previous chapter are now considered as some sorts of discrete energy levels, similar to quantized energy levels in an atom. In other words, a support line for a financial product at a particular moment can become a resistance line at another moment due to the change of *market stimulus*.

6.6.11 M4.3 Channel Trading Strategy—Bollinger Bands

Bollinger Bands technique (BB in short) was developed by Mr. John Bollinger in 1980s for the purpose of factoring in price volatility. Bollinger Bands are effective for virtually any security or market and for any investment time horizon. Rather than placing bands at a certain percentage distance from a moving average line as is done with trading bands, Bollinger Bands are placed two moving standard deviations above and below a simple moving average line.

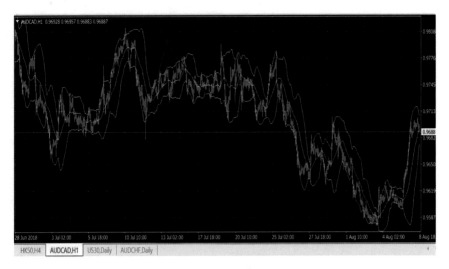

Fig. 6.15 M4.3 Channel trading strategy—Bollinger bands

More specifically, the bands are placed above and below a simple moving average at a distance of 1.5–2.0 times the root mean square of the deviations (σ) from the average. The amount of data used in the calculation is equal to the number of periods used for the simple moving average.

For example, if we use a 20-day simple moving average, all calculations should be based on 20 days of data. Similar to moving average, Bollinger Bands are so important and commonly used that almost all financial trading software/systems are bundled with the BB tools.

In fact, Bollinger Bands are very reliable with effective technical signal and trading tool, especially in day trade of forex products which are usually oscillating between upper–lower *Bollinger Bands* (so-called *Bollinger channel*).

Figure 6.15 is the same H1 chart of AUDCAD with period-20 Bollinger Band.

Note that it provides least three chances (maybe more) to trigger buy/sell signals. But remind that such channel always breakthrough during Europe–US trading period. It is wise to set stop-loss as risk control.

6.6.12 M5 Averaging Trading Strategy

Averaging method (or *averaging* in short) is a very important and commonly used trading strategy for experienced traders and hedge fund traders.

The logic is very simple: instead of putting 100% of investment at a single price (either buy or sell), we divide the investment into at least two (or more, usually three) batches, each of equal amount (or different proportion, according to the trader's preference).

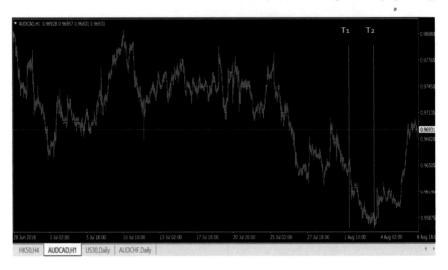

Fig. 6.16 M5 averaging trading strategy

So even when the price moves in reverse direction, the overall execution price will be the average of two (or three) triggering prices.

The merits of averaging trading are threefolds:

1. Provide risk sharing especially during adverse market situation;
2. Provide more chances for investors and traders to enter market at one hand, and locate a good averaging position even the market swings sideway; and
3. Provide a good trading strategy with the integration of S & R lines or/and quantum price levels (we will discuss that in detail in the upcoming section—quantum trading strategy).

Figure 6.16 shows a realistic case of forex trading chart of AUDCAD using averaging trading strategy. As shown in Fig. 6.15, if we use S1 and S2 as support lines, we can execute 50% buy at T1-S2 price. When price reaches S1 at T2, at this "better position", we trigger the other 50% buy. Once it reaches S3, we close the order. In that case, the average gain will be S3 − (S2 + S1)/2, which is much better than solely investing 100% at S2-T1. Same method can be applied and through a combination to any kind of trading strategy.

In fact, professional traders and investors usually combine averaging trading strategy with other trading and hedging methods to provide a more flexible, dynamic trading and risk control strategy.

6.6.13 M6 Stop-Loss Trading Strategy

Stop-loss strategy seems to be very simple, but it is the most important trading strategy especially for derivatives trading and margin trading products such as forex and cryptocurrency trading.

Usually, stop-loss strategy is combined with target profit in which the ratio is normally 1:2. For example, if the target profit is 240 pts, the stop-loss should be around 120–150 pts.

How to set stop-loss?

Traditionally, there are two methods:

1. Fixed stop-loss according to the investor's risk control preference. It varies for different products. For forex trading, normally we set 150–200 points. It can also be the proportion of the average daily range (i.e., range between day-high and day-low) and
2. Dynamic stop-loss according to some technical indices such as S & R lines.

In the next section, we will see quantum trading to provide a third solution for stop-loss strategy using quantum price levels (QPLs).

Using previous example on AUDCAD trading in Fig. 6.15, if we trigger a buy order at time T1 0.96251, we set the stop loss of 125 pts (i.e., 0.96126). So even though price finally drops 464 pts to S1 0.95787 at time T2, the loss is controlled to 125 pts only. Figure 6.17 shows another example based on the trading of Dow Jones Index (DJI) in MT platform. As shown, DJI breakouts at T1-S2 level (23180) and drop 1728 points to S1 (21452) before it finally rebound. So, if we set the stop loss with 200 points, we can control the potential loss to an acceptable level.

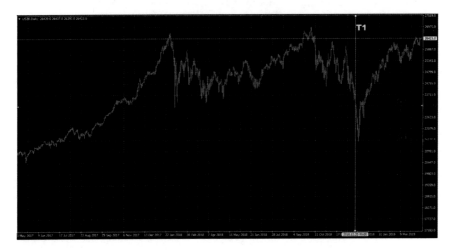

Fig. 6.17 M6 Stop-loss trading strategy

As an old saying in Wall Street: The best and most successful trader is not the one that can always win, but the one that has self-discipline to follow stop-loss strategy in every trading battle.

6.6.14   M7.1 Hedge Trading Strategy—Single-Market Single-Product (SM-SP) Hedge

Hedge trading (or *hedging* in short) is the second kind of risk control-based trading strategy. A hedge is an investment position intended to offset potential losses or gains that may be incurred by a companion investment. A hedge can be constructed from many types of financial instruments including stocks, forex, insurance, forward contracts, swaps, and options.

Technically speaking, *hedging* is the practice of taking a position in one market to offset and balance against the risk adopted by assuming a position in a contrary or opposing market or investment.

It is a common tool for experienced traders as it involves more technical skill to so-called *release the hedge*.

Hedge trading basically is of three types as given below:

1. single-product hedging,
2. multi-product hedging, and
3. cross-trading-platform hedging.

First, let's have a look at a basic case: single-market single-product (SM-SP) hedge.

In a single-product trading, when we execute a buy/sell order at a certain price P, we execute the same amount of reverse order at a price "P" at the stop-loss level to control loss when the market goes in opposite direction.

Same example of AUDCAD trading is used for illustration in Fig. 6.18. Instead of using stop loss, we set a sell hedge at the stop-loss level below S2. When price drops to S1 and rebound, we harvest the hedge order and release the hedge.

6.6.15   M7.2 Hedge Trading Strategy—Single-Platform Multi-product (SP-MP) Hedge

With the advance in computer technology and online trading, real-time calculation of co-relation and co-variance between different financial products under the same trading platform, e.g., forex, is no more a reverie.

Single-platform multi-product (financial products) hedging strategy is a kind of composite hedging strategy with market momentum evaluation of two (or more) similar products in a financial platform. It is commonly used in forex platform which contains over 100+ forex products to study.

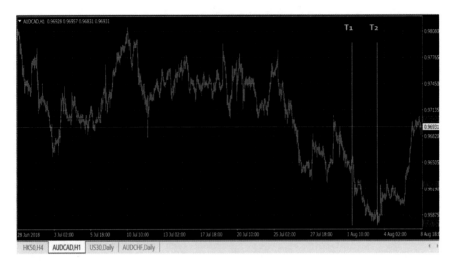

Fig. 6.18   M7.1 Hedge trading strategy—single-market single-product (SM-SP) hedge

SP-MP hedge ranging from simple two products hedge (i.e., traditional *paired-hedge*) to multi-product hedge with a portfolio of products (so-called *portfolio-hedge*) with the help of AI tools such as fuzzy logics, neural networks, and genetic algorithms for portfolio weighting factor "ω" evaluation (we will study all major AI tools in the next chapter).

Figure 6.19 shows a typical five-product cyclic hedge (5PC-hedge) for the cyclic hedge of top five foreign currencies: AUD, CAD, EUR, GBP, and USD. Try to figure out how it works and the logic behind.

Note that most often such kind of high-level and composite hedging strategy can only be valid and effective within a period of time and restricted to number of

Fig. 6.19   Five-product
cyclic hedge (5PCH) in forex

financial products. That's why all major fund houses and laboratories recruited top mathematicians and AI scientists to develop complex trading and hedging algorithms.

In the same way that the author always mentioned in investment talks; nowadays, the design and implementation of trading and hedging algorithms resemble more to mathematical and AI problems than classical financial problems.

6.6.16   M7.3 Hedge Trading Strategy—Multi-platform Multi-product (MP-MP) Hedge

Considering the complexity of world financial market, one of the most important and complex hedging systems is the multi-platform multi-product (MP-MP) hedge which involves multiple-financial products across different financial trading platforms.

The logic behind came from the fundamental concept of financial theory that all financial markets are the currency flow, e.g., USD behind different major worldwide financial markets. So, in the perfect case of hedging for risk control, the best solution should be the hedging across different products and platforms to protect adverse market change of a particular product/financial market.

Traditional wisdom tells us that USD currently remains as major clearance currency for three types of worldwide major financial and commodity markets which include (a) US Index, (b) crude oil, and (c) precious and colored metals gold (or silver) along with agricultural products such as cotton and sugar that they are of great consideration as shown in Fig. 6.20.

Fig. 6.20 MP-MP hedge of world three major markets (gold, crude oil, and US index)

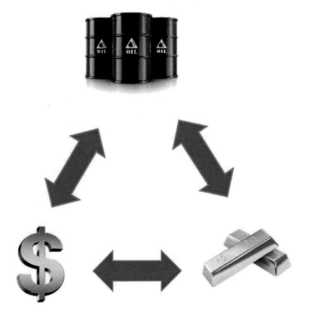

Many fund houses and financial institutions have active research for the *mining* of such *golden combination* using various AI techniques such as fuzzy-neuro networks, hybrid EC (evolutionary computing) systems, deep neural networks, etc. Active R&D on this challenging topic is being carried out at worldwide fund houses, financial institutions, and universities (Iba and Aranha 2012).

6.7 Quantum Trading and Hedging Strategy

6.7.1 Basic Concept

In previous chapters, we have learnt that basic concepts of quantum finance stem from two different aspects:

- Path integral approach and
- Quantum anharmonic oscillator approach.

Path integral approach is based on Feynman's path integral by the modeling and integration of all possible paths of the quantum financial particles. Major applications are exercised on the modeling and primary financial markets evaluation such as forward interest rate and option pricing.

Quantum anharmonic oscillator (QAHO) approach is based on the modeling and numerical approximation of the quantum finance Schrödinger equation (QFSE) into quantum anharmonic oscillation—a quartic polynomial equation that can be solved numerically by simple numerical computational model. Major applications are exercised on the evaluation of quantum finance energy levels (so-called quantum price (energy) levels, QPL) for any financial products in the secondary financial markets (Fig. 6.21).

Fig. 6.21 Illustration of quantum finance energy fields in financial markets

Which one should we employ to design quantum trading and hedging strategy? and why?

6.7.2   Quantum Price Levels Versus S & R Analysis

The answer is clearly to employ quantum anharmonic oscillator (QAHO) approach since all abovementioned trading and hedging strategies fall within the domain of secondary financial markets ranging from stocks and options to online forex and cryptocurrency trading activities.

In addition, one of the major breakthroughs in quantum finance using QAHO model is the quantum price (energy) level (QPL) evaluation, which forms a direct analog to supports and resistances (S & R) in technical analysis.

If we consider the financial world is a quantum finance energy field formed by financial particles, it is a natural logic to presume that these inherited QPL are exactly the same concept of support and resistance we defined in classical technical analysis.

But there is only one difference. If quantum dynamics of the financial particles is strictly followed from quantum theory, there should be no difference between support and resistance levels. They should all be the same as discrete energy levels that the financial particles can jump from one to another (in both directions). In fact, this logic is entirely equivalent to nowadays assumption on support and resistance lines in program trading, which we generally assume that support and resistance lines are simply quantized price levels in price motions.

In other words, once we adopt quantum finance method to evaluate all QPLs for any financial product automatically, we literally evaluate all the possible S & R levels, which can be applied in turn to the seven trading and hedging strategies learnt previously (Fig. 6.22).

6.7.3   Quantum Finance Forecast Center—Daily Forecast and QPLs

Since December 1, 2017 with patented quantum finance technology, quantum finance forecast center (QFFC.org) provides daily (and weekly) free financial forecasts for worldwide 120+ financial products which include the following:

- 9 major cryptocurrencies,
- 84 forex,
- 19 major commodities, and
- 17 worldwide financial indices.

Up to December 31, 2018, QFFC has over 10000+ registered users which include worldwide professional traders and investors.

Figures 6.23 and 6.24 show snapshots of daily forecast results of BTCUSD (Bit-

Fig. 6.22 Support and resistance levels in financial analysis

Fig. 6.23 Quantum finance forecast of BTCUSD on April 15, 2019

Fig. 6.24 Quantum finance forecast of USDCAD on April 15, 2019

coin versus USD) and forex pair USDCAD on April 15, 2019. At 08:00 HKT (UTC 00:00), QFFC calculates the following for 120+ worldwide financial products:

- Forecast high/low and
- Eight closest QPLs.

Professional traders and investors can base on these forecasts with QPLs and integrate with their program trading strategies for intelligent trading.

QFFC members can also make use of QFFC's API to integrate the daily forecasts/QPLs into their MT4 EA programs for intelligent trading.

In the next section, we will study some basic quantum trading and hedging techniques/strategies.

6.7.4 Basic Quantum Trading and Hedging Strategies

In comparison with technical and fundamental analysis used in the past 50 years, the major breakthroughs of quantum finance technology are the scientific computations of

- Daily (weekly) high/low for day-trade and mid-term trading, with
- QPLs to replace subjective evaluation of support and resistance lines that can be used for breakout analysis and mid-term/long-term trading/market trend analysis.

The following quantum trading and hedging techniques/strategies are the theoretical characteristics collection of these quantum financial instruments (i.e., forecast high/low and QPLs) and feedbacks from professional traders and investors using QFFC forecast services in the past 2 years.

As a friendly advice, for any new trading and hedging, please study carefully and thoroughly, together with sufficient simulations and test run (at least 100 trading days as industrial norm) before putting into actual practice.

Different from the major seven trading and hedging strategies, at quantum finance trading, since we already have scientific daily forecasts and QPLs, the quantum trading (includes hedging) techniques are categorized in terms of users' trading experiences, which is also the way we have conducted investment seminars in the financial community nowadays.

6.7.5 QT#1.1 Inexperienced Investors—Quantum Forecast Oscillation Trading Strategy (QFOTS)

Just like driving, financial trading is a kind of *practical activity* that not only require professional knowledge but also practical experience. Any experienced investor with over decades of trading experience must realize that trading (and especially "failure") experiences are very important elements for every investor and trade. Especially nowadays program trading is accompanied by monitoring and handling trading strategies of multiple products in multiple-financial markets.

As an industrial norm, we set 3–5 years (around 1000 transactions) to define whether an investor is *experienced* or *inexperienced*.

For inexperienced investors who don't have sufficient skills on *high-level* trading techniques such as *breakout trading, hedging, and release-hedge* skills, they are suggested to perform basic quantum forecast oscillation trading strategy (QT#1-QFOTS).

QT#1-QFOT are based on daily forecasts, together with daily calculated QPLs as SL (stop loss) and TP (target profit) to perform day trade.

The logic behind is: for any financial market, over 80% of time are undergoing so-called *oscillation* between the day-high and day-low. Since we already have evaluated the forecast high/low, we can simply use them as approximation of the day-high/low

Fig. 6.25 Quantum finance forecast of XAUUSD on April 15, 2019

for reversal trading. That is, set the buy limit at the forecast low and sell limit at the forecast high.

Also, for SL and TP, we can simply use daily QPLs as reference levels as example shown in Fig. 6.25 for XAUUSD on April 15, 2019.

6.7.6 QT#1.2 Inexperienced Investors—Quantum Forecast Oscillation Trading with Averaging Strategy (QFOTAS)

As we have learnt from the beginning of this chapter, *averaging* is a simple and good technique not only as risk control but also for investment diversification.

Naturally, we can adopt *averaging technique* into our quantum forecast oscillation trading strategy.

Just like working S/R lines as averaging levels, we can now employ QPLs for averaging levels, which, in fact, is entirely inherent to quantum theory and quantum field concepts also learnt in previous chapters.

Fig. 6.26 Quantum finance forecast of UK oil on April 15, 2019

Use UK oil forecast on April 15, 2019 as example shown in Fig. 6.26. If we set first buy limit at the forecast low 7149, we can set the second buy limit at the lower QPL, i.e., 7128, and the third one at 7069, each with one-third of the day-trade investment. Of course, remember to set SL, say, at the fourth QPL 6997.

By doing so, we can, on one hand, self-guard investment with a reasonable level of oscillations for the financial product, and on the other hand, provide a reasonable investment diversification with reasonable level of stop-loss mechanism.

In fact, not only inexperienced investors, many professional traders usually perform this strategy when they identify a particular financial market has a very high chance of oscillation that always occurs during Asia trading period.

6.7.7 QT#2.1 Experienced Traders—Quantum Forecast Breakout Trading Strategy (QFBTS)

Among the seven major trading and hedging techniques, *breakout trading* is an important trading skill for experienced traders and investors.

In actual fact, the success in identifying financial market whether it is an *oscillating market* or high chance of *market breakout* defines whether one is an experienced or inexperienced trader.

According to quantum finance theory, if a financial market undergoes severe *excitations*, either by some financial news outbreaks or some important financial figures/indices (e.g., GDP, PMI, etc.) announcement, the financial quantum particle should jump from one QPL to another one, in a discrete manner. In fact, the author customarily observed this phenomenon while working as chief analyst in a fund house and conducting R&D on quantum finance.

The truth is: The more *excited* the stimulus is, the more accurate and precise is the *quantum jump*. That's why *breakout trading* is so important in financial investment. Use UK100 index forecast on April 15, 2019 as example shown in Fig. 6.27. For a day-low breakout, we can use the lower three QPLs (or more) for breakout strategy, i.e., 7395.2, 7364.6, and 7330.5. Similar case for day-high breakout.

One important reminder, according to quantum finance theory, the range between day-high and day-low is the ranging of *normal daily oscillation* of the financial market. We should determine the market as *breakout* if the market jump/fall beyond the oscillating region.

6.7.8 QT#2.2 Experienced Traders—Quantum Forecast Breakout Trading with Averaging Strategy (QFBTAS)

The integration of averaging technique and breakout strategy is a direct enhancement of quantum finance breakout strategy. That is, naturally, the integration of *averaging technique* at every QPL breakouts, such as 30019, 29871 for Forecast Low breakouts and 30482, 30691 for Forecast High breakouts, is shown in Fig. 6.28 for HSI.

It is the usual (in fact standard) technique being adopted by almost every experienced trader. Even if we can identify it is a breakout market, it is impossible for us to predict how much it can breakout.

Thus, averaging is the most viable solution and is also a smart move.

One important reminder is that: as we all know, once the market breakout, it can be very serious.

So, (1) it is important to set (and calculate) as many QPLs as possible and (2) Don't try to do *reverse trading* during a severe market *breakout*. It is a very dangerous act to do so.

Fig. 6.27 Quantum finance forecast of UK100 index on April 15, 2019

The truth is: once market *breakout*, anything can occur. So, for those use breakout strategy, all we need to do (and should) is to choose a good position to harvest the gain. That's all. Nothing else (*traditional wisdom*).

In fact, this also differentiates between an *experienced* versus *inexperienced* trader. For an experienced trader, one will know that statistically speaking it is very difficult to harvest a good breakout gain and followed by a reversal gain. So, one will harvest the big gain and take a rest. But for inexperienced investor, one will make a usual human mistake, *hubris*—thinking one is *invincible* and continue with reverse trading until a huge loss had incurred.

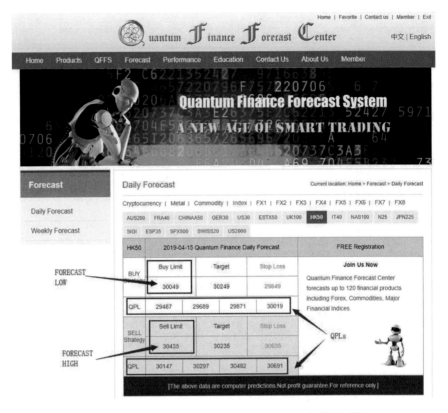

Fig. 6.28 Quantum finance forecast of HSI (Heng Seng Index) on April 15, 2019

6.7.9 QT#3.1 Experienced Traders—Quantum Forecast Single-Product Hedging Strategy (QFSPHS)

As mentioned, *hedging* is one of the most important trading techniques in modern finance. It does not only function as risk control but is also a vital technique in *high-level* trading activity. It is an important component in quantum trading.

According to quantum finance theory and actual practice, 80% of time the market undergoes *oscillation mode* between the forecast high–low region. However, what we are interested in is actually the other 20% *breakout* cases that are usually triggered by *excited events* such as financial news outbreaks or unexpected financial figures announcement, which will push/pull the market from one QPL to another QPL (normally with more than two to three QPL jumps/drops).

So, experienced traders can perform quantum forecast single-product hedging strategy (QFSPHS) like this:

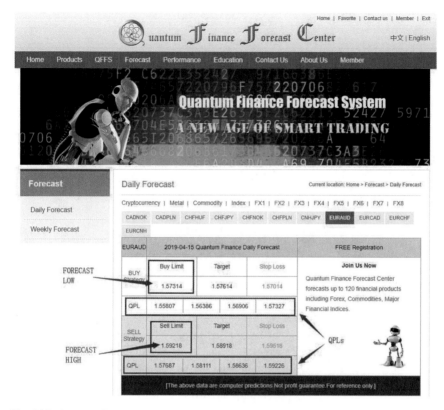

Fig. 6.29 Quantum finance forecast of EURCAD on April 15, 2019

- To capture the 80% *oscillation* market, set buy limit and sell limit at the forecast low and forecast high, respectively, such as 1.57314 and 1.59218, respectively, as shown in Fig. 6.29.
- To cater for the 20% *breakout* scenario, do the SP hedging, i.e., set sell at each QPL beyond the forecast low, combine with averaging and hedging techniques for breakout market. Similar for buy-side breakout.

Depends on the trading skills and/or actual market movement, professional traders can have some variations of QFSPHS. For example, one can release the hedge using QPL(s), i.e., do a hedge at each breakout QPL, and partial cut-loss (release the hedge) simultaneously. That is to conform with quantum finance theory and is a wise move to partial release the hedge.

By using the same technique, one can naturally design and implement more complex, cleverness multi-product and multi-platform quantum hedging strategies.

As an old saying, anything is possible. It all depends on our imagination.

6.8 Conclusion

In this chapter, we have studied seven major trading strategies which can be applied to financial products and trading platform.

Also, we have studied how to apply quantum finance to design quantum trading and hedging strategy.

Note that these methods can be applied effectively only if one has sufficient knowledge and actual experience of technical analysis and quantum finance.

In addition, trading activities can be (and should be) carried out by a combination of these techniques and strategies to yield the best performance.

A final reminder: risk control strategy is the most important strategy that we must always bear in mind. The fact is: The biggest failure is always the one at most confident and lack of risk awareness.

As an old saying in Wall Street:
There is no genius in trading.
The truth is: the common factor for all "successful traders" are "Failures".
Only though numerous Trials-and-Errors one can learnt and perfect one's
trading skills.
So, be patient. A successful trading strategy is the one that can be tested
against TIME.

Problems

6.1   Why *self-discipline* is important to financial trading as mentioned in the opening quotation of this chapter?

6.2   How to maintain *self-discipline* in financial trading? State three techniques or methods and explain how they work.

6.3   Discuss and explain why program trading (or algorithmic trading) becomes so popular nowadays.

6.4   What are the major similarities and differences between cryptocurrency trading versus traditional forex trading?

6.5   Why conventional trading strategies are mainly based on technical analysis tools and techniques? Give any three technical analysis tools and techniques to support your explanation.

6.6   What is the definition of hedging strategy? Describe briefly how it works in a financial market such as forex market.

6.7   What are the major difficulties and challenges to design a good trading/hedging strategy? And how can we overcome them by using quantum finance technology?

6.8   State and explain four commonly used technical analysis tools and indicators for the design of trading and hedging strategies.

6.9   State and explain seven major types of trading and hedging techniques. For each type of technique, use a financial product and its corresponding financial charts as example for explanation.

6.10 Trend trading is the most commonly used trading method, why? What is/are the major limitation(s)?

6.11 Trend trading using multiple MA signal line crossing technique:

   (i)   What is Fibonacci number sequence?

   (ii)   How it works for the design of trading strategy using multiple MA signal lines crossing techniques?

   (iii) If we want to use three MA signal lines crossing technique, what is the best combination of MA number from the Fibonacci to represent: daily, weekly, and monthly pattern crossing? Why?

   (iv)   What are major advantages and shortcomings using Fibonacci number sequence for the selection of MA signal lines?

6.12 Discuss and explain breakout trading strategy. Why it is commonly used by experienced traders and hedgers?

6.13 What are the major challenges in breakout trading strategy and how quantum finance can be helped to solve these problems?

6.14 How can we relate and interpret market breakouts in terms of quantum finance theory? And how we can use QPL to design a scientific breakout strategy?

6.15 Discuss and explain what is reversal trading in finance? And how quantum finance technology can be used for reversal trading?

6.16   Discuss and explain why portfolio trading becomes a popular strategy for the design of trading and hedging strategy. How it works?

6.17   Discuss and explain any three latest R&D of trading and hedging strategies using AI and related technology.

6.18   Why AI technologies such as fuzzy logic and genetic algorithms are commonly used in the design of intelligent trading and hedging systems?

6.19   What are the major limitations of using technical analytical techniques and tools for the design of trading strategies? and how quantum finance provides an alternative solution?

6.20   In the old days, financial trading and hedging are totally two different concepts and methodologies, but now the distinction between them becomes obscure. Why? Is it related to the popularity of program trading?

6.21   What are Bollinger Bands? Discuss and explain how to use Bollinger Bands to design trading strategy. What are the major limitations and how quantum finance can be applied to resolve these problems?

6.22   What is stop-loss strategy? Why it is important especially for the design of program trading strategies?

6.23   State and discuss three major hedging strategies. Give an example of financial product for explanation.

6.24   Discuss and explain the major concepts and approach of quantum trading. Which one is more preferable for the application to program trading? Why?

6.25   Discuss and explain the major similarities and differences between QPL and S & R in technical analysis. How we can use QPL to replace S & R for the design of program trading and hedging algorithms?

6.26   State and explain three levels of quantum finance trading and hedging strategies. Why they are divided into three levels? Each of them uses a financial product as example to explain how it works.

6.27   For inexperienced investors, it is preferable to use daily forecasts together with simple stop-loss strategy instead of hedging technique. Why?

6.28   Quantum trading program.

    (i)   Study MQL workshop from QFFC.org to learn the basic skill of design and writing MQL-based trading programs.

    (ii)   Finish workshop #1 to learn how to design and implement an MQL program to calculate all the QPR and QPL of any financial product based on the past (says) 2000-trading day time series.

    (iii) Finish workshop #2 to learn how to design and implement quantum finance forecast system to calculate the daily forecast high/low of any financial product.

    (iv) Based on the level 1 quantum trading strategies to implement: (1) QT#1.1 quantum forecast oscillation trading strategy (QFOTS) and (2) QT#1.2 quantum forecast oscillation trading with averaging strategy (QFOTAS) using QPL for at least five forex products (e.g., AUDCAD, EURJPY, USDCAD, EURGBP, etc.).

(v) First, use MT4 simulator to test for the trading performances of these five products for at least 1 year.

(vi) Second, use demo account to perform the mock trading for at least 20 trading days (i.e., 1 month) and compare their trading performance and overall monthly returns.

(vii) Once you are experienced in quantum trading, implement the level 2 and level 3 quantum trading strategies and compare their trading performances.

References

Allen, F. and Karjalainen, R. (1999) Using genetic algorithms to find technical trading rules. *Journal of Financial Economics*. 51(2): 245–271.

Byun, H. W. et al. (2015) Using a principal component analysis for multi-currencies-trading in the foreign exchange market. *Intelligent Data Analysis*. 19(3): 683–697.

Chang, Y. and Lee, M. (2017) Incorporating Markov decision process on genetic algorithms to formulate trading strategies for stock markets. *Applied Soft Computing*. 52: 1143–1153.

Evans, C. et al. (2013) Utilizing artificial neural networks and genetic algorithms to build an algo-trading model for intra-day foreign exchange speculation. *Mathematical and Computer Modelling*. 58(5–6): 1249–1266.

Helseth, A. (2016) Stochastic network constrained hydro-thermal scheduling using a linearized progressive hedging algorithm. *Energy Systems*. 7(4): 585–600.

Hsu, Y. et al. (2009) Profit refiner of futures trading using clustering algorithm. *Expert Systems With Applications*. 36(3): 6192–6198.

Iba, H. and Aranha, C. C. (2012) *Practical Applications of Evolutionary Computation to Financial Engineering: Robust Techniques for Forecasting, Trading and Hedging (Adaptation, Learning, and Optimization)*. Springer.

Kaufman, P. J. (2013) *Trading Systems and Methods*. Springer.

Leshik, E. and Cralle, J. (2011) *An introduction to algorithms for stock trading on the NASDAQ and New York Stock Exchange*, Wiley-Blackwell.

Li, Z. et al. (2017) Research on Quantitative Trading Strategy Based on Neural Network Algorithm and Fisher Linear Discriminant. *International Journal of Economics and Finance*. 9(2): 133.

Liu, X. et al. (2017) Quantified moving average strategy of crude oil futures market based on fuzzy logic rules and genetic algorithms. *Physica A: Statistical Mechanics and its Applications*. 482: 444–457.

Murphy, J. J. (1999) *Technical Analysis of the Financial Markets: A Comprehensive Guide to Trading Methods and Applications*. New York Institute of Finance.

Naranjo, R. et al. (2015) An Intelligent Trading System with Fuzzy Rules and Fuzzy Capital Management. *International Journal of Intelligent Systems*. 30(8): 963–983.

Nelken, I. (1999) *Pricing, Hedging, and Trading Exotic Options: Understand the Intricacies of Exotic Options and How to Use Them to Maximum Advantage*. McGraw-Hill.

Pauna, C. (2019) Data Mining Methods on Time Price Series for Algorithmic Trading Systems. *Informatica Economica*. 23(1): 26–36.

Rhoads, R. (2011) *Trading VIX Derivatives: Trading and Hedging Strategies Using VIX Futures, Options, and Exchange-Traded Notes*. Wiley.

Schmidt, G. et al. (2010) Experimental Analysis of an Online Trading Algorithm. *Electronic Notes in Discrete Mathematics*. 36: 519–526.

Shiau, J. (2009) Optimization of Reservoir Hedging Rules Using Multiobjective Genetic Algorithm. *Journal of Water Resources Planning and Management*. 135(5): 355–363.

Tuchong (2019) Intelligent trading and market analysis. http://stock.tuchong.com/image?imageId= 533733773505462352. Accessed 21 Aug 2019.

Vella, V. and Ng, W. L. (2015) A Dynamic Fuzzy Money Management Approach for Controlling the Intraday Risk-Adjusted Performance of AI Trading Algorithms. *Intelligent Systems in Accounting, Finance and Management*. 22(2): 153.

Yang, S. Y. et al. (2018) An investor sentiment reward-based trading system using Gaussian inverse reinforcement learning algorithm. *Expert Systems with Applications*. 114: 388–401.

Zhang, J. and Maringer, D. (2016) Using a Genetic Algorithm to Improve Recurrent Reinforcement Learning for Equity Trading. *Computational Economics*. 47(4): 551–567.

Chapter 7
AI Powerful Tools in Quantum Finance

*A computer would deserve to be called intelligent if it could
deceive a human into believing that it was human.*
Alan Turing (1912–1954)

What is intelligence?
What is artificial intelligence?
What are the core AI technologies?
How can we apply AI to quantum finance?

All who have heard about artificial intelligence (AI) must be familiar with Sir Alan Turing—the father of AI with his famous Turing test. He was an English mathematician, computer scientist, cryptanalyst, and theoretical biologist. He was highly influential in the development of theoretical computer science, provided a formalization on algorithm concepts and computation with his famous invention—*Turing machine* was deemed as original computer prototype. His famous quotation brought out the foundation concept and AI notion—machine or software to mimic and exemplification of human behaviors, which is also the core concept of *Turing test*.

In previous chapters, we have learnt the basic concepts of quantum finance.

More importantly, we have learnt how to make use of time series returns, together with quantum anharmonic oscillator model of quantum finance Schrödinger equation to evaluate quantum price levels (QPLs) on any financial market, which provide us a new way and concrete solution to model financial markets.

We have also learnt how to apply QPLs, together with the seven major financial trading (hedging) techniques to design and implement quantum trading and hedging systems. The only component left behind is: how to make use of time series financial information, together with QPLs to forecast financial markets using state-of-the-art AI techniques?

© Portions of this chapter are reprinted from Lee (2006), with permission of Springer Nature.

In this chapter, we will study the three *pillars* of AI:

- Artificial neural networks,
- Fuzzy logics, and
- Genetic algorithms.

In Chap. 8, we will study other two important computational intelligence (CI) techniques in finance: chaos theory and fractals, how they can be used in quantum finance. Prior to that, let's review a brief history of artificial intelligence.

7.1 An Overview of Artificial Intelligence

7.1.1 From Greek Mythology to Turing Test

Many textbooks on artificial intelligence named Dartmouth conference in 1956 as the *Birth of AI.*

Was it?

The answer is yes and no.

"Yes" in the sense that five AI founders: Professors Allen Newell (CMU), Herbert Simon (CMU), John McCarthy (MIT), Marvin Minsky (MIT), and Arthur Samuel (IBM) coined the word of *artificial intelligence (aka AI)* at this conference. Since then, R&D on AI had commenced officially as a newly emerged discipline. They also became leaders of AI research in different AI domains for the following decades.

"No" in the sense that: the notion of *man-made intelligence*, or what we called *artificial intelligence (AI)* is not a new concept.

According to Greek mythology (Greek epic poet Hesiod's Theogony 700B.C.), Prometheus (Hansen 2005), a titan and culture hero who created first human from clay ("humanoid-robot" in today's term) and defied Zeus to steal fire (intelligence) from the chariot of the sun for further progress and comfort of human civilization.

If we review the formal conceptualization of AI theory, one may think of Sir Alan Turing and his celebrated invention—Turing test.

In 1950, he published *Computing Machinery and Intelligence* (Turing 1950). In his paper, he designed a method to evaluate whether a machine can be characterized as human and pondered the way to measure technology progression. To study the basic problem: *Can machines think?* He initiated a series of tests that were meant to help human answer the question—*Turing test* (at that time he called this test as *imitation game*).

The test comprised of an interrogation room conducted by a *judge*. Figure 7.1 illustrates that there are two unseen subjects in individual terminals being judged: a human and a computer program. Based on the quality of conversation, the judge makes dialogue to ensure both subjects within a designated timeframe and attempts to identify persona between human and computer program. He concluded that if the

Fig. 7.1 Illustration of Turing test

judge is likely to pick out either subject, it showed that the computer has succeeded in revealing to *think* with human intelligence.

Modern version of the *Turing test* has employed more than one human judge to converse and interrogate with both subjects. The test is marked success if more than 30% of the judges contemplated the computer program as human subsequent to five minutes' conversation. The *Loebner Prize* is an annual *Turing test* competition that was started by American inventor and activist Dr. Hugh Loebner (1942–2016) in 1991. He created additional rules to the original Turing test, calling for human and computer program to have 25-min conversation with each of the four judges presiding over the competition. The winner of the competition (not necessarily one that passes the Turing test) is the one whose computer bot received the most votes and highest ranking from the judges, regardless of whether there were 30% judges' votes. Despite Sir Alan Turing predicted that by year 2000 machines would eventually pass the Turing test, this did not occur until 2014. In 2014, Professor Kevin Warwick organized a Turing test competition to mark the 60th anniversary of Sir Alan Turing's death. A computer chatbot called Eugene Goostman with the persona of a 13-year-old boy had passed the Turing test in this competition. He secured 33% judges' votes who were convinced that he was human.

7.1.2 Definition of AI

Although there is no official definition of AI, one refers to definition of AI from Webster's Dictionary and the American Association for Artificial Intelligence (AAAI) as

Fig. 7.2 Illustration of AI (Tuchong 2019a)

> The capacity of computers or programs to operate in ways to mimic human thought processes, such as reasoning and learning.
>
> —Webster's New College Dictionary
>
> The scientific understanding of the mechanisms underlying thought and intelligent behavior and their embodiment in machines.
>
> —AAAI

Professors Stuart Russell and Peter Norvig in their remarkable book: *Artificial Intelligence—A Modern Approach* (2015) provided a substantial survey on AI definition in four major categories:

- Systems that think like human;
- Systems that think rationally;
- Systems that act like human; and
- Systems that act rationally.

In author's book—*Fuzzy-Neuro Approach to Agent Applications* published by Springer in 2006 (Lee 2006) defined AI as (Fig. 7.2)

> The exemplification of human intellectual thoughts, acts and behaviors for the design and implementation of intelligent systems, software objects (agents) and robotic systems.

7.1.3 Strong Versus Weak AI

Since Dartmouth conference, many AI scientists at that time predicted that a machine with human intelligent would exist in no more than a generation and they were given millions of dollars to make this vision come true. However, it became apparent that the difficulties of constructing human-like robot were underestimated.

In 1973, in response to criticism of prolonged progress, shortages of practical and promising AI applications, both U.S. and British Governments terminated funding undirected research on artificial intelligence subsequent to *winter of AI*. Seven years later, a visionary initiative by the Japanese Government inspired governments and industry to provide billions of dollars on AI research, but by the late 80s investors became disillusioned by insufficient computer capacity (hardware) and withdrew funding again.

It was not until the first decades of twenty-first century, investment and interest in AI revived, while machine learning was successfully applied to many problems in academia and industry due to the presence of powerful computer hardware. It was called the second golden age of AI.

One special feature of AI is the conformity with AI. There were debates after Dartmouth conference even ongoing at present, on AI general definition and classification.

Professor Robert Wilensky (1951–2013), an influential AI scholar mentioned in his work—*Planning and Understanding: A Computational Approach to Human Reasoning* stated that:

Artificial Intelligence is a field renowned for its lack of consensus on fundamental issues.

One typical example was the argument between strong (hard) AI versus weak (soft) AI. Some AI scientists believed that AI should focus on the design and systems/programs implementation that mimic and simulate how human think and act— the so-called *weak AI*. Opponents believed that AI systems should not only think and act like human, but rather should *think* and *behave* consciously like human— so-called *strong AI*.

In fact, Turing test is an exemplification example of *strong AI*.

7.1.4   Strong (Hard) AI

Strong AI refers to the study of machine intelligence that could successfully perform any intellectual task as per human. It is also called *artificial general intelligence (AGI)*.

Key characteristics of strong AI include the ability to reason, solve puzzles, make judgments, plan, learn, and communicate. It should have consciousness, objective thoughts, and self-awareness.

Some AI scientists argue that a machine with strong AI should be able to go through the same human development process, starting with a childlike mind through learning in developing an adult mind. It should be able to interact with the world, learn from it, acquire its own common sense, and language. Another argument is that we will not know when we have developed strong AI (if it can indeed be developed) because there is no consensus on what constitutes intelligence.

Core areas of strong AI research include the following:

- Construct intelligent machines, robotic in particular;
- R&D on generalized rule-based and case-based systems, e.g., IBM Deep Blue supercomputer on chess playing; and
- R&D on consciousness, objective thoughts with self-awareness.

Owing to its complexity, the progress of strong AI was prolonged between 1950–1980s.

The celebrate AI versus human contests—Deep Blue versus Mr. Garry Kasparov—six-game chess matches between world chess champion and an IBM supercomputer Deep Blue. The first match was played in Philadelphia and won by Kasparov in 1996. The second was played in New York City and won by Deep Blue in 1997. The 1997 match was the first defeat of a reigning world chess champion by a computer under tournament condition.

7.1.5 Weak (Soft) AI

The emergence of computer technology such as software engineering is the software design (programs) that can be executed by computer (systems) to work out problems. In this respect, we have a new area on AI R&D which is not focused on *general intelligence* implementation, but rather mimic human behaviors, especially on *thinking* and *problem-solving*—so-called weak (soft) artificial intelligence.

Weak AI simulates human cognition and benefits mankind such as data analysis and automates time-consuming functions in ways that human is unable to do on occasions. In order to surpass this limitation, mimic human intelligence grew exponentially in the last 20 years. In fact, weak AI helps big data transform into usable information by detecting patterns and making predictions—data mining. Examples include Facebook's news feed, Amazon's suggested purchases, Apple's Siri, and iPhone's technology that answer users' verbal enquiries.

Major R&D on soft AI include the following:

- Artificial neural networks (ANNs)—using computer models to mimic human brain (neural) structure on thinking and problem-solving.
- Genetic algorithms (GAs)—using computer models to mimic human genetic evolution, also called evolutionary computing (EC) or evolutionary programming (EP).
- Fuzzy logic (or fuzzy logic system, FLS)—using computer models to mimic human imprecise (so-called fuzzy) determination of matters and events, typical applications on real-time control systems.

As we can see, the boundary between hard and soft AI becomes obscure. Modern AI nowadays entails integrated disciplines including cognitive science, neuroscience, biological and artificial neural networks, evolutionary computing technology, robotic (engineering), data mining and deep learning, active vision, natural language processing (NLP), etc. (Fig. 7.3).

Fig. 7.3 Weak AI (Tuchong 2019b)

7.1.6 Classification of Artificial Intelligence

Owing to the advance of computing technology and the diversification of AI application in different areas, the author had defined AI classification into two main categories: macroscopic versus microscopic AI (Fig. 7.4).

Macroscopic AI

- Also called *behavioral approach* (BA) AI refers to the *acts* and *behaviors.*
- Focus on the design and implementation of intelligent machines (systems) that mimic high-level (appears to be seen) human intellectual behavior.
- BA-AI is further sub-divided into two sub-categories: mind-BA—high-level mental activities and body-BA—high-level body and control activities.

Microscopic AI

- Also called *cognitive approach* (CA) AI refers to the *thoughts* and *mental processes.*
- Focus to mimic human low-level mental activities, referred by cognitive scientists as human cognitive mental activities.
- According to various types of mental process, CA-AI approach is sub-divided into four major types: neural network, fuzzy logics, evolutionary computing, and chaotic systems.

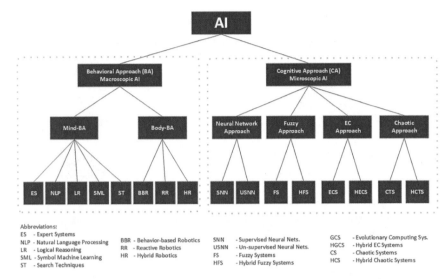

Fig. 7.4 Classification of artificial intelligence

7.2 Neural Networks—The Brain of AI

7.2.1 *Biological Neural Network—Our Brain*

Artificial neural networks (ANNs) (or *neural networks* in short) is one of the major components of microscopic AI, which focus on the study and modeling of intelligent system to mimic one of the most important human organs—the *brain.*

The first scientist to work in the area of *brain science* was Italian physician Professor Camillo Golgi (1843–1926) who invented the *stain method* to investigate neural activities inside the brain (Bentivoglio 2014). By using this method, he proposed that the brain is made up of *syncytium*—a sponge-like tissue that is *activated* by the staining operation. Based on his discovery, Spanish neuroanatomist Professor Santiago Ramón y Cajal (1852–1934) proposed an innovative idea that "*these staining tissues were not sponge-like elements, but rather the collections of the brain cells called neurons*", which were interlinked together to form a complex group—*neural networks* as shown in Fig. 7.5. Both esteemed medical professionals were awarded the Nobel Prize for Physiology or Medicine in 1906.

Figure 7.5 shows each neural cell (neuron) consists of

- nucleus—central body of the neuron;
- axon—prolonged filament which connects to other neurons;
- dendrites—tree-like structures which branch from the neuron; and
- synapse—the axon tips (junctions) that make contact with other neurons by attaching to the dendrites of these neighboring neurons.

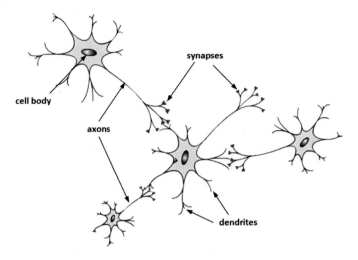

Fig. 7.5 Biological neural network

Latest neuroscience revealed that human brain consists of around 10^{11} neurons. The enigma is: the volume of neurons in human brain is so massive that we can only make use of less than 10% of neurons throughout our lives, which is rather against the basic evolution theory.

In fact, how such biological neural network works was still a mystery at that time.

7.2.2 Integrate-and-Fire Operations in Biological Neural Network

Prior 1943, almost all neuroscientists believed that the sole purpose of neurons was to process energy. Yet the way they work such as how to process information and store memory was unsolved.

In 1943, neurophysiologist Professor Warren McCulloch (1898–1969) and mathematician Professor Walter Pitts (1923–1969) published an influential paper *A logical calculus of the ideas immanent in nervous activity* (McCulloch and Pitts 1943) which set off the birth of artificial neural networks, *ANN* in short (Abraham 2016).

In the paper, they proposed that the main function of neural activities was to *process information*—not energy storage. They maintained that the functions of the neurons were just like *logical switches*. Signal transmission from one neuron to another at synapses is the result of a complex chemical process in which specific transmitter substances are released from the sending points of the junctions. If the potential reaches a certain *threshold*, a pulse will be generated down the axon, known as *firing*, as shown in Fig. 7.6. More importantly, in the paper, they demonstrated

Fig. 7.6 3D illustration of integrate and fire in neural network (Tuchong 2019c)

how their proposed network (now called *artificial neural network*) could be used to perform basic logical operations such as and, or, and not operations.

This breakthrough has not only merely solved the *century mystery* of how *biological neural network* works, but also provided a solid foundation for digital computing technology development.

Although we now understand that the neural activities in our brains are quite different from logical switches such as transistors in digital computer, they behave somewhat nonlinear (or even chaotic) *integrate-and-fire* operators for information process and transmission. This discovery in 1943 coined the *first golden age of artificial neural networks*.

7.2.3   A Neuron Model

As a direct analog of the biological neuron, the schematic diagram of neuron structure can be interpreted as a computational model in which synapses are represented by weights that modulate the effect of associated input signals, the formulation is given by

$$y = f\left[\sum_{i=1}^{n} w_i x_i\right] \tag{7.1}$$

where x is the input signals, w's are the weights, and y is the output.

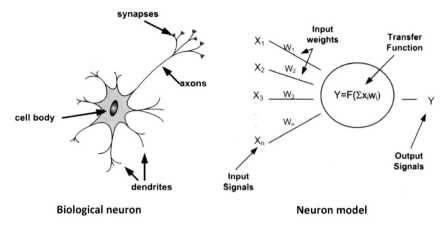

Fig. 7.7   Biological neuron versus neuron model

The nonlinear characteristics exhibited by the neuron is represented by a transfer function $f(x)$ such as a binary or bipolar sigmoid function, given by
Binary sigmoid function:

$$f(x) = \frac{1}{1 + e^{-\sigma x}} \tag{7.2}$$

Bipolar sigmoid function:

$$f(x) = \frac{1 - e^{-\sigma x}}{1 + e^{-\sigma x}} \tag{7.3}$$

in which σ is the *steepness parameter* to control the curvature of the transfer function. The learning capability of an artificial neuron is achieved by adjusting the weights in accordance with a predefined learning algorithm, usually in the form of

$$\Delta w_j = \alpha \sigma x_j \tag{7.4}$$

where α is the *learning rate* and σ is the *learning momentum*.

Figure 7.7 illustrates the analog between the biological neuron and its neuron model.

7.2.4   Single Hidden Layer Artificial Neural Network Model

Artificial neural networks (ANNs, or neural network in short) are the pieces of a computing system designed to simulate the way human brain analyzes and processes information. They are the foundations of artificial intelligence (AI) and solve problems that would prove impossible or difficult by human or statistical standards. ANN has self-learning capabilities that enable them to produce better results as more data become accessible.

A typical artificial neural network consists of intermediate layer(s) known as *hidden layers* to facilitate nonlinear computational capabilities of the network system.

Classical ANNs, such as feedforward neural network (FFNN) shown in Fig. 7.8, allow signals (information) to flow from input units to output units in a forward direction.

Other basic ANNs include classical Kohonen self-organizing map (SOM) and learning vector quantization (LVQ) that are based on competition, along with the adaptive resonance theory (ART), and also our main theme—feedforward backpropagation neural network (FFBPN).

From mathematical perspective, ANNs can be regarded as multivariate nonlinear analytical tools. They are known to be superior at recognizing patterns from noisy, complex data, and estimating their nonlinear relationships.

Many studies revealed that ANNs possess distinguished learning proficiency in the underlying mechanics of time series problems ranging from stocks prediction and foreign exchange rates in various financial markets to weather forecast.

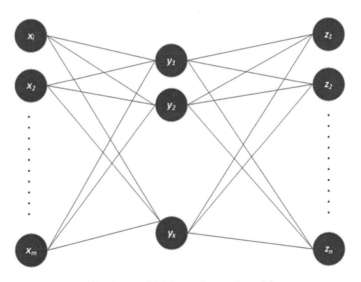

Fig. 7.8 A typical single hidden layer artificial neural network model

7.2.5 Classification of ANNs by Machine Learning Technique

Following half a century of development, numerous neural networks have been proposed with over 20 different types of commonly used ANNs.

Generally, ANNs are commonly classified by (1) machine learning techniques (Fig. 7.9) and (2) areas of application (Fig. 7.10).

In terms of machine learning techniques, ANNs can be categorized into three main types:

1. Supervised-learning (SL) neural networks—network learning (training) based on input–output (target output) pairs. Typical examples include feedforward back-propagation neural network (FFBPN), Hopfield network, support vector machine (SVM), radial basis function (RBF) network, etc.
2. Unsupervised-learning (UL) neural networks—neural networks that do not need any supervised learning and training strategies, include all kinds of self-organizing, self-clustering, and learning networks such as SOM, ART (adaptive resonant theory), etc.
3. Reinforcement-learning (RL) neural networks—different from supervised-learning counterpart with well-defined input–output pairs, reinforcement learning trains the neural network to adopt with feedback signals, namely, reinforcement signal (RS). For the right behavior, the network will respond with a positive RS to *award* the RL network, whereas the wrong behavior, the network will respond with a negative RS to *punish* the RL network. This method is particularity useful to tackle optimization problem without exact target solutions such as trading strategy optimization. We will study the advanced topics in detail on multiagent-based intelligent trading strategies in Chap. 13.

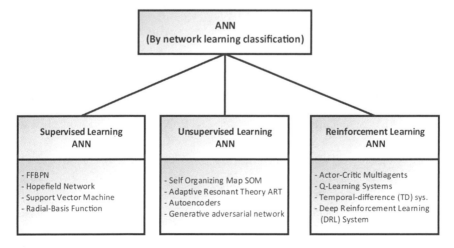

Fig. 7.9 Classification of ANNs by machine learning technique

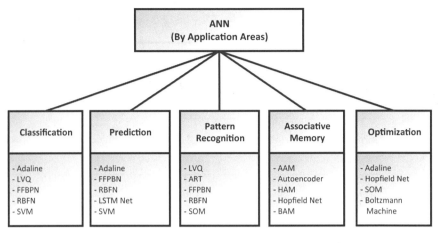

Key:
AAM – Auto-associative Memory
ART – Adaptive Resonant Theory
BAM – Bidirectional Associative Memory
FFBPN – Feedforward Backpropagation Network
HAM – Hetero-associative Memory

LTSM Net. – Long Short-Memory Term Network
RBFN – Radial Basis Function Network
SOM – Self-Organizing Map
SVM – Support Vector Machine

Fig. 7.10 Classification of ANNs by areas of application

7.2.6 *Classification of ANNs by Areas of Application*

In terms of *areas of application*, ANNs can be categorized into five major types:

- classification,
- prediction,
- pattern recognition,
- associative memory, and
- optimization.

Figure 7.10 shows the classification of ANNs by areas of application. As shown in Fig. 7.10, there are certain ANNs such as feedforward backpropagation network (FFBPN) and radial basis function network (RBFN) that can be used on multiple area applications such as classification, prediction, and pattern recognition, whereas other ANNs such as bidirectional associative memory (BAM) and Boltzmann machine are specialized to a particular application area.

In this chapter, we will study three basic and commonly used ANNs, they are as follows:

- Associative network,
- Hopfield network, and
- Feedforward backpropagation network (FFBPN).

7.2.7 *Auto-associative Network—Network Architecture*

Associative learning is one of the major fundamental characteristics of human intellectual behavior.

It is also widely used by human and machines for pattern recognition such as visual pattern identification and recognition include recalling human faces, voices, and music. Significantly, it relates to recalling knowledge and memory association.

Figure 7.11 shows the system architecture of an auto-associative network. It is a single-layered neural network to store a set of patterns for pattern association (or what we call *recalling*).

The training of the associative network is conducted by stored patterns iterative presentation updated by weights in accordance with the training algorithm.

Once the training is completed, the network can associate not only with the stored pattern but can also associate with the correct incomplete or noisy query stored pattern.

Basically, there are two major kinds of associative networks:

1. *auto-associative networks*—in which input (and query) patterns are the same type (and nature) as the associated pattern; and
2. *hetero-associative networks*—in which input (and query) patterns are completely different types (and nature) from the associated patterns.

Fig. 7.11 System architecture of auto-associative network

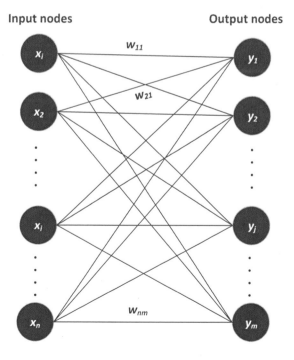

7.2.8 Auto-associative Network—Network Training Algorithm

The Hebb rule (Hebb 1949) is commonly used for network training. Figure 7.12 shows the network training algorithm of auto-associative network (Fausett 1994; Patterson 1996).

Binary and bipolar vectors can be used in associative networks. Both training vectors will be a set of training input and target output pairs (x', y').

Hebb rule is simple to exercise on weight adjustments, and other methods such as the delta rule can also be adopted. In that case, the weight adjustment formula (in Step 2.3) will be replaced by

$$w_{ij}(new) = w_{ij}(old) + \alpha\left(y_j' - y_j\right)x_i \ where \ i \in [1 \ldots n], j \in [1 \ldots m] \qquad (7.5)$$

in which α is the learning rate of the network.

Since associative network architecture is plain to discern complex patterns such as human face association and character recognition, it does provide an innovative means to apply neural networks in association with memory storing and pattern recalling.

Inspired by these discoveries in neuroscience and neurophysiology, the author (Lee, 2004) proposed a chaotic neural associative network—the Lee-associator—to model chaotic and progressive human memory recalling mechanisms.

Training Algorithm – Associative Network
(Fausett, 1994; Patterson, 1996)

Step 1: Network weight initialization
        For all i, j where $i \in [1 \ldots n], j \in [1 \ldots m], n$ and m are the total numbers of neuron input and output nodes respectively.
    Set $w_{ij} = 0$

Step 2: For each training pair (x', y'), perform the following operations:
    Step 2.1:   Set the activation values for the input nodes as the values of the training input.
                i.e. $x_i = x_i'$ $i \in [1 \ldots n]$
    Step 2.2:   Set the activation values for the output nodes as the values of the target output.
                i.e. $y_j = y_j'$ $j \in [1 \ldots m]$
    Step 2.3:   Update ALL the weights in the network
                i.e. $w_{ij}(new) = w_{ij}(old) + x_i y_j$
                $i \in [1 \ldots n], j \in [1 \ldots m]$

Fig. 7.12 Network training algorithm of auto-associative network

Fig. 7.13 System architecture of discrete Hopfield network

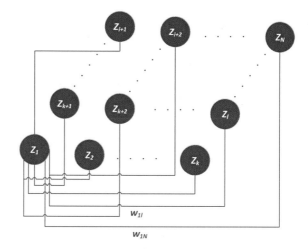

7.2.9 Hopfield Network—Network Training Algorithm

Professor John Hopfield published his influential paper in 1984, *Neurons with graded response have collective computational properties like those of two-state neurons* (Hopfield 1984). In this paper, he described how a simple recurrent auto-associative network can be used for content-addressable memory systems. Hopfield network also provided a model for the understanding of human memory, which can also be used in pattern recognition to tackle complex optimization problems such as the typical traveling salesman problem (TSP).

The architecture of Hopfield network (Fig. 7.13) is similar to a classical auto-associative network but with three basic differences:

1. Hopfield network is a recurrent network in the sense that output nodes in one time step are fed as input in the next time step.
2. In classical associative network, all neurons update their activations simultaneously, whereas in Hopfield network, one neuron selects solely at a time to update its activation and then *broadcast* its new state to other members of the network.
3. Each neuron keeps on receiving the *stimulus* from external signal during the entire auto-association process.

7.2.10 Hopfield Network—Network Training Algorithm

Hopfield network's key aspect is that it demonstrated how a powerful memory storage and retrieval device can be composed of simple auto-associative network modification. Figure 7.14 shows the training algorithm of a classical *discrete Hopfield network*.

Training Algorithm – Discrete Hopfield Network
(Fausett, 1994; Patterson, 1996)

Step 1: Store all the (binary) patterns into the network (using Hebb Rule).
        For each pattern $x'_p = (x'_{p1}, x'_{p2}, \ldots, x'_{pm})$, where p is the pattern number and m
        is the total number of patterns to store, calculate:

$$w_{ij} = \sum_{p=1}^{m}[2x'_{pi} - 1][2x'_{pj} - 1] \qquad for\ i \neq j$$
otherwise $w_{ij} = 0$

Step 2: If neuron activations have not yet converged, do the following:
        Step 2.1: Set the initial activation values for the network as the values of the ex-
              ternal input vector x':
              i.e. $z_i = x'_i$ $i \in [1 \ldots n]$
        Step 2.2: For each unit x, performs Step 2.2.1-2.2.3
              (update unit in random order):
              Step 2.2.1: Calculate the network input:

$$z_{in_i} = x_i + \sum_{j=1}^{n} z_j w_{ji}$$

              Step 2.2.2: Determine the activation value:

$$z_i = \begin{cases} 1 & if\ z_{in_i} > \theta_i \\ z_i & if\ z_{in_i} = \theta_i \\ 0 & if\ z_{in_i} < \theta_i \end{cases}$$

              Step 2.2.3: "Broadcast" the new value of z_i to all neurons in the net-
                      work.
Step 3: Check for the convergence condition.

Fig. 7.14 Network training algorithm of Hopfield Network

In Hopfield network's associative memory, there are two types of association: auto-association and hetero-association. The former is a vector in association with itself, whereas the latter are two different vectors in association with storage. It is important to note that Hopfield's network learning model used the identical to Hebb rule (1949), which basically strived to show that learning occurs as a result of the weights strengthening when the auto-association activity occurs.

In reality, Hopfield network's vast application areas also set off the rebirth of ANNs and explored on how neural networks can be applied to complex problems in daily operations.

Hopfield network also provided a viable model for the understanding and modeling of human memory—a significant component in artificial intelligence.

7.2.11 Feedforward Backpropagation Network (FFBPN)—Network Architecture

Different from the previous two neural networks, feedforward backpropagation networks (FFBPNs) provide a multilayer network architecture (Lee 2006; Silva et al. 2017).

A typical FFBPN consists of an input layer, a hidden layer, and an output layer as shown in Fig. 7.15. Although FFBPN can consist of several hidden layers, one or maximum two hidden layers are usually sufficient in most of the cases.

The network training of FFBPNs consists of three main processes:

1. The *feedforward* process of network training;
2. The *error evaluation* process to calculate the errors between calculated output values and target output values; and
3. The *backpropagation* process of the errors for weight adjustments.

Similar to most other networks, training terminates when errors are bound within the tolerance level.

Note:

- In the network architecture, w's denote network weights between the input and hidden layers, and u's denote network weights between the hidden and output layers.
- The total numbers of neurons located in input (n), hidden (t), and output layers (m), respectively.
- For activation functions, normally a sigmoid function is adopted.

Figure 7.18 shows the training algorithm of a typical FFBPN.

Fig. 7.15 System architecture of FFBPN

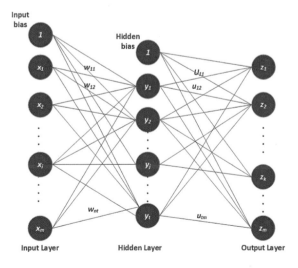

7.2.12   Feedforward Backpropagation Network (FFBPN)—Training Algorithm

An FFBPN can model various kinds of problems such as weather prediction and stock forecasting, pattern recognition such as character recognition, classification, and even optimization. In quantum finance stock (or forex) prediction, after time series financial data (e.g., Daily O, H, L, C, V) are normalized, they are used together with QPLs as input signals for network training in order to calculate next day's H/L forecasts. Figure 7.16 shows the training algorithm of FFBPN.

Training Algorithm – Feedforward Backpropagation Network (FFBPN)
(Fausett, 1994; Patterson, 1996)

Step 1:   Network weight initialization.
Set all network weights w_{ij}, u_{jk} to a small random number between 0 and 1.

Step 2:   While error $\geq$ threshold value, do the following:
Step 2.1   For each training pair (x, z) do Step 2.1.1 – 2.1.7.

Feedforward Operation

Step 2.1.1:   Calculate the input state of each hidden node:
$$y_{in_j} = \sum_{i=0}^{n} x_i w_{ij} \qquad \text{where } x_0 \text{ is the input bias}$$

Step 2.1.2:   Calculate the activation value for the hidden node:
$$y_j = f_y(y_{in_j}) \qquad \text{where } f_y() \text{ is the activation function}$$

Step 2.1.3   Calculate the input state of each output node:
$$z_{in_k} = \sum_{j=0}^{t} y_j u_{jk} \qquad \text{where } y_0 \text{ is the hidden bias}$$

Step 2.1.4:   Calculate the activation value for the output node
$$z_k = f_z(z_{in_k}) \qquad \text{where } f_z() \text{ is the activation function}$$

Backpropagation Operation

Step 2.1.5:   For each output node:
(a)   Calculate the error with the target value
$$\xi_k = (z_k' - z_k)f_z'(z_{in_k}) \qquad \text{where } f_z' \text{ is } df_z/dz$$
(b)   Calculate the correction errors
$$\Delta u_{jk} = \alpha \xi_k y_j \qquad \text{where } \alpha \text{ is the learning rate}$$

Step 2.1.6:   For each hidden node:
(a)   Calculate the accumulated errors in the hidden node
$$\lambda_{in_j} = \sum_{k=1}^{m} \xi_k u_{jk}$$
(b)   Calculate the correction errors in the hidden node
$$\lambda_j = \lambda_{in_j} f_y'(y_{in_k}) \qquad \text{where } f_y' \text{ is } df_y/dy$$
(c)   Calculate the weight adjustments
$$\Delta w_{ij} = \alpha \lambda_j x_i$$

Step 2.1.7:   Update all weights for the two layers simultaneously:
$$w_{ij}(new) = w_{ij}(old) + \Delta w_{ij}$$
$$u_{jk}(new) = u_{jk}(old) + \Delta u_{jk}$$

Step 2.2   Check the stopping criteria

Fig. 7.16   Network training algorithm of FFBPN

However, FFBPN has certain intrinsic limitation such as trapping in local minima along with difficulty in choosing optimal parameter settings (and input vectors). Lengthy convergence rate is also another major concern. A possible solution is the integration with chaotic neural oscillator technique will be studied in Chap. 9.

7.2.13 Neural Network—Where to Go?

In this section, we have studied an overview of neural networks, their basic structure, and mechanisms.

Current trends of neural networks research and development are focused on three major areas:

1. The integration of other AI techniques to remedy certain classical neural networks intrinsic limitations includes fuzzy logic—fuzzy-neuro systems, genetic algorithms (GAs) to overcome parameter selection problem and to fine-tune networks;
2. The investigation and study in neural dynamics, especially in neural oscillators, neural oscillatory models; and
3. The investigation and study in chaotic neural dynamics of chaotic neural networks to model complex AI problems (Figs. 7.17 and 7.18).

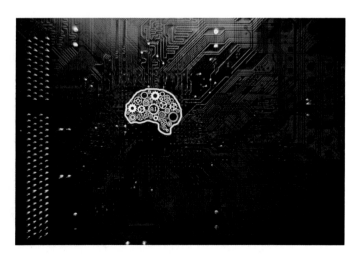

Fig. 7.17 Neural network and AI (Tuchong 2019d)

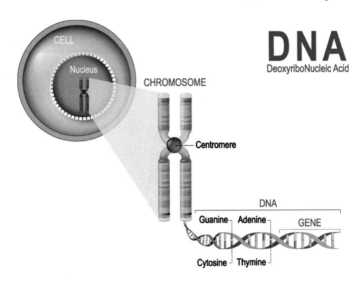

Fig. 7.18 Illustration of DNA and chromosome in a eukaryotic cell

7.3 Genetic Algorithms—The Optimization Engine

7.3.1 Nature of Evolution—Genetic Algorithms

Evolution refers to encode the operations of biological entities (chromosomes) rather than the engineering of the whole living beings themselves.

According to Darwin's evolution theory (Darwin and Huxley 2003), natural selection is based on *survival of the fittest*—in effect, chromosomes with high fitness values will reproduce more than those with low fitness values.

Professor John Holland (1929–2015) introduced genetic algorithms (GAs) in 1960 based on Darwin's concept of evolution theory. In 1975, he wrote a groundbreaking book on genetic algorithms: *Adaptation in Natural and Artificial Systems* (Holland 1992). Afterward, his student Professor David E. Goldberg extended his work on GA and published his distinguished book: *Genetic Algorithms in Search, Optimization and Machine Learning*, which becomes one of the most cited books in computer science and AI (Goldberg 1989).

Basic entity of GAs is the chromosome, which is a sequence of values/states.

Genetic algorithms are commonly used to generate high-quality solutions to optimization and search problems by relying on bio-inspired operators such as mutation, crossover, and selection (Kramer 2017).

7.3.2 Nature of Evolution—Genetic Algorithms

Figure 7.19 shows a typical GA with the following operations:

1. Initialization of the population.
2. Parent selection process.
3. Reproduction process involving *crossover* and *mutation* operations.
4. Fitness value evaluation.
5. Iterative execution on the new population until satisfactory performance is attained.

According to this *evolutionary theory*, an offspring is normally *fitter* if its ancestors are *better*. Also, chromosomes will grow better as the generations grow.

Fig. 7.19 Flowchart of genetic algorithms (GAs)

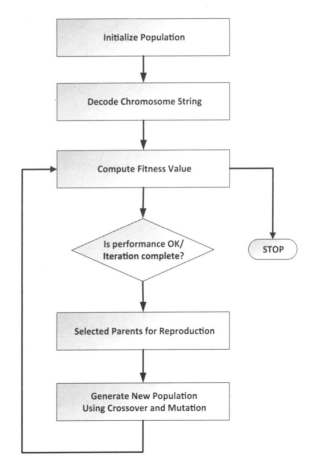

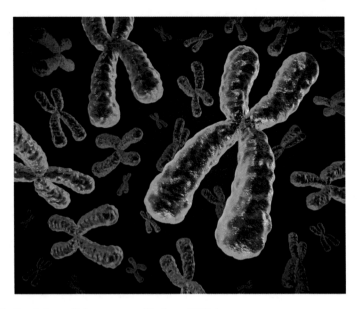

Fig. 7.20 Populations of chromosomes (Tuchong 2019e)

7.3.3 GA—Population Initialization

A *population* is a collection of chromosomes with the representation of a parameter set $\{x_1, x_2, x_3, \ldots, x_m\}$. This parameter set is to be encoded as a finite length string over an alphabet of finite length.

It is usually encoded as a binary value of zeros and ones, or list of *normalized real number between 0 and 1* in parameters optimization problem.

The population size depends on the problem nature, but typically contains several hundreds or thousands of possible solutions. Often, the initial population is generated randomly, allowing the entire range of possible solutions (the search space). Occasionally, the solutions may be *seeded* in areas where optimal solutions are likely to be found.

For a chromosome of length m, the possible number of different chromosome strings is 2^m (Fig. 7.20).

7.3.4 GA—Fitness Evaluation Function

An *evaluation function f (x)* is applied to the population to compute the fitness value of a chromosome.

This evaluation function varies among different problems.

For example, in a parameter optimization problem of a trading system, $f(x)$ might be the accumulated returns in a certain period of time, say, 2-month period.

It is vital that a complete GA remains in a form as it is the only selection criterion for chromosome performance of the entire population and increases the possibility for reproduction.

To justify stopping criterion of the GA, it usually depends on whether the best chromosome in the population has attained a sufficient level or *evolution* (i.e., iteration of reproduction) which exceeded the generation limit (say a maximum of 1000 generations).

The fitness function must not only correlate closely with the designer's goal, but it must also be computed rapidly. Execution speed is very important. A typical genetic algorithm must be iterated many times in order to produce a usable result for a nontrivial problem.

Fitness approximation may be appropriate, especially in the following cases:

- Fitness computation time of a single solution is extremely high.
- Precise model for fitness computation is missing.
- The fitness function is uncertain or noisy.

7.3.5 GA—Parents Selection Scheme

For *parents selection*, a *fitness proportionate selection* (also known as *roulette-wheel selection scheme*) is commonly adopted. Fitness proportionate selection is a genetic operator used in genetic algorithms to select prospective beneficial recombination.

The fitness function assigns a fitness to all chromosomes. This fitness level is used to associate a probability of selection with each individual chromosome.

If f_i is the *fitness value (FV)* of a member i of the population, its probability of being chosen (i.e., reproduction) will be

$$p_i = \frac{f_i}{\sum_j^m f_j} \tag{7.6}$$

Those chromosomes selected for the possibility of reproduction are directly proportional to their fitness value, and they conform to the basic feature of natural selection that: *A fitter organism has a higher chance of survival, hence reproduction.*

7.3.6 GA—Crossover Operations

In GA, there are two main operators for reproduction, namely, *crossover* and *mutation*.

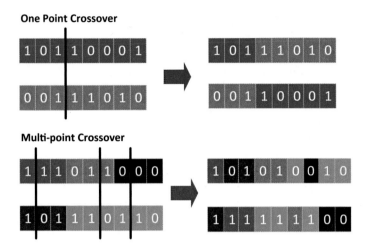

Fig. 7.21 One-point versus multi-point crossover operation

In typical crossover operation, a pair of parent chromosomes is selected from the population.

In *one-point crossover*, a random location is selected from the chromosome strings, and chromosome elements beyond this crossover point are exchanged to form a pair of new offspring according to the crossover rate.

Similarly, for *two-point and uniform crossover*, multiple points are selected for crossover operations.

Figure 7.21 illustrates *one-point versus multi-point crossover* operations.

Different algorithms in evolutionary computation may use different data structures to store genetic information, and each genetic representation can be recombined with different crossover operators.

Typical data structures that can be recombined with crossover are bit arrays, vectors of real numbers, or trees.

7.3.7 GA—Mutation Operations

For *mutation operations*, a single chromosome is selected from the population which will be *screened* throughout the entire list.

Basically, there are five basic types of mutation (Fig. 7.22):

- Deletion,
- Duplication,
- Inversion,
- Insertion, and
- Translocation.

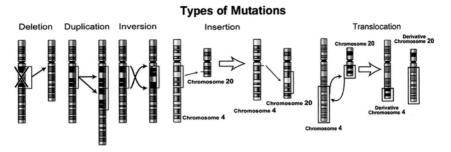

Fig. 7.22 Five basic types of mutation operations

A particular element will be changed according to the mutation rate, which is normally much lower than the crossover rate, say 1–5% of probability.

The classic example of a mutation operator involves a probability that an arbitrary bit in a genetic sequence will be changed from its original state. A common method of implementing the mutation operator involves generating a random variable for each bit in a sequence. This random variable tells whether or not a particular bit will be modified. This mutation procedure, based on the biological point mutation, is called single point mutation. Other types are inversion and floating-point mutation. When the gene encoding is restrictive as in permutation problems, mutations are swaps, inversions, and scrambles.

The main purpose of crossover is to *exchange information* between randomly selected parent chromosomes with the aim of not losing any improvement of information. The purpose of mutation in GAs is preserving and *introducing diversity*. Mutation should allow the algorithm to avoid local minima by preventing the population of chromosomes from becoming too similar to each other, thus slowing or even stopping evolution.

This reasoning also explains the fact that most GA systems avoid only taking the fittest of the population in generating the next but rather a random (or semi-random) selection with a weighting toward those that are fitter.

Although the main objective of mutation is to introduce some *genetic diversity* into the population, it must remain at a slow rate (normally less than 5%) in order not to disrupt the genetic characteristic of the "*good genes*".

7.3.8 GA—Implementation

Based on different parent selection criteria, reproduction scheme, crossover, and mutation methods, there are numerous versions of schema for GAs implementation.

The fundamental one is reproduction to replace all parent population, using one-point crossover and bit mutation.

For parent selection, the roulette-wheel parent selection scheme based on parent fitness value is applied.

In the *elitism scheme*, parents with the highest fitness value will be retained in next generation in order to guarantee the performance of the population at a certain standard.

For crossover operation, a two-point crossover in the other extreme to uniform crossover can be applied for other GA schemes.

For setting GA parameters, besides using a fixed crossover and mutation rates throughout the whole evolutionary process, a dynamic crossover and mutation rate assignment scheme can also be used.

The ratio of mutation rate will normally be set to a higher value when the number of generations increases to a higher level, such as 500 iterations.

The main reason for this is to induce a higher diversity of the chromosomes when the entire population evolves to a more mature stage, whereas a higher mutation rate can bring more *"freshness"* to the population.

7.3.9 GA—Applications

As one can see, genetic algorithm (GA) is based on the foundation of evolution theory that focuses on target system improvement after generations of *evolution*.

Naturally speaking, GA is particularly useful to tackle two main categories of problem:

- Optimization problem and
- Scheduling problem.

Besides, GAs have been widely used with neural networks in two specific areas:

- Topology optimization
- Genetic training algorithms

In topology optimization, GAs are used to select the optimal topology (or parameter setting) for neural network which in turn is trained using some fixed training scheme such as FFBP training.

In genetic training algorithms, the learning of a neural network is formulated as a weight optimization problem, usually using the inverse mean square error as the fitness evaluation scheme.

In Part II of the book—applications of quantum finance, we will study how GA can be used for the selection of top-10 fuzzy financial signals in a quantum finance transient-fuzzy DNN (deep neural network) for financial prediction as in Fig. 7.23.

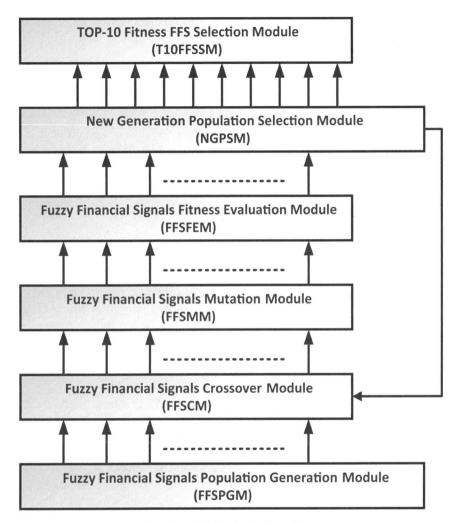

Fig. 7.23 GA-based top-10 fuzzy financial signals selection scheme

7.4 Fuzzy Logic—The Fuzzification Engine

7.4.1 The World of Fuzziness, Chaos, and Uncertainty

In our daily lives, we have to handle numerous things all the time—solving problems, asking and answering questions, etc.

At the end of the day, if we sit down and recall how many things were handled that day, we will find that this number is surprisingly large.

If you take a *closer look* at all these things, we will find that most (or almost all) of them have the following three basic properties:

- They are fuzzy and highly uncertain in nature.
- We need our knowledge and/or experience to handle them.
- The methods and/or solutions to handle them are usually highly dynamic (or sometimes chaotic) in nature.

In fact, we might be aware that things that happen always seem to be highly *chaotic* and *unpredictable*.

Does this mean that we cannot do anything to predict them, or cannot model all these phenomena to be solved and handled automatically?

7.4.2 The Birth of Fuzzy Logic

The notion of *fuzzy* is not a new idea. The first scientific and mathematical interpretation for the *phenomenon of fuzzy* (or one can say the *fuzzy property of matter*) attracted little attention from scientists until the feverish study of Heisenberg Uncertainty Principle of matter in the late 1920s and 1930s.

In 1937, Professor Max Black (1909–1988) published his paper *Vagueness: an exercise in logical analysis* in the distinguished journal *Philosophy of Science* proposed to use continuous logic component-wise techniques in sets and lists of elements (Black 1937).

However, instead of using the name fuzzy sets to describe these *vague sets*, he referred to these structures as *vagueness*. Nevertheless, he was considered to be the forerunner to propose and define the first fuzzy membership function.

The term fuzzy set was coined by Professor Lotfi A. Zadeh (1921–2017) in 1965 in his influential paper *fuzzy sets* published in the journal Information & Control (Zadeh 1965). In this paper, he proposed a modified new kind of set theory (so-called *fuzzy set theory*), which was used to describe the belonging degree of an individual member that can exist as any real value between 0 and 1, instead of just either 0 or 1 (Lee 2006; Ross 2016). Later, in his distinguished 3-Part journal papers *The Concept of a Linguistic Variable and Its Application to Approximate Reasoning* published by *Information Sciences* in 1975, he established a concrete theoretical and mathematical foundation on *type-1* and *type-n fuzzy logics* which began the age of fuzzy logic in the following half-century (Zadeh 1975a, b, c).

Moreover, he also demonstrated how these fuzzy sets could be operated, including fuzzy set union and intersection, and also developed a consistent framework for dealing with all of these structures and operations.

He believed that fuzzy sets, instead of crisp sets, should intuitively exist in our world of experience (Zadeh and Aliev 2019).

More importantly, these kinds of existence were even more naturally accepted by us and consistent with what we understood as our world of existence.

7.4.3 Fuzzy Theory and the Uncertainty Principle

In fact, Professor Zadeh's observations were very sensible.

Most people might think that a fuzzy set tries to turn everything in our perceived world *blurry*, *unclear*, or *uncertain*, while a *crisp set* is the reality.

Actually, it is just the opposite situation.

Lack of crispness should instead be the truth face of a real world.

According to the Heisenberg uncertainty principle learnt in previous chapter, we cannot exactly (100%) determine (measure) the position and momentum of an object at the same moment.

From fuzzy theory point of view, what Heisenberg's uncertainty principle tells us is that all measured qualities appearing in our world of perception are fuzzy in the sense that it is not the quality itself that is fuzzy, but rather that we cannot always measure these physical qualities with 100% certainty.

In fact, there are three types of uncertainties that are now recognized:

1. *fuzziness* (or *vagueness*), which results from the imprecise boundaries of fuzzy sets;
2. *non-specificity* (or *imprecision*), which is connected with sizes (cardinalities) of relevant sets of alternatives; and
3. *strife* (or *discord*), which expresses conflicts among the various sets of alternatives.

Figure 7.24 shows the taxonomy of uncertainty and decisions. *Vagueness* is a form of uncertainty where the analyst is unable to clearly differentiate between two different classes, such as *person of average height* and *tall person*. This form of vagueness can be modeled by some variation on fuzzy logic. We will return to this interesting topic again when we study chaos theory in Chap. 8.

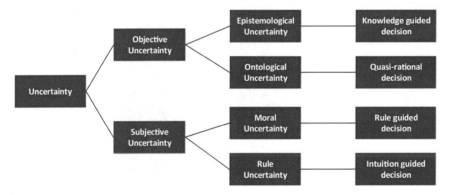

Fig. 7.24 A taxonomy of uncertainty

7.4.4   ICMR Model—Fuzzy Identification Module

Structurally speaking, fuzzy logic can be decomposed into four basic components—
the so-called *ICMR structural framework of fuzzy logic* as shown in Fig. 7.25.

Fuzzy Identification Module
This module focuses on determination and identification of *fuzzy variable(s)* (*FV*)
being used in the system.

It is the most important and foremost stage in the entire system design, because an
incorrect (or inappropriate) identification of the FV(s) will defeat the whole purpose
of system modeling.

One important fact is that the same *quality* (*variable*) can play totally different
roles in different scenarios for the design of fuzzy systems.

For example, if one wants to use fuzzy logic to represent the following statement:
John is tall—the fuzzy variable being used is the quality *height*. Although this quality
does not appear in the statement, conceptually speaking it is exactly what we mean
by common sense, or what we called *priori knowledge*.

However, consider another statement: *Most of the students in the class are tall*.

In this case, if we want to represent this statement using fuzzy logic, the FV being
used should not be the quality *height*, but rather it should be the quality *most*—it
depicts the quality of students' proportion who are tall in this class! In other words,
we should be very careful in selecting the FV(s) being used in a fuzzy system.

Fig. 7.25 Fuzzy
identification module of
ICMR model

7.4.5 ICMR Model—Fuzzy Categorization Module

Based on FV(s) being identified in the last module, this module focuses on how to categorize the selected FV(s) into different fuzzy sets.

By using the example in previous case, for FV *height*, the possible fuzzy sets will be *short*, *average*, and *tall*—in which each of them is a fuzzy set with a range of values:

- *short*—for student height below 1.55 m.
- *average*—for student height between 1.5 and 1.65 m.
- *tall*—for student height over 1.6 m.

As shown in the above example, one major characteristic of fuzzy sets is that there exists *fuzziness* in their boundaries.

In other words, there is not always a clear boundary between one *fuzzy set* and another, which is totally unacceptable in a traditional *crisp set*.

Of course, the categorization of fuzzy sets is up to system designer in the sense that there are no strict criteria for one to decide the number, the type, and nature of fuzzy sets being used. But it should be logical and make sense.

For instance, instead of using three fuzzy sets, one can further categorize *height* into five categories: *very short*, *short*, *average*, *tall*, and *very tall* (Fig. 7.26).

Fig. 7.26 Fuzzy categorization module of ICMR model

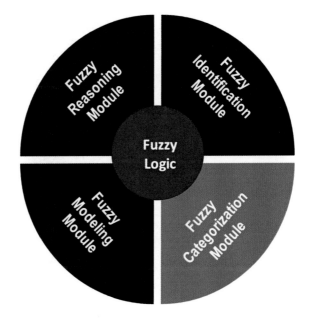

7.4.6   ICMR Model—Fuzzy Modeling Module

For each fuzzy set, the designer has to define its *membership function (MF)*. The membership function is one of the most important components in fuzzy logic. It defines fuzzy mapping—the fuzzy belongings of each element within its fuzzy set.

From the mathematical point of view, let X be a collection of objects denoted by $\{x|x \in X\}$ which is the so-called *universe of discourse.*

A fuzzy set A in this universe of discourse X is characterized by a membership function P A(x) which takes values in the interval [0, 1].

For instance, if we define the fuzzy set A as

$$A = \{(x, \mu_A(x)|x \in X)\} \tag{7.7}$$

If X is continuous, the fuzzy set A will be written as

$$A = \int_X \mu_A(x)/x \tag{7.8}$$

Otherwise, if X is discrete, the fuzzy set A will be written as

$$A = \sum_N \mu_A(x_i)/x_i \tag{7.9}$$

where N is the total number of possible discrete values in A.

By using the above example, one can model the fuzzy membership functions of each fuzzy element for the FV *height* as shown in Fig. 7.27.

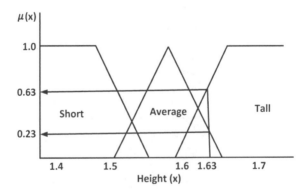

Fig. 7.27 Fuzzy membership function. *Note* According to the above fuzzy membership functions. If we say: *the height of a student (Jack) is 1.63*, the membership functions for fuzzy sets *average* and *tall* will be 0.23 and 0.63, respectively. The implications for these fuzzy values are: *Jack is tall, but he is not among the tallest ones in the class. But overall speaking, he is taller than the average students in the class*

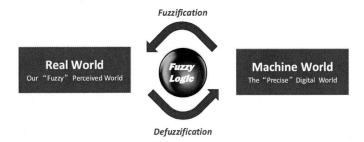

Fig. 7.28 Fuzzification versus defuzzification

7.4.7 ICMR Model—Fuzzification Versus Defuzzification

The above example also demonstrates a very important concept in fuzzy logic—the *fuzzification* versus *defuzzification* scheme. As said, the main concept of fuzzy logic is to reflect the reality experience of our world (our perceived world), which is uncertain and fuzzy.

In other words, the main job of fuzzy logic is to act as a bridge (or a *crystal ball* as shown) for us to convert *precise* values (and measurements) from machine world (the so-called digital world) to *imprecise* values in real world (the so-called *perceived world*) which is full of *fuzziness* and *uncertainty*.

From the operational point of view, we denote the forward operation, which is the conversion of *digital measurements* from machine world to *fuzzy values* in the real world as the fuzzification scheme, while the reverse one is the defuzzification scheme. From the mathematical point of view, *fuzzification* is the forward mapping of membership function, whereas *defuzzification* is reverse mapping (inverse function) of the membership function as shown in Fig. 7.28.

7.4.8 ICMR Model—Fuzzy Reasoning Module

This module is the *brain* of the entire fuzzy system that aims to explore system's heuristics by applying the so-called fuzzy reasoning technique.

This module consists of three main processes:

- fuzzy rules construction process (FRCP),
- fuzzy knowledge-based construction process (FKCP), and
- fuzzy inference process (FIP).

Fuzzy Rules Construction Process—Fuzzy Relations (FRs)
During the construction process, it is essential to define a set of rules. Fuzzy rules are basically set of *fuzzy relations (FRs)* that normally appear in the form of *if-then* propositions. They are also called fuzzy implication functions (FIFs) (Fig. 7.29).

Fig. 7.29 Fuzzy reasoning
module of ICMR model

Mathematically speaking, a FIF is denoted in the following form:

$$\text{If } x \text{ is } A \text{ then } y \text{ is } B, \text{ where } x, A \in X, y, B \in Y \tag{7.10}$$

or written as

$$R = A \rightarrow B \text{ where } R \text{ is the fuzzy rule} \tag{7.11}$$

Fuzzy Knowledge-Based Construction Process—Fuzzy Knowledge Base (FKB)
The *fuzzy knowledge base (FKB)* is the *knowledge center* of a fuzzy system.

In a typical fuzzy system, the FKB normally appears as a collection of fuzzy rules
to represent a particular domain (or domains) of knowledge.

Following the FIF's notation used in Table 7.1, the FKB of a fuzzy system is
denoted by a collection of fuzzy rules:

$$R_k = A_k \rightarrow B_k, \ k \in [1 \ldots N] \tag{7.12}$$

where N is the total number of fuzzy rules used in the FKB.

From the implementation point of view, there are various approaches to implement
the fuzzy relation operations.

It is up to the system developers to select the most suitable one for their applications.

*The following figure shows the four commonly used fuzzy rule implementation
schemes.*

Table 7.1 Fuzzy rules and methods being used

Methods used	Fuzzy rule "If x is A then y is B" where $x, A \in X, y, B \in Y$
Mini rule	$R_{mini} = A \times B = \int_{X \times Y} \mu_A(x) \wedge \mu_B(y)/(x, y)$
Product rule	$R_{product} = A \times B = \int_{X \times Y} \mu_A(x) \cdot \mu_B(y)/(x, y)$
Max-min rule	$R_{max-min} = (A \times B) \cup (not\ A \times Y)$ $$= \int_{X \times Y} (\mu_A(x) \wedge \mu_B(y)) \vee (1 - \mu_A(x))/(x, y)$$
Arithmetic rule	$R_{arith} = (not\ A \times Y) \oplus (X \times B)$ $$= \int_{X \times Y} 1 \wedge (1 - \mu_A(x) + \mu_B(y))/(x, y)$$

Fuzzy Inference Process (FIP)

Fuzzy inference refers to the *fuzzy reasoning (heuristic)* operations based on facets (or fuzzy facets) and knowledge (FKB) in the fuzzy systems.

There are two kinds of fuzzy inference methods:

1. The generalized modus ponens (GMP) and
2. The generalized modus tollens (GMT).

Owing to their characteristics of inference operations, GMP and GMT are also called direct reasoning and indirect reasoning, respectively.

For example, given two groups of fuzzy set A, A' ∈ X while B, B' ∈ Y:

GMP—For any two fuzzy variables x and y, suppose we are given:

(a) *Implication (the fuzzy rule): if x is A then y is B.*
(b) *Premise: x is A'.*
(c) *Conclusion y is B'.*
    which can be denoted by

$$B' = A' \circ R \ (where\ R\ is\ the\ relation). \tag{7.13}$$

GMT—For any two fuzzy variables x and y, suppose we are given:

(a) *Implication (the fuzzy rule): if x is A then y is B.*
(b) *Premise: y is B'.*
(c) *Conclusion x is A'.*
    which can be denoted by

$$A' = R \circ B' \ (where\ R\ is\ the\ relation). \tag{7.14}$$

Fuzzy Compositional Rule

One of the most important and powerful functions of fuzzy reasoning is the *fuzzy compositional rule (FCR)* at fuzzy inference.

This situation arises when there is a complex knowledge base containing a list of *fuzzy rules (R_k)* that correspond to the measured input signals A'_k for each fuzzy set A_k.

That is,

$$x = (A'_1, A'_2, \ldots A'_k, \ldots A'_n), \quad y = B'_j, \ j = [1 \ldots n] \tag{7.15}$$

Based on the GMP defined in (7.13), the fuzzy variable B' can be inferred as the composition operation between the fuzzy set A_k and the *fuzzy relation R*:

$$B' = (A'_1, A'_2, \ldots A'_k, \ldots A'_n) \circ R \tag{7.16}$$

From the implementation point of view, two composition schemes are commonly adopted: (1) the sup-min operation and (2) the sup-product operation.

In the sup-min operation, the membership function value $\mu_{B'}$ after the composition operations, is given by

$$\mu_{B'} = \bigcup_{k=1}^{n} \left[supr_x \left(\bigwedge_{i=1}^{n} \mu_{A'_i} \right) \bigwedge \left(\bigwedge_{i=1}^{n} \mu_{A_{ki}} \to \mu_{B_k} \right) \right] \tag{7.17}$$

Basically, a composite fuzzy reasoning process consists of the following steps:

1. Identify each fuzzy rule and determine the *connectives* for each condition in the rule base, in which an *OR* clause represents a union operation and an *AND* clause represents an intersection operation.
2. For each fuzzy rule (R_k), locate the *fire strength α_k*.
3. Use either the sup-min or sup-product method and infer the overall defuzzification control action.

7.4.9   Case Study—Fuzzy Air-Conditioning Control Systems, FACS

In this section, we will use a simple fuzzy air-conditioning control system (FACS) to illustrate how fuzzy composite reasoning works.

First of all, suppose the FACS can vary its power control (Power) by two environmental factors, namely, air *temperature (Temp)* and *relative humidity (RH)*, which are given by two sets of fuzzy variables:

Temp = {Very Cold, Cold, Cool, Mild, Warm, Hot, Very Hot}
RH = {Very Dry, Dry, Humid, Very Humid}

and ranges of *temperatures* and *relative humidity* for each fuzzy variable are given by Table 7.2. The membership function of *Temp (x)* and *Relative Humidity (y)* are shown in Figs. 7.30 and 7.31.

Suppose that the power setting of FACS has three different fuzzy states, given by

Power = {Low, Medium, High}

with the membership function shown in Fig. 7.32.

Table 7.2 Ranges of temperatures and relative humidity for each fuzzy variable

Temperature description (Fuzzy variables)	Fuzzy temp range (°C)	Relative humidity (Fuzzy variable)	Fuzzy RH range (%)
Mild	17–25	Dry	≤ 60
Warm	23–31	Moderate	55–85
Hot	25–33	Humid	65–95
Very hot	≥ 28	Very humid	>75

Fig. 7.30 Fuzzy membership function of *temperature*

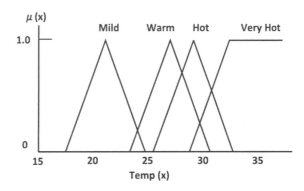

Fig. 7.31 Fuzzy membership function of *relative humidity*

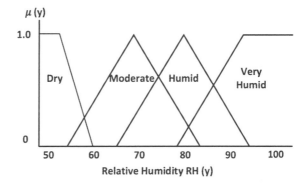

Fig. 7.32 Fuzzy
membership function of
power

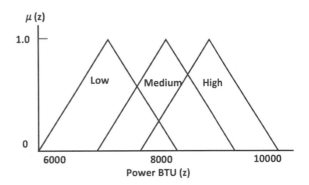

Assume this power switch is controlled by the following three fuzzy rules:

Rule 1: If Temp is *very hot* and RH is *very humid* then switch to *high.*
Rule 2: If Temp is *hot* and RH is *humid* then switch to *medium.*
Rule 3: If Temp is *warm* or RH is *moderate* then switch to *low.*

Figure 7.33 illustrates a conceptual diagram of a fuzzy logic air-conditioning system.

Now, given that the temperature is 30 °C and the relative humidity is 80%, the defuzzification schemes of the power by using the sup-min method and sup-product method are illustrated in Figs. 7.34 and 7.35.

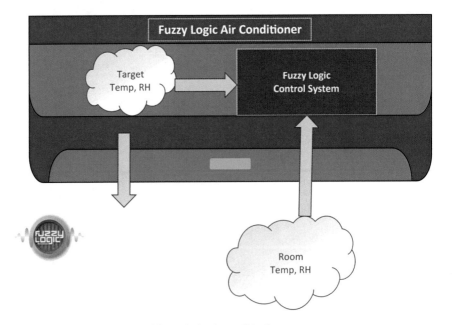

Fig. 7.33 Conceptual diagram of fuzzy logic air-conditioning system

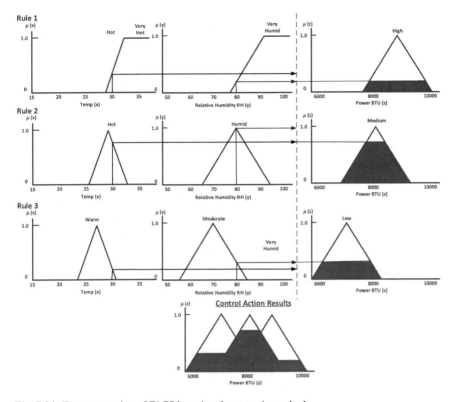

Fig. 7.34 Fuzzy reasoning of FACS by using the sup-min method

7.4.10   Applications of Fuzzy Logic in Daily Life

Nowadays, fuzzy logic (FL) is widely adopted in many real-world applications rang-ing from fuzzy electronic appliances such as rice cookers, air conditioners, and washing machines as in Fig. 7.36 to sophisticated control systems such as robotics and fuzzy-based ABS (automatic braking systems).

In reality, many Japanese car manufacturers incorporate fuzzy systems for antilock braking, active suspension systems, automatic transmissions, and engine emission controls into their automobiles.

Fuzzy systems are easy to set up, typically require less processing power than alternative approaches and provide robust performance.

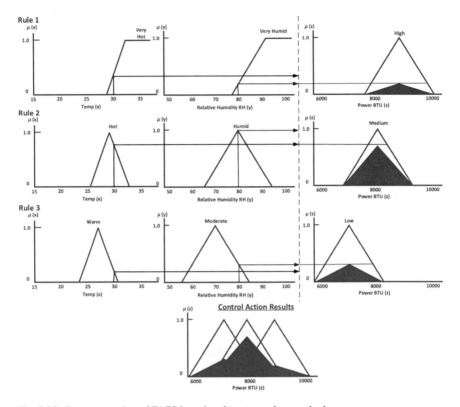

Fig. 7.35 Fuzzy reasoning of FACS by using the sup-product method

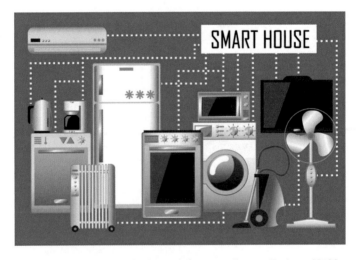

Fig. 7.36 Smart home with fuzzy logic electronic home appliances (Tuchong 2019f)

7.4.11   Fuzzy Expert Systems on Financial Trading

From the implementation point of view, current fuzzy logic is mainly integrated with other AI technologies in order to solve complex problems and produce more sophisticated and comprehensive systems (Lee 2006; Siler and Buckley 2004). This type of hybrid system integration technique is so-called "hybridization".

Basically, there are two major kinds:

- Fuzzy expert systems (or fuzzy-ES for short) and
- Fuzzy-neuro systems.

Fuzzy Expert Systems on Financial Trading
Owing to the capability of fuzzy systems that can handle imprecise concepts and values, fuzzy logic has been widely used in many practical commercial and industrial applications. Fuzzy-ES focus on the integration of fuzzy with expert knowledge in classical expert systems.

When fuzzy logic applies in financial trading, a typical schematic diagram of a fuzzy expert system is required to incorporate a fuzzy knowledge base (KBS) together with the fuzzification of technical indicators, in order to drive expert advice and perform day trade.

Figure 7.37 illustrates the conceptual diagram of fuzzy expert systems on financial trading.

7.4.12   Fuzzy-neuro System in Quantum Finance

The major shortcoming of the fuzzy system is the lack of learning capability.

The design of fuzzy sets and the assignment of all fuzzy relations are done by system designers (or experts) without any way of acquiring membership functions and inference rules automatically.

To solve this problem, hybridization with other AI models that provide automatic machine learning such as neural networks—the so-called fuzzy-neuro systems were developed.

Nowadays, fuzzy-neuro systems are widely adopted for solving complex problems such as stock prediction, severe weather forecasting (e.g., rainstorm prediction), adaptive tracking systems (e.g., tropical cyclone (TC) tracking systems, missile tracking systems), robot vision, etc.

Figure 7.38 shows the integration of fuzzy logic with chaotic neural network for stock prediction, and we will study quantum finance advance applications in Chap. 12.

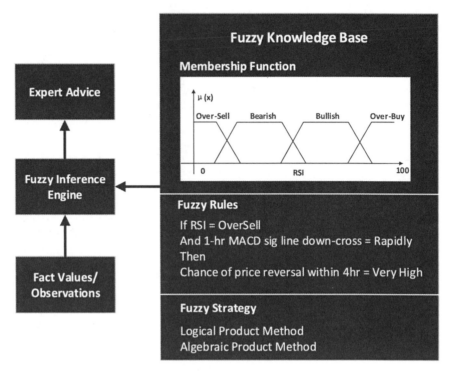

Fig. 7.37 Conceptual diagram of fuzzy expert systems on financial trading

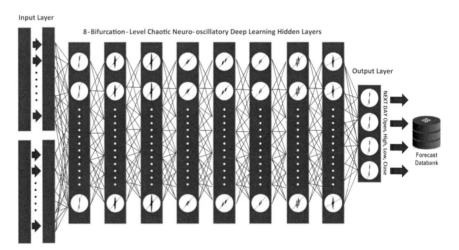

Fig. 7.38 System architecture of chaotic type-2 fuzzy deep neural network for financial prediction (CT2TFDNN)

7.5 Conclusion

In this chapter, we have learnt the basic concept of artificial intelligence, what it is and how it works.

We had also studied three major components of modern AI technology:

- Artificial neural networks,
- Genetic algorithms, and
- Fuzzy logic.

As mentioned, owing to the advancement of computer technology, the popularity of Internet technology along with social medium and mobile computing, the needs of AI and intelligent systems are more than anytime in this new era.

In fact, there are many new AI technologies, algorithms, techniques, and applications emerging on daily basis.

This chapter is only an introductory course for the path of AI.

The truth is: Different from other disciplines, AI is an integrated and cross-discipline doctrine ranging from neuroscience and visual psychology to quantum computing without any boundary. The only limitation is imagination and thoughts.

In next chapter, we will discuss two innovative AI technologies which are related to financial engineering, and also important components of quantum finance—chaos theory and fractals.

Problems

7.1 What is artificial intelligence (AI)? Give any three applications of AI in finance industry and explain how they work.

7.2 What is Turing test? Discuss and explain how it works?

7.3 What discipline of AI does Turing test focus on? Why such test is believed to be the best method to evaluate a machine (robot) is truly intelligence or not? Give a live example in daily human interactions and communication to support your explanation.

7.4 John Searle in 1980 proposed a thought experiment so-called "Chinese room experiment" to challenge the validity of Turing test for the evaluation of machine intelligence.

> **Chinese Room Thought Experiment**
>
> Similar to the setting of Turing test, but this time all Q&A between the invigilator and the challenger are in Chinese in which the challenger (human) nothing about Chinese. So, the challenger in the Chinese room follows English instructions in his computer program for manipulating Chinese symbols, where a computer "follows" a program written in a computing language to reply with Chinese symbols. The challenger produces the appearance of understanding Chinese by following the symbol manipulating instructions but does not thereby come to understand Chinese. Since a computer just does what the human does—manipulate symbols on the basis of their syntax alone, merely by following a program, comes to genuinely understand Chinese to pass the test.

(i)   What is the main argument of John Searle's Chinese room experiment to challenge the validity of Turing test for the evaluation of machine intelligence in AI?

(ii) Does the argument and challenge by John Searle's Chinese room experiment against Turing test is valid or not? Why?

7.5 What is an *intelligent agent* in AI? State and explain the three main difference between an intelligent agent versus traditional computer program (system)? Give two examples of how intelligent agents can be applied to finance.

7.6 Discuss and explain the major differences between strong AI versus weak AI. Give two examples of strong AI and weak AI applications and explain how they work.

7.7 In 1997, IBM Deep Blue defeated world chess champion Mr. Garry Kasparov. It is also the first time AI computer program defeat world chess champion.

(i)   Why this is an important event and milestone in AI history? In other words, why chess playing with chess master is an important challenge in AI community?

(ii) In fact, IBM Deep Blue system at that time is a classical AI-based expert system tailored for chess playing. Discuss and explain what is an expert system. How it works?

(iii) However, after the world chess contest in 1997, Kasparov in the TV interview complained that it was not a fair game and not really a kind of human versus machine intelligence. What was his argument behind? Do you agree? How this event influences the future development of AI?

7.8 Why was the past 20 years can be claimed as the second golden age of AI? and what kind of AI is focused on—strong AI or weak AI? What are the logics behind?

7.9 State and explain the three main disciplines of weak (soft) AI. For each disciple, give one real-world application of soft AI and explain how it works.

7.10 What is soft AI? Discuss and explain why soft AI is also known as cognitive approach (CA) AI technology. Name the major four soft AI technologies and briefly explain how they work.

7.11 What is artificial neural network (ANN)? What is the main difference between ANN and contemporary computer system in terms of (1) memory storage and (2) machine learning and computational methods.

7.12 What are integrate-and-fire operations in biological neural networks? Why it is important in machine learning? And how artificial neural networks model these operations?

7.13 Describe and state the roles and importance of transfer function in typical artificial neural networks (ANNs). How can we model transfer function in traditional ANN?

7.14 Describe and explain the major pros versus cons of using sigmoid function as transfer function in typical ANN. Give a real-time example in finance prediction to support your explanation.

7.15 What is a multilayer feedforward backpropagation network (FFBPN)? Describe and explain how it works? State and explain two examples for the application of FFBPN in finance engineering.

7.16 In a typical multilayer FFBPN, how can we determine the: (1) number of input and output nodes? (2) number of hidden layers and the number of neurons in each hidden layer?

7.17 Describe and explain the three major categories of artificial neural networks (ANNs) in terms of machine learning technique. For each category of ANN, give one example of application in financial engineering and describe briefly how it works.

7.18 In machine learning, pattern recognition and time series prediction are sides of the same coin. Discuss and explain why. Give an example in finance engineering to support your explanation.

7.19 Programming exercise I

  (i)   Write the flowchart and system training algorithm of feedforward back-propagation network (FFBPN).
  (ii) Write a MATLAB program to implement the FFBPN training algorithm by using the time series of any financial product (e.g., DJI) exported from MT4/R platform for short-term financial prediction.
  (iii) First, using 5 days to predict next day open/high/low/close.
  (iv) Second, using 10 days to predict next day open/high/low/close.
  (v)   Third, using 20 days to predict next day open/high/low/close.
  (vi) Compare their performance in terms of (1) training performance and (2) forecast performance.
  (vii) Which one is better? Why?

7.20 Describe and explain what is genetic algorithm (GA)? What kind(s) of AI applications and problems GA tackles with? Why? Draw the flowchart of GA and explain how it works.

7.21   What is the core idea and innovation of genetic algorithms (GA) in machine learning? Give two examples of how we can apply GA on finance engineering and hence quantum finance.

7.22   Discuss and explain the roles and functions of crossover and mutations in GA. As compared with crossover, explain why mutation must keep at a much lower rate. What is the logic behind?

7.23   State and explain two typical applications for the integration of ANN with GA in financial engineering.

7.24   Discuss and explain the relationship between fuzzy logic and uncertainty principle in quantum theory. Use two financial markets as example to support your explanation.

7.25   Discuss and explain how fuzzification process works in fuzzy logic? How can we apply fuzzification in financial engineering? Give two examples to support your explanation.

7.26   State and discuss two main types of fuzzy expert systems. For each type of fuzzy expert system, give one potential application on (1) finance engineering and (2) quantum finance and explain how they work.

7.27   Discuss and explain why the integration of fuzzy logic with artificial neural networks (or recurrent neural networks) is commonly used in AI applications. Give two examples of how fuzzy-neuro systems can be used in quantum finance.

7.28   Programming exercise II

   (i)   Repeat 7.19, but this time using MQL to implement the FFBPN in MT4 platform.

   (ii)   For the input nodes, in addition to the time series information, add 10 commonly used technical indicator as input nodes, which include iAlligator (alligator oscillator), iMA (moving average), iBands (Bollinger bands), iMomentum (momentum indicator), iMACD (MACD oscillator), iRSI (RSI indicator), iRVI (RVI indicator), iStdDev (standard deviation indicator), iStochastic (stochastic oscillator), and iWPR (Williams' percent range indicator).

   (iii) Apply fuzzy logic to fuzzification of these 10 technical indices, using the fuzzification scheme for RSI discussed in this chapter as example.

   (iv)   Apply genetic algorithm (GA) to choose the TOP 5 technical indicators for live application.

   (v)   Implement this GA-based fuzzy-neuro network and compare its performance with the traditional FFBPN implemented in (i).

(Note: For detail training workshop on MQL, please visit QFFC.org official site.)

References

Abraham, T. Rebel (2016) Genius: Warren S. McCulloch's Transdisciplinary Life in Science. MIT Press, Cambridge.

Black, M. (1937) Vagueness: An Exercise in Logical Analysis. *Philosophy of Science* 4(4): 427–455.

Bentivoglio, M. (2014) "Golgi, Camillo". In Daroff, Robert B.; Aminoff, Michael J. *Encyclopedia of the Neurological Sciences* (Second edition. ed.). Burlington: Elsevier Science. 464–466.

Darwin, D. and Huxley, J. (2003) The Origin of Species: 150th Anniversary Edition. Signet; Reprint, Anniversary edition.

Fausett, L. (1994) Fundamentals of Neural Networks Architectures Algorithms and Applications. Prentice Hall.

Goldberg, D. E. (1989) Genetic Algorithms in Search, Optimization, and Machine Learning. Addison-Wesley Professional.

Hansen, W. *Classical Mythology: A Guide to the Mythical World of the Greeks and Romans*. Oxford University Press, 2005.

Hebb, D.O. (1949) *The Organization of Behavior*. New York: Wiley & Sons.

Holland, J. H. (1992) *Adaptation in Natural and Artificial Systems: An Introductory Analysis with Applications to Biology, Control, and Artificial Intelligence*. A Bradford Book; Reprint edition.

Hopfield, J.J. (1984) Neurons with Graded Response Have Collective Computational Properties like Those of Two-State Neurons. *Proceedings of the National Academy of Sciences of the United States of America*. 81(10): 3088–3092.

Kramer, O. (2017) *Genetic Algorithm Essentials (Studies in Computational Intelligence)*. Springer.

Lee, R. S. T. *Fuzzy-Neuro Approach to Agent Applications (From the AI Perspective to Modern Ontology)*. Springer-Verlag, Heidelberg Germany, 2006.

Lee, R. S. T. A Transient-chaotic Auto-associative Network (TCAN) based on LEE-oscillators. *IEEE Transactions on Neural Networks*. 15(5): 1228–1243, 2004.

McCulloch, W. and Walter, P. (1943). A Logical Calculus of Ideas Immanent in Nervous Activity. *Bulletin of Mathematical Biophysics*. 5 (4): 115–133, 1943.

Patterson, D. W. (1996) *Artificial Neural Networks*. Prentice Hall.

Ross, T. J. (2016) *Fuzzy Logic with Engineering Applications*. Wiley.

Russell, S. and Norvig, P. (2015) *Artificial Intelligence: A Modern Approach,* 3rd edition. Pearson Education India.

Siler, W. and Buckley, J. J. (2004) *Fuzzy Expert Systems and Fuzzy Reasoning*. Wiley-Interscience.

Silva, et al. (2017) *Artificial Neural Networks: A Practical Course*. Springer.

Tuchong (2019a) Illustration of AI. http://stock.tuchong.com/image?imageId= 286594758192922638. Accessed 21 Aug 2019.

Tuchong (2019b) Weak AI. http://stock.tuchong.com/image?imageId=467459610627670027. Accessed 21 Aug 2019.

Tuchong (2019c) 3D illustration of integrate-and-fire in neural network. http://stock.tuchong.com/ image?imageId=483388184041619483. Accessed 21 Aug 2019.

Tuchong (2019d) Neural network and AI. https://stock.tuchong.com/image?imageId= 490703320800559171. Accessed 21 Aug 2019.

Tuchong (2019e) Populations of chromosomes. http://stock.tuchong.com/image?imageId= 453246275369566360. Accessed 21 Aug 2019.

Tuchong (2019f) Smart Home with Fuzzy Logic Electronic Home Appliances. http://stock.tuchong. com/image?imageId=267053275471413372. Accessed 21 Aug 2019.

Turing, A. (1950) Computing Machinery and Intelligence. *Mind, LIX* (236): 433–460.

Zadeh, L. A. and Aliev, R. A. (2019) Fuzzy Logic Theory and Applications: Part I and Part II. World Scientific Publishing Company.

Zadeh, L. A. (1975a) The Concept of a Linguistic Variable and Its Application to Approximate Reasoning-I. *Information Sciences*. 8(1): 199–249.

Zadeh, L. A. (1975b) The Concept of a Linguistic Variable and its Application to Approximate Reasoning-II. *Information Sciences*. 8(4): 301–357.

Zadeh, L. A. (1975c) The Concept of a Linguistic Variable and its Application to Approximate Reasoning-III. *Information Sciences*. 9(1): 43–80.

Zadeh, L. A. (1965) Fuzzy sets. *Information and Control*. 8 (3): 338–353.

Chapter 8
Chaos and Fractals in Quantum Finance

> The theory of chaos and theory of fractals are separate but have
> very strong intersections. That is one part of chaos theory is
> geometrically expressed by fractal shapes.
> Benoit Mandelbrot (1924–2010)

What is chaos theory?
What are fractals?
What is the relationship between chaos and fractals?
How to apply chaos theory and fractals in quantum finance?

In previous chapters, we have learnt the basic concepts of quantum theory.

Wave–particle duality—as the foundation and basics of quantum theory—relates to the nature of all fundamental particles (quantum particles) that exist in two kinds of dynamics—*particles* (motions and dynamics) and *waves* (energy field and patterns) can be narrated agreeably by Schrödinger equation.

In modern mathematics and computational physics, we also have two mathematical models that are agreeably analog to this unique phenomenon—*chaos* and *fractals*, like a coin with both sides, such as Professor Benoit Mandelbrot—the father of fractal said in his remarkable quotation.

More importantly, these two mathematical models have important implications in computational finance and modern AI, which are classified as a branch of computational intelligence (CI) and a major component in quantum finance model.

As mentioned in Chap. 1, financial markets refer to a range from stock markets to cryptocurrency markets. If we look closely at their market dynamics, it is not difficult to discover that they exist in both *particle* (market price movements) and *wave* (market pattern movements) nature.

By applying numerical computation method in quantum finance Schrödinger equation (QFSE), we have successfully evaluated all the quantum energy levels (quantum price levels—QPLs).

So, can we model the *particle* and *wave* dynamics as well, without the need to solve the complex path integral formulation?

The answer is yes—*chaos* and *fractals*.

© Springer Nature Singapore Pte Ltd. 2020
R. S. T. Lee, *Quantum Finance*,
https://doi.org/10.1007/978-981-32-9796-8_8

8.1 Basic Concepts and Formulation on Chaos Theory

8.1.1 The World of Nonlinear Dynamics

What is the nature of our world?

This problem has puzzled philosophers, scientists, poets, and scholars of various disciplines for centuries.

What is the importance of knowing the dynamics of all matter in the world (and universe)?

To understand this question, let us have a look at this topic of *dynamics* which include the following:

- The dynamics of free-falling objects—governed by Newton's laws of motion.
- The formation and dynamics of black holes—governed by Einstein's theory of general relativity.
- The dynamics of subatomic particles such as neutrinos and positrons—governed by Heisenberg's quantum mechanisms.
- The formation and dynamics of tropical cyclones, hurricanes, tornadoes, and earthquakes.
- The fluctuation of stocks.
- The chance of catching the bus this morning.
- The theory of causation.
- Our fate.
- Our future.

The fact is that this world of dynamics covers not only the *physical* motion of matter in the universe, but also the *nonphysical* occurrence of *events, notions, and ideas*.

Chaos theory is the study of such *dynamics*, which is believed to be highly nonlinear—chaotic. In other words, if one wants to explore the nature of the dynamics governing matter (and notions) in the universe, one must *enter* the world of chaos theory.

Figure 8.1 shows the famous analog in chaos theory—the *butterfly effect*—the phrase was first coined by Professor Edward Lorenz (1917–2008) the father of chaos theory gave a talk at a scientific meeting in 1972 (Krishnamurthy 2015; Lorenz 1995) on his work regarding the chaotic phenomena found from his weather prediction model—the *Lorenz equations*. The phrase suggests that the flap of a butterfly's wings in Africa could create a small change in the atmosphere that might eventually lead to a hurricane in Florida. The butterfly effect concept has become important in the finance world due to the continuous interdependence between capital markets subsequent to increased globalization of the world economy. The volatility of a minor sector at international markets can dilate rapidly and bleed into other markets. Broadened Internet access and technology improvements also contribute to episodes of extreme market volatility.

Fig. 8.1 The butterfly effect in chaos theory

8.1.2 Deterministic World Versus Probabilistic World

As everything that happens in the world is highly *dynamical* (or *chaotic*, as we said) and uncertain, does this mean that chaos theory is a study of probability?

The answer is absolutely *not*.

The truth is: chaos theory has a basis that is entirely different from probability.

Although chaos theory is the study of the world of dynamics, it is not a world of uncertainty or a matter of chance.

Strictly speaking, chaos theory deals with the world of *determinism* rather than *probabilism*.

Chaos theory maintains that all the dynamics in the world, no matter the complexities involved, can be (and must be) somehow modeled deterministically by certain chaotic motions. But whether we can find these chaotic motions is another question.

In fact, the formal name for chaos is *deterministic chaos* in order to reflect this characteristic (Schuste and Just 2005; Lee 2005).

From the mathematical point of view, no matter the complexity of a physical phenomenon, the challenge of chaos theory is to model its *dynamics* by a set of equations (of time) as follows:

In discrete format:

$$\vec{x_{t+1}} = \vec{f}\left(\vec{x_t}\right) \tag{8.1}$$

Fig. 8.2 Lorenz attractor (upper right) and the butterfly flapping wings (lower left)

where $\vec{x}_t$ is n-dim vectors (with n-variables) and $\vec{f}$ is its dynamics.

In continuous format:

$$\dot{\vec{x}} = \vec{f}(\vec{x}, t) \tag{8.2}$$

where $\vec{x}$ is n-dim vectors of variable x and $\vec{f}$ is its dynamics.

Figure 8.2 shows the *Lorenz's attractor* and *butterfly effect*—the second implementation of chaos theory, in which the chaos dynamics of classical Lorenz's attractor resembles the flipping wings of a butterfly.

8.1.3   Chaotic Flow Versus Chaotic Maps

In chaos theory, the discrete chaotic motions are also called *chaotic maps*, and the continuous motions are called *chaotic flows*.

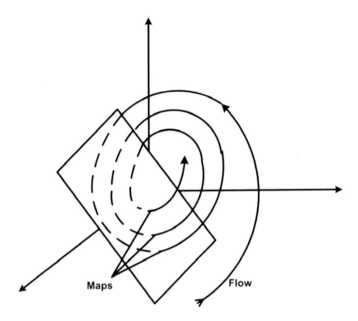

Fig. 8.3 Chaotic flow versus chaotic map

From the physical point of view, the discrete maps can be regarded as taking *snapshots* (or *cross sections*) of the continuous flow on a regular (or timely) basis as shown in Fig. 8.3.

What is the significance of the findings of maps and flows?

The answer is: It is extremely important!

For example, if the variables under consideration are weather elements such as air temperature or rainfall, and the time step (say for a discrete map) is 5 days, then if we can find such chaotic function(s), what we have really done is solving a weather forecasting problem!

Imagine if functions are used to model the next-day stock market or the 15-day financial pattern!

We can expect that what chaos theory is telling us is not something uncertain (or chaotic), but rather how to *identify* the uncertain event in a chaotic way!

8.1.4 Origin of Chaos Theory—Lorenz Attractor

Chaotic phenomena—the first time they were noticed by scientists was in the late nineteenth century by Professor Henri Poincaré (1854–1912) during his exploration of planetary motions in 1886.

However, the major breakthrough in chaos theory was made by mathematician and research meteorologist Professor Edward Lorenz (1917–2008) during his experiment on a *miniature* weather modeling system (Lorenz 1963), using three simple differential equations (the so-called *Lorenz equations*) instead of using complex statistical models which were commonly used by meteorologists at that time.

The Lorenz equations are

$$\dot{x} = -\sigma x + \sigma y \qquad (8.3a)$$

$$\dot{y} = -xz + rx - y \qquad (8.3b)$$

$$\dot{z} = xy - bz \qquad (8.3c)$$

These equations were based on the fundamental *Navier–Stokes equations* of fluid dynamics, which describe weather formation as the fluid circulation of air upon distant heating by the sun.

In Lorenz equations, x is the fluid stream function, whose value is proportional to speed of motion of the fluid due to convection. y is proportional to the temperature gradient between rising and falling parts of the fluid at a given height—the so-called *horizontal flow*. z is proportional to the deviation from temperature linearity as a function of vertical position—the so-called *vertical flow*.

In such a highly simplified weather model, he unexpectedly discovered two important things about this nonlinear model:

1. The nature of dynamics for the system itself can be entirely altered with an adjustment in parameters. In his original work, he discovered that when parameters σ and b are set to 10 and 8/3, respectively, the entire system will turn out to be highly chaotic without any steady states (and limit cycles), whenever the *Rayleigh number* (r) exceeds critical value of 24.74. In other words, all the solutions would evolve to be neither periodic nor convergent to any solution or equilibrium state.
2. The weather solution of this model, under such chaotic conditions, will be extremely sensitive to its initial conditions. In other words, within the chaotic region, only a very small deviation of initial conditions, say minor variations (or a *mistake*) in inputting the initial weather conditions (even of the order of 10^{-5}), will result in an entirely different weather situation in the following days, as shown in Fig. 8.4.

Such discovery is essential to the scientific world because for any system with dynamic equations $y = f(x, a)$, *where x is the variable and a is the parameter*. *Common sense* tells us that only variable x will change the dynamics of the system (that's why we called it *variable*), the changes of parameter a will only cause some minor changes of the overall system, the overall physical behavior of system will not be changed. Besides, traditional experimental science tells us that the overall experimental errors of a system (e.g., prediction system) should be proportional to the errors of variables. However, his discovery told us another story of reality. In which for some systems, just like Lorenz's equations, if we change the parameters to certain range or values, the overall physical behavior and dynamics of the system

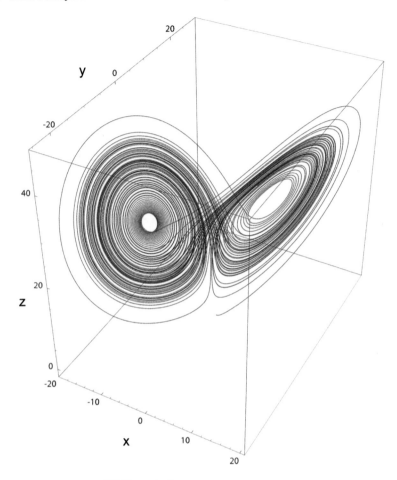

Fig. 8.4 Lorenz attractor (3D illustration)

will be changed entirely into what we called *deterministic chaos*; or *chaos systems*. In that situation, merely a very minor change on initial condition of the system will cause a huge difference of experimental results in future.

8.2 Characteristics of Chaotic Systems

In summary, there are several features and characteristics that can be found in a typical chaotic system.

They are

1. sensitivity to initial condition;
2. highly nonlinear but deterministic in nature;

3. consisted of bifurcation points that lead to chaos; and
4. consisted of self-similarity and fractals.

8.2.1   Sensitive Dependence on Initial Conditions

It is one of the most remarkable features of chaotic system, but it alone does not mean that a system is chaotic, but every chaotic system must possess this property.

Dependence of initial conditions means that for a very slight change in initial conditions or measurements (observations) of a chaotic system, the result produced differ tremendously after exhibiting for a period of time.

People like to quote *butterfly effect* to express such characteristics.

In layman's term, it states that a butterfly wings' flapping in Africa could result in hurricane in Florida.

Figure 8.5 illustrates the sensitivity to initial condition of Lorenz's equation of two systems, one with $x(0) = 1.00001$ and the other with $x(0) = 1.00000$, given $\sigma = 10; r = 28; b = 8/3$.

One should note that the sensitivity to initial conditions, although are very similar to the concept of *% errors of measurements* in experimental physics, they are two totally different concepts.

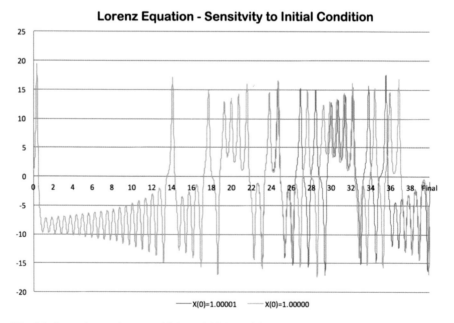

Fig. 8.5 Lorenz's equation—sensitivity to initial conditions.

In classical *% error of measurements* problem, the resulted % errors of a dynamical system are normally proportional to the *degree-of-discrepancy* for the measurement of observations. However, in chaos theory, the sensitivity dependence nature is unrelated to such discrepancy, it is the intrinsic nature of chaotic system itself.

8.2.2 Highly Nonlinear but Deterministic in Nature

Even though a slightest difference in initial conditions can reach extremely different results, all chaotic systems are *deterministic* in nature.

Deterministic means that *equal causes having equal effects.*

Or, from another viewpoint, *equal effects have equal causes.*

Equal causes having equal effects means *predictability*. That is, given e1 is the effect of cause c1, if there is another cause c2 which is equal to c1, we can *predict* that c2 will have an effect e2 equals to e1.

Equal effects have equal causes means *reversibility*. That is, given e2 and e1 are two equal effects, if e1 is caused by c1, then we can say the cause c2 equals to c1 of e2. Figure 8.6 illustrates two time series: one is *deterministic chaos* and the other is a *random walk*. As shown, although they look alike, however, they are totally different in terms of dynamics behavior.

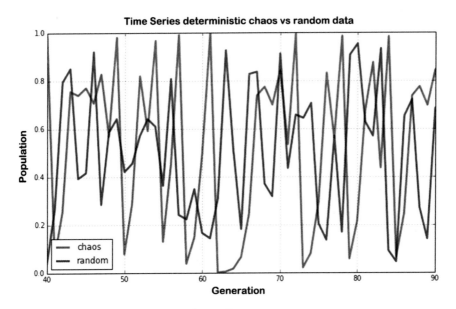

Fig. 8.6 Deterministic chaos versus random walk

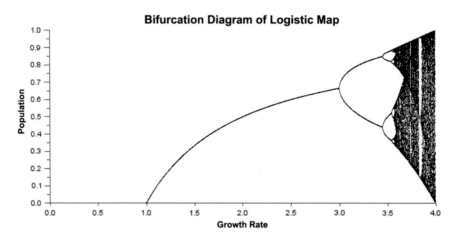

Fig. 8.7 Bifurcation diagram of a logistic map

8.2.3   Bifurcation Phenomenon

Does it mean that only complex nonlinear equations such as the Lorenz equations have chaotic effect?

The answer is *no*.

As mentioned, chaotic phenomena are so common and usual that even simple nonlinear dynamics as in the logistic Eq. (8.4) can have this chaotic phenomenon.

$$x_{t+1} = rx_t(1 - x_t) \tag{8.4}$$

Figure 8.7 depicts a typical bifurcation diagram for the logistic equation with control parameter r ranging from 0 to 4 (Kapitaniak 2000; Lee 2005).

Bifurcation is a common phenomenon found in chaotic systems which indicates sudden, qualitative changes in the dynamics of the system under investigation either:

1. From one kind of periodic case (with limit cycle(s) and fixed point(s)) to another kind of periodic situation, such as the change in bifurcation of the logistic map from r = 3 (from one fixed point to two fixed points) or
2. From a periodic stage to a chaotic stage, such as dramatic changes in the dynamics of the logistic map for r increasing from 3.5 to 4.

8.2.4   Applications of Chaos Theory

Although chaos theory is rather new and its technology yet emerging, there are already numerous applications in various industries which adopted this innovative technology.

Table 8.1 Applications of chaos theory

Application areas	Potential applications
Consumer electronics (Chau and Wang 2011)	• Chaotic kerosene fan heater (Katayama 1993) • Chaotic dishwasher • Chaotic air conditioners • Chaotic mixer
Computer systems	• Lee-associator for progressive memory recalling (Lee 2006) • Temporal pattern search (Davis 1990) • Chaotic mobile robot navigation (Sambas et al. 2016) • Chaotic network design and implementation (Soriano-Sanchez et al. 2018) • Chaotic cryptosystems (Lee 2005)
Health care and life science	• EEG analysis of heart rhythms (Freeman 2001) • Chaos-aware defibrillator (Munakata 1998) • Chaotic brain wave analysis (Freeman 2001)
Engineering (Kapitaniak 2000)	• Circuit stability analysis • Vibration control and study • Heat combustion
Business and finance	• Fractal market pattern analysis (Mandelbrot 1997) • Financial prediction (Lee 2005; Kazem et al. 2013) • Market prediction (Peters 1994)

- consumer electronics,
- computer systems,
- health care,
- life sciences,
- engineering, and
- business and finance.

Table 8.1 summarizes the potential applications derived from this fascinating technology.

8.2.5 Chaos Theory Versus Uncertainty Principle

In Chap. 7, we have learnt the implication of Heisenberg's uncertainty principle on fuzzy logic. In fact, the principle is also closely related to chaos theory.

Professor Heisenberg mentioned from his work *Physics and Philosophy* (Heisenberg 2007) that:

The Revolution in Modern Science, quantum mechanisms and the Uncertainty Principle should not be unusual or unique phenomena, but rather should be universal phenomena that exist in all matter (at the subatomic level).

If this is the case, the subatomic phenomena somehow might be coherent with chaos theory.

Taking weather forecasts as an example, we know that most of the international weather centers use numerical weather prediction (NWP) techniques to perform forecasts. For every single forecast, we need to know the so-called *boundary conditions,* that is, grid-point weather observations include wet- and dry-bulb temperatures, mean sea-level pressure (MSLP), upper level wind speed and direction, and so on. According to chaos theory, slight differences (errors) in the measured values may result in completely different weather situations in the future.

Moreover, according to the Heisenberg's uncertainty principle, it is quite *impossible* for us to obtain 100% accuracy in measuring all different weather conditions at the same time.

The same situation applies to finance, uncertainty principle in quantum finance tells us that we cannot be 100% sure about the quantum finance pair: return (r) and return-rate $(\dot{r})$ given by

$$\Delta r \cdot \Delta \dot{r} \geq \hbar/2 \tag{8.5}$$

Does this really mean that we can never predict our future accuracy?

The answer is *yes* and *no.*

Yes, in the sense that if Heisenberg's uncertainty principle is really true, we should have no way to "measure" the quantum variables for any chaotic system 100% correctly that will lead to tremendous deviation according to first rule of chaotic system.

No, in the sense that although we cannot be 100% sure about the initial condition, we have one mathematical model that can help us to solve this forecast problem, one model that can predict the future without knowing the exact initial condition, which is

Neural network, time series neural network to be exact.

In the coming chapters on quantum finance application, we will study a special kind of neural network, so-called *chaotic neural oscillator,* that can solve this problem agreeably with the integration of chaos theory and neural network in quantum finance prediction.

8.3 Fractals in Quantum Finance

8.3.1 Fractal Versus Chaos—Two Sides of a Coin

The bifurcation diagram of logistic map learnt in last section also revealed other interesting phenomena of a typical chaotic system—namely, the existence of self-similarity in chaotic dynamics.

If we take a closer look at the bifurcation pattern in the first chaotic region (i.e., $3.5 < r < 4$), we will discover that similar bifurcation patterns can be found within the original pattern, but in a smaller or scaled-down size. Figure 8.8 shows self-similarity between bifurcation at μ_1, μ_2, and μ_3 with same pattern but decrease in scale to infinity.

More surprisingly, if we continue to *zoom into* these bifurcation patterns, we will discover that they, in turn, contain even more scaled-down features with completely same shapes and patterns—what we called the *self-similarity* nature of chaotic systems.

Based on these interesting findings, a group of scientists established a new doctrine of applied mathematical geometry—*fractal geometry* (or simply *fractals*).

In fact, fractals are the graphical (visual) presentations of chaos while chaos is the physical dynamics of fractals.

In this section, we will explore what fractals are, their formulations, and how they can be applied to quantum finance.

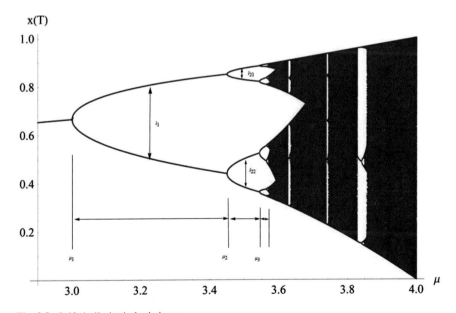

Fig. 8.8 Self-similarity in logistic map

8.3.2   A Brief History of Fractals

The study of self-similarity in graphs and patterns is not a new thing. German polymath and philosopher Gottfried Leibniz (1646–1716) studied recursive self-similarity and used the term *fractional exponent* to describe such phenomena.

In 1872, German mathematician Professor Karl Weierstrass (1815–1897) presented the first definition of a function with a graph that would be considered as fractal today.

In 1883, German mathematician Professor Georg Cantor (1845–1918) who attended lectures by Weierstrass, published examples of subsets of the real line known as *Cantor sets*—one of the fundamental fractals.

At the same period, German mathematician Professor Felix Klein (1849–1925) and French mathematician Professor Henri Poincaré (1854–1912) introduced a category of fractal that has come to be called *self-inverse fractals*.

In 1904, Swedish mathematician Professor Helge von Koch (1870–1924) extended Poincaré's ideas and gave a more geometric definition of fractal geometric with famous fractal discovery known as the *Koch snowflake*.

By 1918, two French mathematicians, Professors Pierre Fatou (1878–1929) and Gaston Julia (1893–1978) discovered fractal behavior associated with mapping complex numbers and iterative functions leading to further ideas about *attractors* and *repellers*.

In 1967, Professor Benoit Mandelbrot (1924–2010) published a paper about self-similarity: *How Long Is the Coast of Britain?* (Mandelbrot 1967) aroused public interests in fractal geometry in nature. Later in 1985, he published the remarkable work in *New Scientist: Fractals—Geometry Between Dimensions* and coined the term *fractals* and illustrated his mathematical definition with striking computer-constructed visualizations—the *Mandelbrot set* (Mandelbrot 1982).

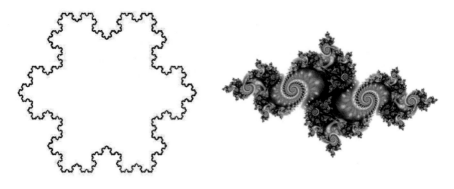

(a) Koch snowflake fractals   (b) Julia set fractals

Fig. 8.9 Koch snowflake and Julia set fractals

Figure 8.9a and b illustrates the *Koch snowflake fractals* and *Julia set fractals,* respectively.

8.3.3 Self-similarity in Mandelbrot Set Fractals

Four main properties of fractals are as follows:

- self-similarity;
- simple but non-differentiable;
- fractal dimension; and
- one-to-many, many-to-one.

Figure 8.10a–c shows the self-similarity of Mandelbrot set fractals under different scales.

The Mandelbrot set is the set of c values in the complex plane for which the orbit of $z_0 = 0$ under iteration of the quadratic map remain bounded:

$$z_{n+1} = z_n + c, c \in M \Leftrightarrow \lim \sup_{n \to \infty} |z_{n+1}| \leq 2 \tag{8.6}$$

That is, a complex number c is part of the Mandelbrot set if and only if when starting with $z_0 = 0$, and applying the iteration repeatedly, the absolute value of z_n remains bounded by a number, say, 2.

If $c = 1$, it will give the sequence 0, 1, 2, 5, 26, ..., which tends to infinity. As this sequence is unbounded, 1 is not an element of the Mandelbrot set.

If $c = -1$ gives the sequence 0, −1, 0, −1, 0, ..., which is bounded, and so −1 belongs to the Mandelbrot set.

By using this simple method, we can construct 2D Mandelbrot fractals by plotting all possible c with x–y-axis as the real–imaginary components (1 belongs to Mandelbrot, whereas 0 does not belong).

8.3.4 Fractal Dimension

A fractal dimension is a ratio providing a statistical index of complexity comparing how detail in a pattern (fractal pattern) changes with the scale at which it is measured.

Counter-intuitive is that a fractal dimension does not have to be an integer.

The essential idea of *fractured* dimensions has a long history in mathematics, but the term itself was brought to the fore by Professor Benoit Mandelbrot based on his 1967 paper on self-similarity in which he discussed fractional dimensions.

The concept of a fractal dimension rests in unconventional views of scaling and dimension.

For example, in traditional (discrete) dimension, given N is the number of sticks, ε is the scaling factor, and D is the dimension, using *box-counting* technique, we have (Fig. 8.11)

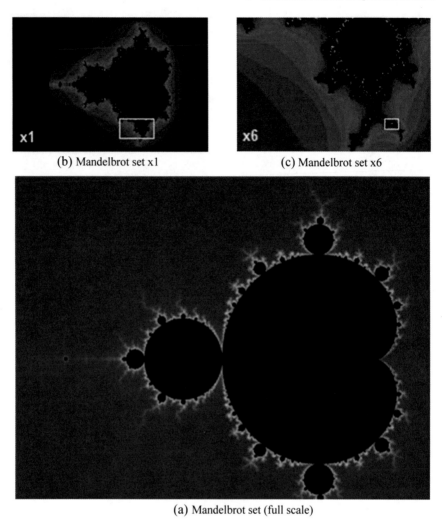

(b) Mandelbrot set x1   (c) Mandelbrot set x6

(a) Mandelbrot set (full scale)

Fig. 8.10 Mandelbrot set fractals (self-similarity under different scales)

$$N = \frac{1}{\varepsilon^D} \quad \text{or} \quad D = -\log(N)/\log(\varepsilon) \tag{8.7}$$

As shown in Fig. 8.12, when
$\varepsilon = 1/3, N = 27 => D = 3$.
Using this method to calculate the fractal dimension of Koch fractal, we have
For i = 2:
$D_{koch} = \frac{-\log(N)}{\log(\varepsilon)} = \frac{-\log(4)}{\log(\frac{1}{3})} = 1.261859507$
Similarly, $D_{koch} = \frac{-\log(16)}{\log(\frac{1}{9})} = 1.261859507$

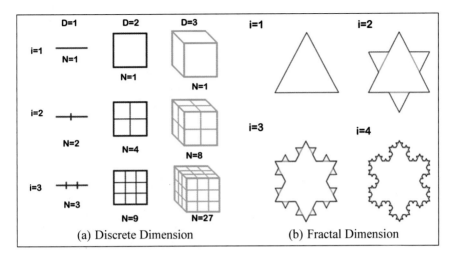

(a) Discrete Dimension   (b) Fractal Dimension

Fig. 8.11 Discrete versus fractal dimension

8.3.5 Natural Phenomena with Fractal Features

How many natural phenomena can be found in fractals?

The answer is … many.

It is similar to the golden ratio on computational finance mentioned in Chap. 6, fractals can be found anywhere in Mother Nature.

• Algae	• Blood vessels	• Brain waves
• Brownian motions	• Cloud formation	• Corals
• Coastlines	• DNA and RNA	• Crystals
• Financial patterns	• Heartbeats	• Lightning bolts
• River networks	• Mountain ranges	• Sand and soil
• Seashells	• Snowflakes	• Saturn rings
• Trees and branches	• Turbulent flows	• Tornadoes and hurricanes

Why?

The answer is obvious, if we are clever enough ….

Let's come back to our old friend … quantum theory and subatomic world

Such one-to-many and many-to-one phenomena not only trigger modern graphics and CGI such as the fantasy in sci-fi motion pictures, but also …

Worlds of simulation ….!

Fig. 8.12 Fractals in nature (Tuchong 2019a, b, c, d, e, f, g, h, i, j)

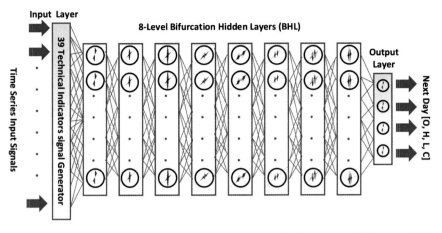

Fig. 8.13 System architecture of CDNOMS chaotic deep supervised-learning (CDSL) network for time series financial prediction

8.4 Applications of Chaos Theory and Fractals in Quantum Finance

8.4.1 Chaos Theory in Quantum Finance

Owing to the chaotic property of financial market, chaos theory becomes one of the hottest topics in financial engineering, especially on financial prediction.

In the second part of this book, we will study quantum finance application on worldwide financial prediction, explore how to integrate QPL; chaos theory, and neural oscillators for real-time financial prediction.

Figure 8.13 shows a snapshot of the chaotic deep neural network for time series financial prediction.

8.4.2 Fractal Finance

Self-similarity and recursive property of fractals led to active study application in finance engineering since late 1980s.

As said, although fractals can generate very complex pattern, the interesting thing is, it can be (should be) originated from very simple pattern (equation).

Let's review a simple example of how fractals can be applied to technical finance.

In the most fundamental case, basic financial fractals are composed of five or more bars as shown in Fig. 8.14.

The rules for identifying fractals are as follows:

Fig. 8.14 Bearish versus
bullish fractals

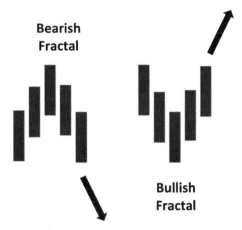

1. A bearish turning point occurs when there is a pattern with the highest high in the middle and two lower highs on each side.
2. A bullish turning point occurs when there is a pattern with the lowest low in the middle and two higher lows on each side.

Actually, in standard MT platform, MQL already provides such indicator *iFractals* for real-time detection of bearish/bullish fractals.

Fractals are best used in conjunction with other indicators or forms of analysis such as alligators or MAs.

Figure 8.15 shows a long-term uptrend with price staying predominantly above the alligator's teeth (middle moving average). Since the trend is up, bullish signals can be used to generate buy signals.

In addition to identify buying and selling signals, can we use fractals to forecast financial market, say, stock markets?

The answer is yes!

Fig. 8.15 Fractals in financial pattern

Current research in financial forecast using fractals in two directions:

1. Integration of fractals (as technical signals) with chaotic neural network for financial prediction and
2. Simulation of multidimensional market pattern by fractal equations, if one may do this. We can probably predict the future. Why?

8.5 Conclusion

In this chapter, we have learnt chaos and fractals—the two contemporary mathematical models with vast application on nonlinear dynamics and financial engineering.

As mentioned, chaos and fractals are sides of the same coin. They show us two ways of looking (modeling) the world's dynamics and all physical phenomena.

Just like the particle–wave duality in quantum theory, we are talking about the same scenario in different (mathematical) languages.

But if we look at them from different perspective, chaos and fractals, in fact, can be considered as mathematical (and hence computational) way to describe the particle–wave duality in terms of nonlinear equations, which is totally coherent with our computational model of quantum finance and price level (QPL) evaluation to form an important component of quantum finance.

In fact, the entire *puzzle* of quantum finance is almost complete, but with one piece of component—how to integrate chaos theory into neural network for chaotic modeling of financial time series?

In next chapter, we will study the last component of quantum finance—chaotic neural oscillator.

Problems

8.1 Why scientists always say chaos and fractals are two sides of the same coin? What is the logic behind it? Use two financial markets as examples to support your explanation.

8.2 What is the meaning of deterministic chaos? Is it equivalent to random event? Why or Why not?

8.3 What kind of systems or phenomena in nature are categories as chaotic systems? State five examples and explain why they are considered as chaotic systems.

8.4 What is the relationship between chaotic flow versus chaotic map?

8.5 What is butterfly effect in chaos theory? Can we find any examples of such phenomena in finance industry? Please give three typical examples of finance and explain how they work.

8.6 Programing Exercise I—Lorenz attractor

(i)   Write a MATLAB program to implement the Lorenz attractor by using the three Lorenz Eqs. (8.3a–8.3c);

(ii) Plot the 3D plot of the butterfly effect, as below.

(iii) What are the key parameters and their values in Lorenz equation in order to generate the 3D chart of butterfly effect in chaos?

(iv) Explain why such findings are important in the modeling of complex systems such as the modeling of weather or financial markets.

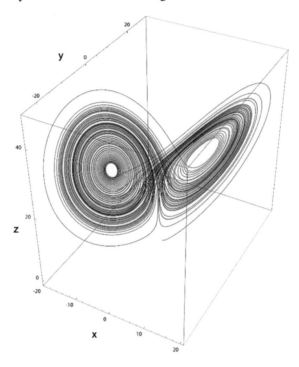

8.7   State and explain the four major characteristics of chaotic systems. Use financial market such as forex market to explain these four characteristics in financial engineering.

8.8   Programming exercise II—sensitivity to initial conditions of Lorenz system

(i)   First, implement the MATLAB system in 8.6.

(ii) Make a very minor change of initial condition x say (a minor difference of 0.00001 of x), plot the result with the x-axis as time step and Y-axis as future value. See what happen.

(iii) Change the initial conditions of y and z and check the results.

(iv) Any interesting findings and explain why.

8.9   What is bifurcation phenomenon in chaos theory? What is the importance of bifurcation in a complex system?

8.10 Programming exercise III—logistic equation

(i)   Write a MATLAB program to implement the logistic map using Eq. (8.4).
(ii) Plot the logistic map for x between 0 and 4.0, see whether you can obtain the result below.

Bifurcation Diagram of Logistic Map

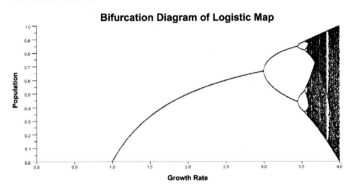

(iii) Locate all the bifurcation zones.
(iv) What is the implication and importance for the discovery of chaotic phenomena and bifurcation zones in this simple logistic map equation?

8.11 State any five applications of chaos theory in consumer electronics and health care. Explain how they work.

8.12 State three applications of chaos theory in finance industry and explain how they work.

8.13 What is the relationship between chaos theory and Heisenberg's uncertainty principle in quantum theory? And discuss how such findings are important to quantum finance.

8.14 What are fractals? State any five examples of fractals in nature.

8.15 Programming Exercise III—fractals

(i)   Write MATLAB programs to implement: (1) Koch snowflake fractals; (2) Julia set fractals; and (3) Mandelbrot set fractals.
(ii) What are the main properties and characteristics of fractals?
(iii) Calculate the fractal dimension for these three fractals.

8.16 Programming exercise IV—fractals in MQL

(i)   State and explain fractals in MQL.
(ii) In order to implement fractals in MQL, at least one other technical indicator is needed. What is this technical indicator? How it works.
(iii) Write a simple trading program using MQL on three financial products: (1) Dow Jones Index (DJI); forex product such as AUDCAD; gold or silver (i.e., XAUUSD or XAGUSD).
(iv) Compare their trading performances.

(v)   State and design how can we integrate quantum finance with this fractal-based trading program.

(vi) Implement this hybrid system and compare it with fractal-based trading system.

References

Chau, K.T. & Wang, Z. (2011) Application of Chaotic Motion in *Chaos in Electric Drive Systems.* Wiley, Singapore, 283–314.

Davis, P. (1990) Application of optical chaos to temporal pattern search in a nonlinear optical resonator. *Japanese Journal of Applied Physics.* 29(7): 1228–1240.

Freeman, W. J. (2001) *How brains make up their mind.* Columbia University Press.

Heisenberg, W. (2007) Physics and Philosophy: The Revolution in Modern Science. Harper Perennial Modern Classics.

Kapitaniak, T. (2000) *Chaos for engineers: theory, applications, and control.* Springer.

Katayama, R. (1993) Developing tools and methods for applications incorporating neuro, fuzzy and chaos technology. Computers and Industrial Engineering. 24(4): 579–592.

Krishnamurthy, V. (2015) Edward Norton Lorenz: Discoverer of chaos. *Resonance.* 20(3): 191–197.

Kazem, A. et al. (2013) Support vector regression with chaos-based firefly algorithm for stock market price forecasting. *Applied Soft Computing Journal.* 13(2): 947–958.

Lee, R. S. T. (2006) LEE-Associator – A Transient Chaotic Autoassociative Network for Progressive Memory Recalling. *Neural Networks.* 19(5): 644–666.

Lee, R. S. T. (2005) *Advanced Paradigms in Artificial Intelligence From Neural Oscillators, Chaos Theory to Chaotic Neural Networks.* Advanced Knowledge International, Australia.

Lorenz, E. (1995) *The Essence of Chaos.* University of Washington Press.

Lorenz, E. (1963) Deterministic nonperiodic flow. *Journal of the Atmospheric Sciences.* 20: 130–141.

Mandelbrot, B. (1997) *Fractals and scaling in finance: discontinuity and concentration,* Springer.

Mandelbrot, B. (1982) *The fractal geometry of nature.* Times Books.

Mandelbrot, B. (1967) How Long Is the Coast of Britain? Statistical Self-Similarity and Fractional Dimension. *Science, New Series.* 156(3775): 636–638.

Munakata, T. (1998) *Fundamentals of the new artificial intelligence: beyond traditional paradigm.* Springer.

Peters, E. E. (1994) *Fractal market analysis: applying chaos theory to investment and economics.* Wiley.

Sambas, A. et al. (2016) A 3-D novel jerk chaotic system and its application in secure communication system and mobile robot navigation. *Studies in Computational Intelligence.* 636: 283–310.

Soriano-Sanchez et al. (2018) Synchronization and FPGA realization of complex networks with fractional–order Liu chaotic oscillators. *Applied Mathematics and Computation.* 332: 250–262.

Schuste, H. G. and Just, W. (2005) *Deterministic Chaos: An Introduction.* Wiley-VCH.

Tuchong (2019a) Fractals in Nature - Snowflakes. http://stock.tuchong.com/image?imageId= 79052163044817970. Accessed 21 Aug 2019.

Tuchong (2019b) Fractals in Nature - Blood vessels. http://stock.tuchong.com/image?imageId= 452726017391067492. Accessed 21 Aug 2019.

Tuchong (2019c) Fractals in Nature - Landscape. http://stock.tuchong.com/image?imageId= 479820371888701550. Accessed 21 Aug 2019.

Tuchong (2019d) Fractals in Nature - Saturn ring. http://stock.tuchong.com/image?imageId= 429030579742310411. Accessed 21 Aug 2019.

Tuchong (2019e) Fractals in Nature - Trees. http://stock.tuchong.com/image?imageId= 130438294248095746. Accessed 21 Aug 2019.

Tuchong (2019f) Fractals in Nature - Sea shell. http://stock.tuchong.com/image?imageId= 459360616581955957. Accessed 21 Aug 2019.

Tuchong (2019g) Fractals in Nature - Corals. http://stock.tuchong.com/image?imageId= 411274248616935479. Accessed 21 Aug 2019.

Tuchong (2019h) Fractals in Nature - Cloud formation. http://stock.tuchong.com/image?imageId= 238389471191761069. Accessed 21 Aug 2019.

Tuchong (2019i) Fractals in Nature - Financial patterns. http://stock.tuchong.com/image?imageId= 481643508196442202. Accessed 21 Aug 2019.

Tuchong (2019j) Fractals in Nature - Lightning bolts. http://stock.tuchong.com/image?imageId= 540477009592057896. Accessed 21 Aug 2019.

Chapter 9
Chaotic Neural Networks in Quantum Finance

*If we look at the way the universe behaves, quantum mechanics
gives us fundamental, unavoidable indeterminacy, so that
alternative histories of the universe can be assigned probability.*
Murray Gell-Mann (1929–2019)

What would happen if time stopped?
Would they like movies and TV shows: everything freezes when time stopped?

In 1952, Professor Erwin Schrödinger addressed at a lecture with a witty remark to his audience what he was about to say might "seem lunatic" (Fig. 9.1).

He stated that when his equations seemed to narrate several different histories, these were *not alternatives, but all really happen simultaneously*—as we are now so-called *parallel universe/multiverse theory*. The author was really and truly, enticed by this statement to take up theoretical physics especially on quantum mechanics and general relativity courses at Hong Kong University in 1986–89.

Current *cosmology* (*superstring and M-theory*) even tells us that we live in one of the many so-called *realities*, as Professor Murray Gell-Mann, Nobel Prize winner for physics in 1969 mentioned in his famous quotation.

More importantly, is that all of these are coexisting realities and vibrating in different frequencies.

If we are clever enough, we should realize why Professor Schrödinger said so.

If it is true. If the entire universe with everything inside it is under continuous vibrations and oscillations in every single moment. If why we can percept the reality and our surroundings is because of these continuous vibrations and oscillations between our sensory organs and the external world—which means when time is really stopped, everything is stopped; or should it be rather disappeared and/or cannot be precepted instead?

In this chapter, we will study an interesting topic—chaotic neural oscillators in quantum finance. First, we will explore the quantum world of chaotic oscillation from the multiverse in quantum theory, chaos in the brain to our sensory of oscillations. Next, we will review chaotic neural oscillators, neuroscience motivation, Wang-

© Springer Nature Singapore Pte Ltd. 2020   235
R. S. T. Lee, *Quantum Finance*,
https://doi.org/10.1007/978-981-32-9796-8_9

Fig. 9.1 What would
happen if time stopped?
(Tuchong 2019a)

oscillators, and the author's latest research on Lee-oscillators; its neural dynamics and applications. After that, we will study quantum finance oscillators (QFO) using Lee-oscillators and different application of QFO in quantum finance including quantum financial prediction using chaotic neural networks, chaotic deep neural networks, and chaotic intelligent multiagent-based trading systems.

9.1 Quantum World of Chaotic Oscillation

9.1.1 Multiverse and Quantum Theory

Cosmologist Professor Max Tegmark has provided a taxonomy of universes beyond the familiar observable universe (Tegmark 2014). In Tegmark's chaotic inflation theory—*bubble universes* in Fig. 9.2—every disk represents a bubble universe.

According to Tegmark's bubble universe theory, our universe is just one of the many disks of universes. Universes 1–6 represent bubble universes. Five of them have different physical constants than our universe.

The four levels of the multiverse are the following:

Fig. 9.2 Illustration of
Tegmark's bubble universe

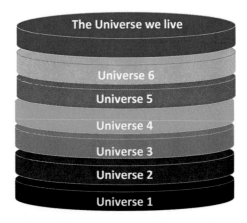

1. Level I: an extension of our universe
2. Level II: universes with different physical constants
3. Level III: many-worlds interpretation of quantum mechanics
4. Level IV: ultimate ensemble.

Related to the many-worlds idea were originated from Professor Richard Feynman's multiple histories theory, who used the theory to explain the existence of infinite path integrals of realities.

9.1.2 Chaos in Brain—Brainwave

Current neuroscience tells us that oscillations can not only be found in the physical world but also can be found at the core of our body—our brain.

If one takes a closer look at neural dynamics in our brain—the original of neural networks: What triggers the entire neural dynamics?

The stimulus, maybe, but how?

Current neurophysiology also tells us that all neurons in our brain exist in the state of *oscillations, chaotic oscillation* to be exact.

It is that such oscillations trigger the entire *thinking* and *learning* processes.

The oscillations of these neurons within our brain are collectively called *brain waves* in layman's terms as shown in Fig. 9.3.

- Awake, conscious state—beta wave
- Insight and peak focus conscious state—gamma wave
- Relaxed, calm, lucid dream (but not thinking)—alpha wave
- REM dream state, deep relaxation and mediation—theta wave
- Deep, dreamless sleep—delta wave.

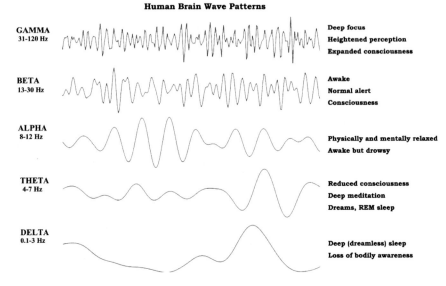

Fig. 9.3 Illustration of human brain waves patterns

In fact, not only from our brain, chaotic oscillations can be easily found in the human body including our five senses:

- eye (sight)—visual nerves;
- nose (smell)—olfactory nerves;
- ear (sound)—auditory nerves;
- tongue (taste)—gustatory nerves;
- skin (touch)—somatosensory nerves;
- our heartbeats and others.

9.2   Chaotic Neural Oscillators

9.2.1   An Introduction

We learnt from Chap. 7 that classical artificial neural networks, ANN in Fig. 9.4 are composed of simple artificial neurons that emulate biological neural activities.

These kinds of neural models have been strongly criticized as being far simpler than real neural models.

The latest studies in neuroscience and neurophysiology have examined such issues as functional properties of the hippocampus, neural activities in pyloric CPG (central pattern generator) of the lobster, and other human brain activities. For instance, hippocampus—a major component of the brain as shown in Fig. 9.5, plays important

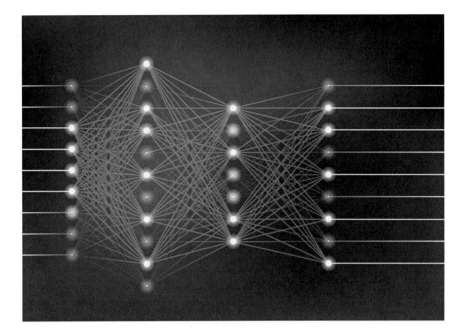

Fig. 9.4   Artificial neural networks (Tuchong 2019b)

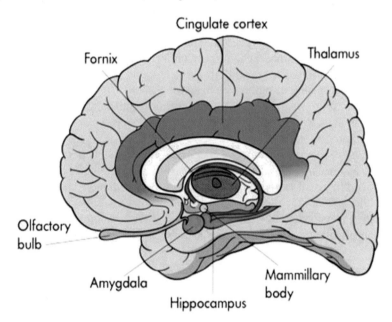

Fig. 9.5   Hippocampus of human brain

roles in the consolidation of information from short-term to long-term memory, and in spatial memory that enables navigation.

In Alzheimer's disease, the hippocampus is one of the first regions of the brain to suffer damage with short-term memory loss and disorientation. Studies have provided strong evidence of chaotic neural activities in these complex neural behaviors.

Researchers have also proposed various chaotic neural models over the past decades. Latest research include: chaotic oscillators proposed by Falcke et al. (2000) to model pyloric CPG neurons, cortical networks proposed by Hoshino et al. (2003) for recalling LTM (long-term memory), TCNN (transient chaotic neural network) proposed by Chen and Aihara (1995) for handling combinational optimization problems, and Zhou's work (Zhou and Chen 2000) based on chaotic annealing technique for dynamic pattern retrieval.

9.2.2 Wang-Oscillator—The Structure

Most of the contemporary neural oscillators that developed theoretically from Wilson–Cowan model are focused on the time-continuous framework, which is in fact too complex for network modeling, let alone for actual computer implementation on actual applications.

Wang (1991) in his Ph.D. thesis, proposed a simple time-discrete neural oscillator model (namely, the Wang-oscillator).

Figure 9.6 shows the neural structure of Wang-oscillator.

Unlike its continuous model counterpart, the Wang-oscillator provided simple but remarkable neural dynamics ranging from fixed points (through quasi-periodicity) to chaos, as revealed by its bifurcation diagram.

This bifurcation diagram can be used as computational elements (called the BTU—bifurcation transfer unit) to replace traditional *sigmoid function* for temporal information processing.

The Wang-oscillator is a neural oscillatory model consisting of two neurons, one *excitatory* and one *inhibitory* neuron.

Just like biological neurons, the excitatory neurons *promote* chances of neural *firing* while the inhibitory neurons *hinder* neural activities.

Together, they form a simple but powerful *oscillatory couple*.

In Wang-oscillator, the time-discrete neural dynamics are given by

$$E(t+1) = Sig[\omega_{EE} \cdot E(t) - \omega_{EI} \cdot I(t) + S_E(t) - \xi_E] \tag{9.1}$$

$$I(t+1) = Sig[\omega_{IE} \cdot E(t) - \omega_{II} \cdot I(t) + S_I(t) - \xi_I] \tag{9.2}$$

$$W(t) = E(t) - I(t) \tag{9.3}$$

Fig. 9.6 Wang-oscillator

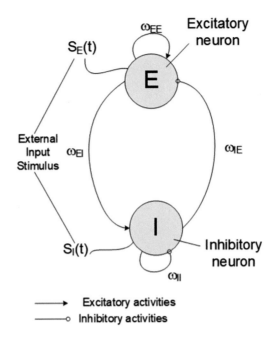

where S_E and S_I are input stimulus; ω are weights; ξ_E and ξ_I are threshold values; the sigmoid function Sig() is given by

$$\text{Sig}(k) = {}^{1}/1 + e^{-k} \qquad (9.4)$$

One important finding of Wang-oscillator is the bifurcation behavior of Wang-oscillator with the variation of input stimulus I.

9.2.3 Wang-Oscillator—Bifurcation Diagram

Figure 9.7 shows the bifurcation diagram of Wang-oscillator.

As shown, the bifurcation diagram of a typical Wang-oscillator consists of *sigmoid* and *bifurcation zones*.

The original idea is that the *sigmoid zone* imitates the classical sigmoid function and *bifurcation zone* imitates the chaotic property of a complex system during the sigmoid-transition period.

However, there are two intrinsic problems:

1. The existence of an extra *bifurcation zone I* destructs the continuity of sigmoid curve in the initial period;
2. *The bifurcation zone II* is "too chaotic" to model the chaotic-transition region in real-world problems.

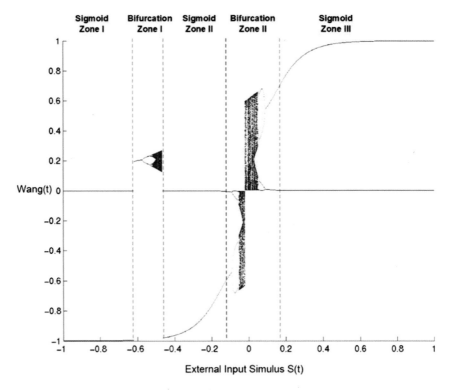

Fig. 9.7 Bifurcation diagram of Wang-oscillator

A more transient progressive growth in terms of neural dynamics is needed to be achieved.

9.2.4   Wang-Oscillator—Major Contributions and Limitations

One important finding and contribution of Wang-oscillator is the property of change in neural dynamics according to the input stimulus.

In most classical neural network models for information encoding and association (e.g., Hopfield Network), the network model can be generalized as a nonlinear function operator which, based on various input stimuli, alters its internal states and *fires* output according to the nonlinear transfer function.

The Wang-oscillator, on the other hand, encodes information (i.e., the input stimulus) and gives responses by altering the behavior of neural dynamics (from chaotic states to sigmoid growth), which is totally consistent with the latest findings on how the brain processes information (Freeman 2000, 2008, 2013).

However, the Wang-oscillator has several major limitations in the bifurcation behavior of the model, preventing it from being used as an effective BTU or temporal information processing model.

As mentioned previously, the *undesirable* chaotic region exists in *bifurcation zone I* which not only affects the continuity of the BTU but also violates the original function of BTU, namely that of stimulating the temporal information processing for the brain.

So, we need to look for a *total solution*.

9.3 Lee-Oscillator

9.3.1 The Motivation

After over 5 years of research and testing, the author created an ideal chaotic neural oscillator, now known as *Lee-oscillator* in 2004 and published in two major neural network journals *IEEE Transactions on Neural Networks* (Lee 2004) for the introduction of *Lee-oscillator* and its transient chaotic behavior; and *Journal of Neural Networks* (Lee 2006) for the exploration of *Lee-oscillator* on progressive memory recalling capability in visual psychology.

Different from *Wang-oscillator*, *Lee-oscillator* successfully emulates the transient chaotic progressive growth in its neural dynamics, which helps to shed light on acting as a perfect CTU to model complex and chaotic problems.

Figure 9.8 shows the neural structure of Lee-oscillator.

Basically, Lee-oscillator consists of 4 neural elements: E, I, Ω, and L, which corresponds to *excitatory, inhibitory, input,* and *output* neurons.

The neural dynamics are given by

$$E(t+1) = Sig[e_1 \cdot E(t) - e_2 \cdot I(t) + S(t) - \xi_E] \qquad (9.5)$$

$$I(t+1) = Sig[i_1 \cdot E(t) - i_2 \cdot I(t) - \xi_I] \qquad (9.6)$$

$$\Omega(t+1) = Sig[S(t)] \qquad (9.7)$$

$$L(t) = [E(t) - I(t)] \cdot e^{-kS^2(t)} + \Omega(t) \qquad (9.8)$$

where e_1, e_2, i_1, and i_2 are weights; ξ_E and ξ_I are threshold values and S(t) is an external input stimulus.

The bifurcation diagram of Lee-oscillator is illustrated in Fig. 9.9.

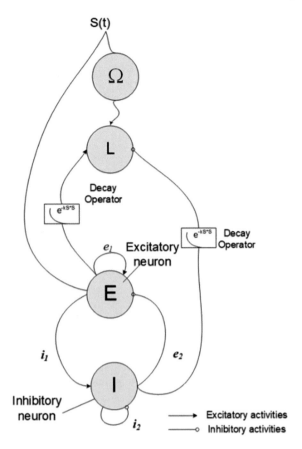

Fig. 9.8 Lee-oscillator

9.3.2   Lee-Oscillator—Bifurcation Diagram

Unlike *Wang-oscillator*, the *Lee-oscillator* bifurcation diagram in Fig. 9.9 is composed of three main regions, *sigmoid zones I and II* and single *bifurcation zone I*.

From neural dynamics point of view, *sigmoid zones I and II* denote sigmoid-shape region, which corresponds to nonchaotic neural activities in the oscillators; and *bifurcation zone I* is *hysteresis* region; which corresponds to the area of chaotic behavior that results when a weak external input stimulus is received.

Such neural dynamics can be perfectly served as *chaotic transfer unit* (CTU) to model complex and chaotic systems such as severe weather prediction, complex scene analysis, and real-time financial prediction such as quantum finance.

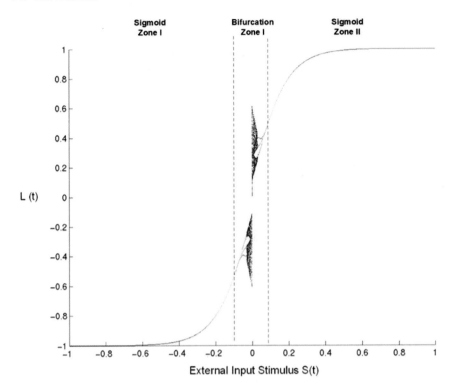

Fig. 9.9 Bifurcation diagram of Lee-oscillator

9.3.3 Lee-Oscillator—Potential Applications

Potential applications of *Lee-oscillator* include the following:

1. Basic chaotic neural elements for temporal information processing

   - Fundamentally, information processing is the overall process of encoding, recognizing, and discriminating various types of information (e.g., images, patterns, etc.).
   - Since a single Lee-oscillator can provide a two-state transient chaotic attraction from input space, it is suitable to serve as the basic element for the information processing unit for quantum computing.
   - In addition, owing to its unique chaotic features where neural dynamics change with variations in the input signal, the Lee-oscillator provides a good analog for simulating chaotic and temporal information processing behavior in brain science (Freeman 2000, 2008; Malsburg 1995; Fernandes and Malsburg 2015).

2. Transient chaotic auto-associator

- As an extension and generalization of using Lee-oscillators for information processing, a two-dimensional (2D) layer of Lee-oscillators can be adopted as a pattern associator (Lee 2006).
- As an analog of the classical *Hopfield Network* (Hopfield 1984) as an auto-associator, a transient chaotic autoassociative network based on Lee-oscillators as its constituting neuron elements can be used to provide an innovative progressive memory association and recalling scheme (Lee 2004; 2006) to expand how human can recall memory and to explain an important phenomena—Gestalt psychology.

3. Chaotic neural oscillatory units for advanced applications

- In fact, Lee-oscillators can be adopted and integrated with each other to form a complex chaotic neural oscillatory model to tackle problems such as scene analyses, robot vision, navigation, and of course, real-time financial prediction in quantum finance.
- We will study this in detail in Part II of this book.

9.4 Quantum Finance Oscillators (QFO)

9.4.1 The Concept

In Chap. 2, quantum field theory, we mentioned that the quantum world can be modeled as quantum oscillators, influenced and oscillating like a ball over a trampoline as shown in Fig. 9.10a.

So, in terms of quantum finance, it is natural to analog the dynamic of every financial market as quantum oscillations of quantum financial particles influenced by quantum energy (price) field as shown in Fig. 9.10b.

But instead of simple quantum harmonic oscillators, we are talking about quantum anharmonic oscillator counterparts.

Such notion, in fact, is totally equivalent and coherent with our modeling of quantum dynamics as quantum anharmonic oscillations studied in Chap. 4.

These quantum anharmonic oscillators are known as *quantum finance oscillators (OFO)*.

9.4.2 The Model

One of the most direct adoptions of the chaotic oscillator is the introduction of quantum finance oscillator—QFO by the replacement of Lee-oscillator with financial particle.

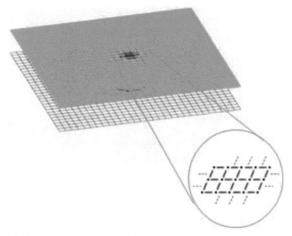

(a) Quantum price oscillator

(b) Quantum energy field of quantum finance particles

Fig. 9.10 Quantum finance oscillators (QFO)

Figure 9.11 shows a typical USDX (USD Index) QFO model in the forex market. The neural dynamics is given by

$$E(t+1) = Sig[e_1 \cdot E(t) - e_2 \cdot I(t) + S(t) - \xi_E] \tag{9.5}$$

$$I(t+1) = Sig[i_1 \cdot E(t) - i_2 \cdot I(t) - \xi_I] \tag{9.6}$$

$$\Omega(t+1) = Sig[S(t)] \tag{9.7}$$

$$QFO|_{USDX}(t) = [E(t) - I(t)] \cdot e^{-kS^2(t)} + \Omega(t) \tag{9.9}$$

Fig. 9.11 Quantum finance oscillator model

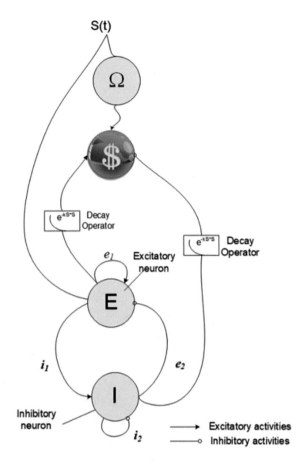

By using the same interpretation, practically speaking we can model any financial market using QFOs and investigate the chaotic oscillation dynamics of these financial markets.

More importantly, such modeling technique can be integrated with AI tools and technologies mentioned in Chap. 7 to complex and highly chaotic phenomena, financial fluctuation in particular.

9.5   Applications of QFO in Quantum Finance

9.5.1   The Concept

In Chaps. 7 and 8, we have learnt 3 basic AI tools/technologies that include the following:

- artificial neural network (ANN),
- fuzzy logic and fuzzy logic systems (FLS),
- genetic algorithms (GA),
- chaos theory, and
- fractals.

In fact, QFO (quantum finance oscillators) can be integrated into all these tools/technologies to tackle highly complex and chaotic phenomena—financial markets and forex markets in particular.

This section will study some typical QFO applications in quantum finance applications.

They are the following:

- QFO for real-time financial prediction using chaotic neural networks,
- chaotic fuzzy-neuro oscillators—quantum finance signals (QFS),
- chaotic deep neural networks in quantum finance, and
- chaotic multiagent trading system (Fig. 9.12).

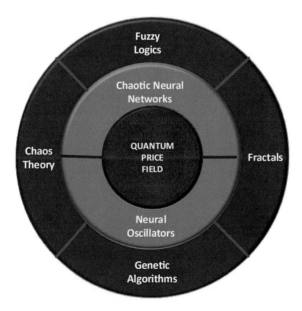

Fig. 9.12 Concentric sphere model of quantum finance

9.5.2   QFO for Real-Time Financial Prediction Using Chaotic Neural Networks

As a direct adoption in artificial neural networks, QFO can be directly integrated with classical neural networks such as feedforward backpropagation networks (FFBPN) by replacing all simple neurons in input/hidden/output layers by Lee-oscillations.

By doing so, technically speaking, we convert a simple neural network system into a multilayer chaotic neural network as shown in Fig. 9.13.

In the figure, two inputs types (time series input signals and technical index signals) are replaced by Lee-oscillators.

From neural dynamics point of view, such changes bring (1) chaotic oscillation dynamics during network training; (2) chaotic bifurcation transfer function (CBTF) into forecast system to resolve the network training *deadlock* and *trapped by local minimum* problems which usually occur during network training of highly complex patterns (e.g., financial patterns) using classical FFPBN.

Detail implementation will be studied in Chap. 11.

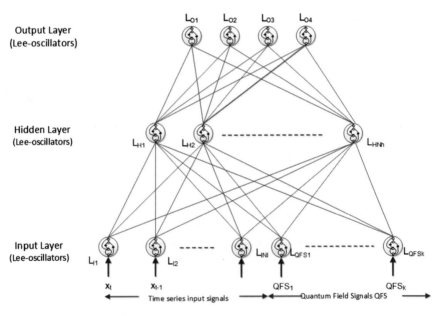

Fig. 9.13   QFO for real-time financial prediction using chaotic neural networks

9.5.3 *Chaotic Fuzzy-neuro Oscillators—Quantum Finance Signals (QFS)*

In Chap. 7, we have studied fuzzy logic (FL) and fuzzy logic system (FLS).

A simple fuzzy logic membership function (also known as *type-1 fuzzy logic*) of financial technical signal (RSI) is shown in Fig. 9.14a.

Different from *type-1 FL*, a *type-2 FL* provides a higher degree of fuzziness by the adoption of *second-tier fuzziness* in overlapping regions of the primary fuzzy membership function.

Figure 9.14b shows a typical so-called *interval type-2 MF f* or RSI.

By QFO adoption, we technically can generate a chaotic overlapping region for any technical financial signals—quantum finance signals (QFS).

Figure 9.14c shows a typical chaotic type-2 fuzzy logic MF (membership function) for RSI QFO.

Detailed implementation will be studied in Chap. 12.

(a) Type-1 fuzzy logic MF on RSI   (b) Interval type-2 fuzzy logic MF on RSI

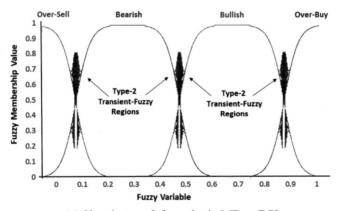

(c) Chaotic type-2 fuzzy logic MF on RSI

Fig. 9.14 Chaotic fuzzy-neuro oscillators—quantum finance signals

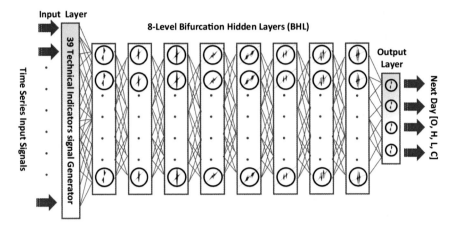

Fig. 9.15 Chaotic deep neural networks in quantum finance

9.5.4   Chaotic Deep Neural Networks in Quantum Finance

Latest neuroscience research discovered that neural dynamics for memory recall and learning existed as the so-called *retrograding signaling (RS)*, a kind of delayed and feedback mechanisms of information processing.

Such an important finding leads to the latest development of LORS (LEE-oscillators with Retrograde Signals).

Between 2012 and 2017, the author had categorized 8 major types of LORS bifurcation signals (LORS#0-#7), which corresponded to 8 different kinds of chaotic bifurcations and neural dynamics.

More importantly, such findings can be adopted to DNN (deep neural network) technology for the conversion of single hidden layer FFBPN into CDNN (chaotic deep neural networks) with 8-level bifurcation hidden layers (BHL).

Figure 9.15 shows the system architecture of CDNN on quantum finance. Detail implementation will be studied in Chap. 12.

9.5.5   Chaotic Multiagent Trading System

Neural networks can be used in *supervised learning (SL)* such as financial prediction, it can also be used on *unsupervised learning (UL)* and *reinforcement learning (RL)* as we learnt from Chap. 7.

A multiagent financial trading system is a typical reinforcement-learning problem in which there are no well-defined output/target pairs in a typical SL system, e.g., financial forecast.

All we have (and need to do) is to *optimize* the net-profit (or returns) at (fixed) period of time and *minimize* the potential risks.

In that case, we can make use of reinforcement-learning (RL) neural networks such as actor-critic (AC) RBF-based RLN.

Again, by replacing simple neurons in the RBF-layer of the AC-RBFN, we technically convert this RLN into chaotic AC-RBFN which can significantly improve the RL efficiency.

Figure 9.16 shows the system architecture of chaotic AC RBF-based RLN.

Detail implementation will be studied in Chap. 13.

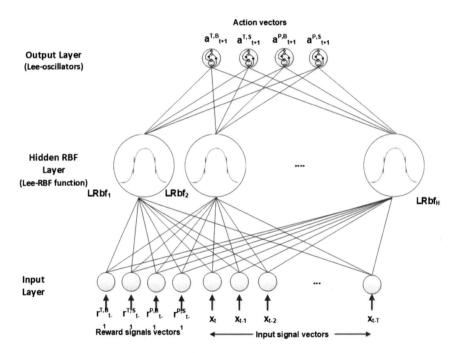

Fig. 9.16 Chaotic multiagent-based trading system

9.6 Conclusion

In this chapter, we have learnt the basic theory of chaotic neural oscillators.

Also, we have learnt how to integrate chaotic neural oscillators (Lee-oscillator) to effectively convert a typical neural network into chaotic neural networks to handle highly complex and chaotic phenomena—quantum finance in particular.

As we can see, chaotic neural oscillators can be adopted into almost any AI tools and technologies to improve efficiency and performance.

How can we do that?

As mentioned, all these various kinds of AI tools and technologies are just different tools and methods to model and handle real-world problems, finance in our case, which is highly complex and chaotic in nature.

More importantly, when we mentioned a (financial) market is highly chaotic or fuzzy in nature, we are just *seeing* (*modeling* in technical perspective) the problem with a particular *perspective*, the problem never changes itself.

Just like we are talking about the particle–wave duality property of financial market in quantum finance.

The financial market always *as it is*, we are just using different *angles* and *perspectives* to observe and investigate it.

In Part II of this book, we will study four major applications of quantum finance in financial prediction and intelligent trading systems.

Problems

9.1 Discuss and explain why Schrödinger in his 1952 lecture interpreted the quantum events such as Schrödinger's cat as different coexisting realities instead of different alternative events, and what such notion is critical and related to Heisenberg's uncertainty principle.

9.2 Based on the brain waves shown in Fig. 9.3, please discuss and explain why our *perception of time* is different in various brain wave states, and how it can be related to the understanding of time in: (1) human perception; and (2) financial markets.

9.3 If all human sensory schemes are products of sensory oscillators and their interactions with neural networks in the brain, is it possible to have what we called *sensory interchange* in theoretical perspective? And what is its implication?

9.4 What is the importance of converting a traditional artificial neural network (ANN) into chaotic neural networks in terms of (1) network training; and (2) network performance ?

9.5 What are the physical evidence and explanations for the existence of chaotic neural dynamics in humans? And how such discoveries are vital to the future development of AI and intelligent systems?

9.6 Based on Eqs. (9.1)–(9.4) and Eqs. (9.5)–(9.8), write two MATLAB programs to plot the normalized bifurcation diagrams of Wang-oscillator and Lee-oscillator (i.e., input and output in the range of [0 … 1]) and discuss how Lee-oscillator can be used to convert ANY artificial neural networks (ANN) or recurrent neural networks (RNN) into chaotic neural networks (CNN).

9.7 In both Wang-oscillator and Lee-oscillator, there are excitatory and inhibitory neurons in their neural models. Discuss and explain the biological and mathematical meanings of these neurons in a typical chaotic neural network, as compared with traditional ANN.

9.8 For the modeling of forex market (say AUDCAD) using quantum finance oscillator (QFO):

   (i)   Draw the quantum finance oscillator model of $QFO|_{AUDCAD}$.
   (ii) Write down the formulation of $QFO|_{AUDCAD}$.
   (iii) What are the physical meanings of excitatory (E), inhibitory (I), and input stimulus (S) in $QFO|_{AUDCAD}$ model?
   (iv) Discuss how $QFO|_{AUDCAD}$ can be integrated with chaotic neural networks for (1) short-term forex price prediction; (2) long-term trend pattern recognition.

9.9 Programming Exercise I (bifurcation diagrams of Lee-oscillator and Wang-oscillator)

Below MATLAB program implements the Wang-oscillator and generates its bifurcation diagram.

```
%%%%%%%%%%%%%%%%%%%%%%%%%%%%%%%%%%%%%%%%%%%%%%%%%%%%%%%%%%%%%%%%%%%%%
%
% Wang-Oscillator
%
% Created By Dr. Raymond Lee on May 2019
%
%%%%%%%%%%%%%%%%%%%%%%%%%%%%%%%%%%%%%%%%%%%%%%%%%%%%%%%%%%%%%%%%%%%%%
% Define parameters
%
  N = 600;      % n = no. of time step default is 600
  s = 5;        % parameter for tanh function
  a1 = 5;       % default is 5
  a2 = 5;
  b1 = 1;       % default is 5
  b2 = 1;
  eu = 0;       % u threshold default is 0
  ev = 0;       % v threshold default is 0
  e  = 0.02;

%%%%%%%%%%%
% Define and initialize u(t) and v(t)
%
  u = zeros(1,N);
  v = zeros(1,N);
  z = zeros(1,N);
  z(1) = 0.2;   % IC of z set to 0.2
  u(1) = 0.2;

  hold on;

  for i=-1:0.001:1  % 1000 i values
    valueOf_stimulus_i = sprintf('%0.5g',i)
    sim = i + e*sign(i);
    for t = 1:N-1
        tempu = a1*u(t) - a2*v(t) + sim - eu;
        tempv = b1*u(t) - b2*v(t) - ev;

        u(t+1) = (exp(s*tempu) - exp((-1)*s*tempu))
                 /(exp(s*tempu) + exp((-1)*s*tempu));
        v(t+1) = (exp(s*tempv) - exp((-1)*s*tempv))
                 /(exp(s*tempv) + exp((-1)*s*tempv));
        z(t+1) = u(t+1) - v(t+1);

        if (t > 500)
          plot(i, z(t+1));
        end
    end
  end
```

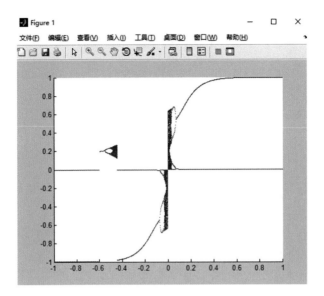

(i)   Copy this program script into MATLAB and run the program, you should see the bifurcation diagram as above.

(ii) As shown in the MATLAB code, we only plot the output z when t > 500. Why?

(iii) Modify this MATLAB to implement the Lee-oscillator using Eqs. (9.5)–(9.8). You should obtain the below bifurcation diagram.

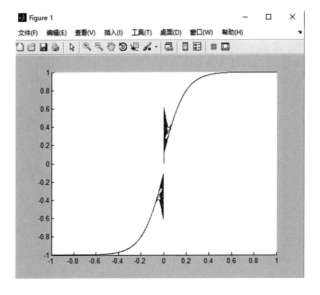

(iv) From the bifurcation diagram of Lee-oscillator, identify the sigmoid zone and the bifurcation zone.

(v) As shown in the bifurcation diagram, the x-axis and y-axis are both between
[−1 ... +1]. Modify the MATLAB program to convert the bifurcation diagram
of Lee-oscillator to *Lee chaotic transfer function (LCTF)* with normalized value
between [0 ... 1], as shown in the below figure.

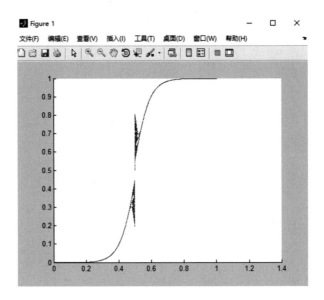

(vi) Modify the MATLAB program by creating a 2-D array Lee() which contains
the output values of Lee-oscillator for 1000 values of i between [1 ... 1] and
100 time-step values of Lee-oscillator between 500–600. The definition of
Lee() array as shown below:
Lee = zeros(1001,100);

(vii) Store the LCTF-Lee() table array into a CSV data file for later use.

9.10 Programming exercise II

(i) Start with 7.28 with the implementation of MQL-based FFPBN for the
next-day forecast of at least 5 Forex products (e.g., AUDCAD, USD-
CAD, GBPUSD, USDJPY, EURUSD, etc.).

(ii) Implement the MATLAB program in 9.16 to obtain the CSV datafile of
LCTF (Lee chaotic transfer function).

(iii) Modify the MQL FFBPN program in (i) by: (1) Read the LCTF array in
the program initialization process; (2) Replace sigmoid function for the
FFPBN program with the LCTF function array.

(iv) Compare these two neural networks in terms of: (1) training performance; and (2) forecast performance using five forex products mentioned
in (i) by plotting the following charts:
• Mean RMSE (root-mean-square error) in network training for 1000
iterations;

- Mean standard deviation of RMSE in network training for 1000 iterations;
- Mean forecast % Error for a test run (or simulation) of 100 trading days.
- Mean standard deviation of forecast % Error for a test run (or simulation) of 100 trading days.

References

Chen, L. and Aihara, K. (1995) Chaotic stimulated annealing by a neural model with transient chaos. *Neural Networks*. 8(6): 915–930.

Falcke, M. et al. (2000) Modeling observed chaotic oscillators in bursting neurons: The role of calcium dynamics and IP3. *Biol. Cybern*. 82: 517–527.

Fernandes, T. and Malsburg, C. (2015) Self-organization of control circuits for invariant fiber projections. *Neural Computation*. 27(5): 1005–1032.

Freeman, W. J. (2013) *Imaging Brain Function With EEG: Advanced Temporal and Spatial Analysis of Electroencephalographic Signals*. Springer.

Freeman, W. J. (2008) A pseudo-equilibrium thermodynamic model of information processing in nonlinear brain dynamics. *Neural Network*. 21(2–3): 257–265.

Freeman, W. J. (2000) *Neurodynamics: An Exploration in Mesoscopic Brain Dynamics (Perspectives in Neural Computing)*. Springer.

Hopfield, J. J. (1984) Neurons with Graded Response Have Collective Computational Properties like Those of Two-State Neurons. *Proceedings of the National Academy of Sciences of the United States of America*. 81(10): 3088–3092.

Hoshino, O. et al. (2003) Roles of dynamic linkage of stable attractors across cortical networks in recalling long-term memory. *Biol. Cybern*., 88(3): 163–176.

Lee, R. S. T. (2006) LEE-Associator – A Transient Chaotic Autoassociative Network for Progressive Memory Recalling. *Neural Networks*. 19(5): 644–666.

Lee, R. S. T. (2004) A Transient-chaotic Auto-associative Network (TCAN) based on LEE-oscillators. *IEEE Trans. Neural Networks*. 15(5): 1228–1243.

Malsburg, C. (1995) Binding in models of perception and brain function. *Current Opinion in Neurobiology*. 5(4): 520–526.

Tegmark, M. (2014) *Our Mathematical Universe: My Quest for the Ultimate Nature of Reality*. Vintage.

Tuchong (2019a) What would happen if time stopped? http://stock.tuchong.com/image?imageId= 446677423538045208. Accessed 21 Aug 2019.

Tuchong (2019b) Artificial Neural Networks. http://stock.tuchong.com/image?imageId= 492056055110172742. Accessed 21 Aug 2019.

Wang, X. (1991) Period-doublings to chaos in a simple neural network: An analytic proof. *Complex Syst*. 5: 425–441.

Zhou, C. and Chen, T. (2000) Chaotic neural networks and chaotic annealing. *Neurocomputing*. 30(1): 293–300.

Part II
Quantum Finance Applications

Chapter 10
Quantum Price Levels for Worldwide Financial Products

> *The markets represent the aggregate interaction of many investors. Their attitudes, philosophies, and behavioral patterns on many levels are predictable... and repetitive.*
> Rick Santelli (Born in 1956)

Mr. Rick Santelli gave a vivid description of financial markets. As an old saying: *History repeats itself.*

As mentioned in Chap. 1, humans are a species of *habit*. Under the same conditions, we will always respond in the same ways, no matter the consequences are. If we believe financial markets are the consequences of investors' *collective conscious* for market responses in every single moment. *Patterns* exist will be a natural cause.

If we believe every single phenomenon (including financial market) happened among us no matter how *chaotic* or *unpredictable* at the very basic and microscopic level, are all governed by quantum mechanics under the influence of quantum energy levels generated by their own existence with neighboring quantum particles; thus, the existence of *financial energy levels*; or what we called *quantum price levels (QPL)* should be a natural cause too.

In Part I of this book, we have studied the basic concept and theory of quantum finance with methods for the modeling and *quantum price levels (QPL)* evaluation. In Part II, this chapter will begin on how to apply the theory and formulation of QPLs evaluation on worldwide *secondary financial markets* which include: cryptocurrency, forex, commodity, and worldwide financial indices.

First, we will introduce financial market types and their importance to the worldwide economy.

Second, we will give a brief overview of traditional concepts and beliefs of *quantized market price levels*—the *support and resistance (S&R) levels*, and how they are related to our understanding of *quantum price levels (QPLs)* in quantum finance.

Third, we will introduce the MetaTrader (MT) system and MQL (MetaQuotes Language)—the real-time worldwide financial program trading and system development platform.

After that, we will study the system architecture of QPL evaluation system and introduce two major MT4 platforms: Forex.com (one of the biggest worldwide forex

© Springer Nature Singapore Pte Ltd. 2020
R. S. T. Lee, *Quantum Finance*,
https://doi.org/10.1007/978-981-32-9796-8_10

trading platforms) and AvaTrade.com (one of the major cryptocurrency trading plat-forms).

For the system implementation section, we will study in detail the MT4 program on automatic evaluation of worldwide 120+ financial products, based on the numerical formulations revealed in Chap. 5.

In the last section, we will study QPL system implementation results; conclude its related works including the design; implementation of real-time financial prediction, and multi-agent-based trading system in Chaps. 11–13.

10.1 Financial Markets

10.1.1 An Overview

A financial market is a broad term describing any marketplace where trading of securities including equities, bonds, currencies, and derivatives occurs.

Some financial markets are small scale with little activities, while some financial markets like the New York Stock Exchange (NYSE) trade trillions of dollars of securities daily.

Financial market prices may not indicate the true intrinsic value of stock due to macroeconomic forces like taxes. In addition, the prices of securities are heavily reliant on informational transparency to ensure efficient and appropriate prices are set by the market.

The stock market is a financial market that enables investors to buy and sell shares of publicly traded companies.

The primary stock market is where new issues of stocks are first offered. Any subsequent trading of stock securities occurs in the secondary market.

In terms of market pricing—*primary market prices* are often set beforehand, while prices in the *secondary market* are determined by the basic forces of supply and demand. If the majority of investors believe a stock will increase in value and rush to buy it, the stock's price rises typically. If a company loses favor with investors or fails to post sufficient earnings, its stock price declines as demand for that security dwindles.

In *secondary markets*, investors exchange with each other rather than with the issuing entity. Through a massive series of independent yet interconnected trades, the secondary market drives the price of securities toward their actual value. Moreover, the secondary markets create additional economic value by allowing more beneficial transactions to occur. The net result is that almost all market prices—interest rates, debt, houses, and the values of businesses and entrepreneurs—are more efficiently allocated because of secondary market activity.

10.1.2 Primary Versus Secondary Market

It is important to understand the distinction between the *primary market* and the *secondary market*. When a company issues stock or bonds for the first time and sells those securities directly to investors, that transaction occurs on the *primary market*. Some of the most common and well-publicized primary market transactions are IPOs, or initial public offerings. During an IPO, a primary market transaction occurs between the purchasing investor and the investment bank underwriting the IPO. Any proceeds from the sale of shares of stock on the primary market go to the company that issued the stock, after accounting for the bank's administrative fees.

If these initial investors later decide to sell their stakes of the company, they can do so on the *secondary market*. Any transactions on the *secondary market* occur between investors, the proceeds of each sale go to the selling investor, not to the company that issued the stock or to the underwriting bank. The *secondary market* is where securities are traded after the company has sold its offering on the *primary market*. It is also referred to as the *stock market*. The New York Stock Exchange (NYSE), London Stock Exchange, and Nasdaq are *secondary markets*.

Small investors have a much better chance of trading securities on the secondary market since they are excluded from IPOs. Anyone can purchase securities on the secondary market as long as they are willing to pay the asking price per share.

A broker typically purchases the securities on behalf of an investor in the secondary market. Unlike the primary market, where prices are set before an IPO takes place, prices on the secondary market fluctuate with demand. Investors will also have to pay a commission to the broker for carrying out the trade.

The volume of securities traded varies from day to day, as supply and demand for the security fluctuates. This has a great effect on the security's price.

When an initial offering is completed, the issuing company is no longer a party to any sale between two investors, except in the case of a company stock buyback. For example, after December 12, 1980, Apple's IPO on the primary market, individual investors were able to purchase Apple stock on the secondary market. Because Apple is no longer involved in the issue of its stock, investors would, essentially, deal with one another when they trade shares in the company.

The *secondary market* (Kuznetsov 2006; Drayer et al. 2008) has two different categories: the *auction* and the *dealer markets*. The *auction market* is home to the open outcry system, where buyers and sellers congregate in one location and announce the prices at which they are willing to buy and sell their securities. The NYSE is one such example. In *dealer markets*, though, people trade through electronic networks. Most small investors trade through *dealer markets*.

10.1.3 Types of Financial Markets

Over-The-Counter Markets

The *over-the-counter* (OTC) market is an example of a *secondary market*. An OTC market handles the exchanging of public stocks not listed on the Nasdaq, New York Stock Exchange, or American Stock Exchange. Companies with stocks trading on the *OTC market* are usually smaller organizations, as this financial market requires less regulation and trading is less expensive.

Financial Markets for Bonds

A bond is a security in which an investor loans money for a defined period of time at a preestablished rate of interest. Bonds are not only issued by corporations but may also be issued by municipalities, states, and federal governments from around the world. They are also referred to as debt, credit, or fixed-income market. For example, the bond market sells securities such as notes and bills issued by the United States' Treasury.

Money Markets

A money market is a portion of the financial market that trades highly liquid and short-term maturities. The intention of the money market is for short-term borrowing and lending of securities with maturity typically less than 1 year. This financial market trades certificates of deposit, banker's acceptances, certain bills, notes, and commercial paper.

Derivatives Market

The derivatives market is a financial market that trades securities that derive its value from its underlying asset. The value of a derivative contract is determined by the market price of the underlying item. This financial market trades derivatives include forward contracts, futures, options, swaps, and contracts-for-difference.

Forex Market

The forex market is a financial market, where currencies are traded. This financial market is the most liquid market in the world, as cash is the most liquid assets. The interbank market is the financial system that trades currency between banks.

10.1.4 OTC Market and Market Makers

Over-the-counter (OTC) or off-exchange trading is done directly between two parties, without the supervision of an exchange (Arnett 2011; Atkeson et al. 2015; Huber et al. 2012).

It is contrasted with exchange trading, which occurs via *exchanges*.

In traditional security trading such as stock trading, a *stock exchange* has the benefit of facilitating liquidity, providing transparency, and maintaining the current market price.

In an OTC trade, the price is not necessarily published for the public.

OTC trading, like exchange trading, occurs with commodities, financial instruments (including stocks), and derivatives of such products.

In 2008, approximately 16% of all U.S. stock trades were OTC trading, by April 2014 it has increased to about 40%.

A *market maker* or *liquidity provider* is a company or an individual that quotes both a buy and a sell price in a financial instrument or commodity held in inventory, hoping to make a profit on the bid-offer spread, or turn.

The U.S. Securities and Exchange Commission defines a market maker as a firm that stands ready to buy and sell stock on a regular and continuous basis at a publicly quoted price.

Most foreign exchange trading firms are market makers and so are many banks.

Market makers both sell to and also buy from its clients and is compensated by means of price differentials for the services of providing liquidity, reducing transaction costs and facilitating trade.

They make a profit by the difference between the price at which a market maker is willing to buy a stock (the bid price) and the price that the firm is willing to sell it (the ask price) is known as the *market maker spread*, or *bid–ask spread.*

Market makers also provide liquidity to their own firm's clients for which they earn a commission.

10.1.5 Forex, OTC, and Market Maker

The foreign exchange market (forex, FX, or currency market) is a global decentralized or OTC market for currencies trading.

This market determines the foreign exchange rates.

It includes all aspects of buying, selling, and exchanging currencies at current or determined prices.

In terms of trading volume, it is by far the largest market in the world.

In retail forex trading, most foreign exchange trading centers/firms are market makers and so are many banks.

Unlike a stock market, the foreign exchange market is divided into levels of access.

At the top is the interbank foreign exchange market, which is made up of the largest commercial banks and securities dealers.

Key participants include the following:

- central banks,
- commercial banks,
- foreign exchange brokers,
- investment management firms,
- retail foreign exchange traders, and
- money transfer/remittance companies.

The main participants in this market are the larger international banks.

Financial centers around the world function as anchors of trading between a wide range of multiple types of buyers and sellers around the clock, with the exception of weekends.

Currencies are traded against one another in pairs.

Each currency pair thus constitutes an individual trading product and is traditionally noted XXXYYY or XXX/YYY, where XXX and YYY are the ISO 4217 international three-letter code of the currencies involved.

For instance, the quotation EURUSD (EUR/USD) 1.5465 is the price of the Euro expressed in US dollars, meaning 1 euro = 1.5465 US dollars.

Table 10.1 shows the list of the top 35 most traded currencies in 2016.

10.1.6 Forex Trading Characteristics

Forex trading is a huge trading volume, representing the largest asset class in the world leading to high liquidity due to its geographical dispersion.

Its continuous operation is 24 h a day except weekends, i.e., trading from 22:00 GMT on Sunday (Sydney) until 22:00 GMT Friday (New York) as Asia trading session ends, Europe session begins, followed by North America session and then back to the Asia session.

There are a variety of factors that can affect exchange rates.

The margins of relative profit are moderate as compared with other markets of fixed income.

There is no unified or centrally cleared market for the majority of trades, and there is very little cross-border regulation.

The main trading centers are London, New York, Tokyo, Hong Kong, and other prominent trading centers throughout the world.

The top 3 most heavily traded bilateral currency pairs were the following:

1. EURUSD: 23.0%;
2. USDJPY: 17.7%;
3. GBPUSD: 9.2%.

Different from stocks which are localized to the company's financial situation, fluctuations in exchange rates are usually caused by actual worldwide monetary flows as well as by expectations of changes in monetary flows.

These are caused by changes in worldwide gross domestic product (GDP) growth, inflation (purchasing power parity theory), interest rates, budget and trade deficits or surpluses, large cross-border M&A deals, and other macroeconomic conditions.

Major financial reports, news, and indices from major financial/government institutions such as Federal Reserve Board (USA) and Organization of the Petroleum Exporting Countries (OPEC) are important factors, which are released publicly, often on scheduled dates.

Political conditions including internal, regional, and international political conditions and events can have a profound effect on currency markets.

Supply and demand is also a major factor.

Table 10.1 Top 35 most traded currencies 2016

Rank	Country	Currency	Symbol	% of daily trades
1.		United States dollar	USD (US$)	87.60
2.		Euro	EUR (€)	31.40
3.		Japanese yen	JPY (¥)	21.60
4.		Pound sterling	GBP (£)	12.80
5.		Australian dollar	AUD (A$)	6.90
6.		Canadian dollar	CAD (C$)	5.10
7.		Swiss franc	CHF (Fr)	4.80
8.		Renminbi	CNY (元)	4.00
9.		Swedish krona	SEK (kr)	2.20
10.		New Zealand dollar	NZD (NZ$)	2.10
11.		Mexican peso	MXN ($)	1.90
12.		Singapore dollar	SGD (S$)	1.80
13.		Hong Kong dollar	HKD (HK$)	1.70
14.		Norwegian krone	NOK (kr)	1.70
15.		South Korean won	KRW (₩)	1.70
16.		Turkish lira	TRY (₺)	1.40
17.		Russian ruble	RUB (₽)	1.10
18.		Indian rupee	INR (₹)	1.10
19.		Brazilian real	BRL (R$)	1.00
20.		South African rand	ZAR (R)	1.00
21.		Danish krone	DKK (kr)	0.80

(continued)

Table 10.1 (continued)

Rank	Country	Currency	Symbol	% of daily trades
22.		Polish złoty	PLN (zł)	0.70
23.		New Taiwan dollar	TWD (NT$)	0.60
24.		Thai baht	THB (฿)	0.40
25.		Malaysian ringgit	MYR (RM)	0.40
26.		Hungarian forint	HUF (Ft)	0.30
27.		Saudi riyal	SAR (ريال)	0.30
28.		Czech koruna	CZK (Kč)	0.30
29.		Israeli shekel	ILS (₪)	0.30
30.		Chilean peso	CLP (CLP$)	0.20
31.		Indonesian rupiah	IDR (Rp)	0.20
32.		Colombian peso	COP (COL$)	0.20
33.		Philippine peso	PHP (₱)	0.10
34.		Romanian leu	RON (L)	0.10
35.		Peruvian sol	PEN (S/)	0.10
Other				2.10
Total				200.00

Note
(1) *Source* Triennial Central Bank Survey Foreign exchange turnover in April 2016. Triennial Central Bank Survey. Basel, Switzerland: Bank for International Settlements. 11 December 2016. p. 7
(2) The total sum is 200% because each currency trade always involves a currency pair; one currency is sold (e.g., US$) and another bought (€). Therefore, each trade is counted twice, once under the sold currency (US$) and once under the bought currency (€). The percentages above are the percent of trades involving that currency regardless of whether it is bought or sold, e.g., the U.S. Dollar is bought or sold in 87% of all trades, whereas the Euro is 31.4%

10.1.7 *Retail Forex Trading Platforms*

Retail foreign exchange trading is a small segment of the larger foreign exchange market, where individuals speculate on the exchange rate between different currencies.

This segment has developed with the advent of dedicated electronic trading platforms and the internet, which allows individuals to access the global currency markets.

In 2016, it was reported that retail foreign exchange trading represented 5.5% of the entire foreign exchange market ($282 billion in daily trading turnover).

Prior to the development of forex trading platforms in the late 90s, forex trading was restricted to large financial institutions.

Nowadays, traders are able to trade spot currencies with market makers on margin. This means they need to put down only a small percentage of the trade size and can buy and sell currencies in seconds.

As both profit and loss will be *amplified* by margin trading, risk management such as *stop-loss* strategy must execute in every trading transaction.

MetaTrader 4 (MT4) platform is the most popular retail forex trading platform.

By using *MetaQuotes Language* (MQL), a trader can develop one's own trading program for automatic program trading (Young 2015; Walker 2018).

Details of MT4 program trading will be studied in the next section.

Figure 10.1 shows a snapshot for forex trading over MT4 provided by Forex.com (www.Forex.com), one of the largest forex trading platforms in the world.

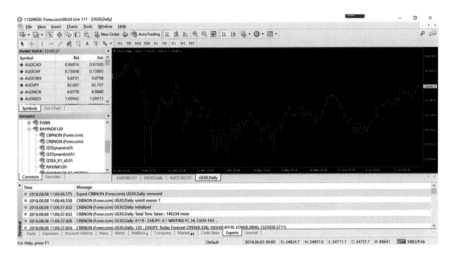

Fig. 10.1 Snapshot for forex trading over MT4 platform provided by Forex.com

10.2   Supports and Resistances

10.2.1   Introduction

The concepts of *support* and *resistance* are undoubtedly two of the most highly studied attributes of technical analysis. Part of analyzing chart patterns, these terms are used by traders to refer price levels on charts that tend to act as barriers, preventing the price of an asset from getting pushed in a certain direction (Bulkowski 2005; Kirkpatrick and Dahlquist 2015; Murphy 1999; De Angelis and Peskir 2016; Gomes and Waelbroeck 2010; Zapranis and Tsinaslanidis 2012).

In the Wall Street environment, the terms *support and resistance* are almost synonymous with *demand and supply* respectively.

Support is a price level at which there is adequate demand for security to stop its downward price movement and, normally, turn prices upward.

Typically, *support* occurs at reaction lows.

Resistance is a price level at which there is a significant supply of a stock causing prices to halt an upward move and turn prices down.

Typically, *resistance* occurs at reaction highs.

Once an area or *zone* of *support or resistance* has been identified, it provides valuable potential trade entry or exit points. This is because, as a price reaches a point of support or resistance, it will do one of two things—bounce back away from the support or resistance level or violate the price level and continue in its direction—until it hits the next support or resistance level.

Most forms of trades are based on the belief that support and resistance zones will not be broken. Whether price is halted by the support or resistance level, or it breaks through, traders can *bet* on the direction and can quickly determine if they are correct. If the price moves in the wrong direction, the position can be closed at a small loss. If the price moves in the right direction, however, the loss may be substantial. Figure 10.2 shows typical support and resistance lines.

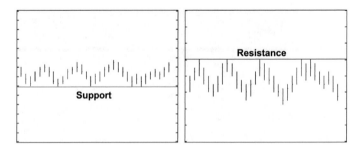

Fig. 10.2 Typical supports and resistances

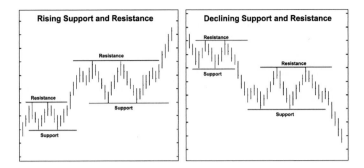

Fig. 10.3 Rising and declining supports and resistances

10.2.2 Rising and Declining Supports and Resistances

In an uptrend, both support and resistance levels rise as illustrated in Fig. 10.3.

Typically, support levels hold while resistance offers temporary halts to upward movements in prices. *Resistance levels* are repeatedly broken until the uptrend is reversed. In a downtrend, both support and resistance levels move lower as illustrated in the right figure. Typically, *resistance levels* hold while support levels temporarily stop downward price movements. *Support levels* are repeatedly broken until the downtrend is reversed.

10.2.3 Role Reversal on Support and Resistance

In an uptrend, both support and resistance levels rise as illustrated in Fig. 10.4.

Typically, support levels hold while resistance offers temporary halts to upward movements in prices.

Resistance levels are repeatedly broken until the uptrend is reversed.

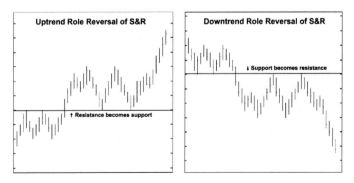

Fig. 10.4 Role reversals in supports and resistance

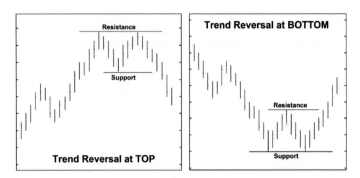

Fig. 10.5 Trend reversals in supports and resistance

In a downtrend, both support and resistance levels move lower as illustrated in the right figure.

Typically, resistance levels hold while support levels temporarily stop downward price movements.

Support levels are repeatedly broken until the downtrend is reversed.

10.2.4   Trend Reversal on Support and Resistance

Trend reversal at top: In an uptrend, a trend reversal occurs when prices are held at a resistance level. A double top or some other reversal formation develops at that point, and the trend changes direction.

Trend reversal at the bottom: A trend reversal occurs in a downtrend when prices are unable to penetrate a support level. In this case, a bottom reversal pattern is formed, and the trend changes direction to the upside.

Figure 10.5 shows typical trend reversal on support and resistance.

Keep in mind that a trend reversal is not signaled by the first failure to break through a resistance level (in an uptrend) or a support level (in a downtrend). A reversal pattern must fully develop before one gets the signal that the trend has changed. In other words, a trader should not rush to sell all of one's securities or sell short just because prices have held at a resistance level.

10.2.5   Support and Resistance Versus QPLs

As mentioned in Chap. 5, although we don't know whether quantum price level (QPL) whether exists or not in financial market. But for many experienced analysts and traders, we all believe that *quantized price levels* definitely exist in all financial markets such as *support (S) and resistance (R)*.

In fact, the basis of *technical analysis* in the past 50 years is based on this fundamental concept.

The difference is that, with the exponential growth of program trading, especially HFPT (high-frequency-program-trading) (Durbin 2010). The financial markets become so complex and chaotic that we don't believe one can observe (or locate) all these *hidden-force-levels* manually.

However, different from the traditional understanding of support and resistance (S&R) levels, there is one major difference between them.

In traditional technical analysis, S&R levels are normally totally different price levels as they play different roles in the supply and demand *forces* for market movements. However, in quantum finance, *quantum price levels (QPLs)* are interpreted as discrete energy levels for the movements of the financial particles from one energy state to another, so basically there is no concepts either *support* or *resistance*, but just *energy states*. In other words, one QPL can be acted as a support level for a moment and then acted as *resistance level* at another moment.

The truth is, such interpretation becomes more realistic and practical in nowadays financial markets, as we normally see a particular price level can be acted as both a support and resistance levels all the times within a day or a period of time, especially for highly fluctuated financial markets such as forex and cryptocurrency. The main reason is maybe due to the popularity of program trading and especially HFPT (Durbin 2010) in which trading programs normally define the same set of price levels as the thresholds to trigger *buying or selling* signals, and seldom evaluate them as two separated price levels for automatic program trading. That's why the frequent interchange of S&R roles is commonly found in nowadays financial markets include stock, forex, and cryptocurrency markets (Fig. 10.6).

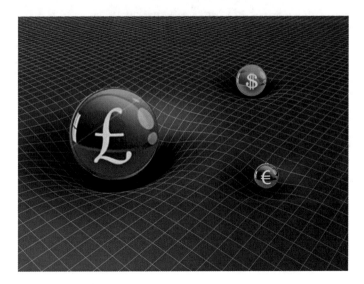

Fig. 10.6 QPLs in financial markets

10.3   The MetaTrader (MT) Platform and MQL

10.3.1   An Introduction

MetaTrader 4, also known as MT4, is an electronic trading platform widely used by online retail foreign exchange speculative traders (Brown 2016; Walker 2018; Young 2015).

It was developed by *MetaQuotes software* and released in 2005.

The software is licensed to foreign exchange brokers, who provide the software to their clients.

The software consists of both a client and server component. The server component is run by the broker and the client software is provided to the broker's customers, who use it to see live streaming prices and charts; to place orders; to manage their accounts.

The client is a Microsoft windows-based application that became popular mainly due to the ability to end-users to write their own trading scripts and robots that could automate trading. Figure 10.7 shows a snapshot of MT4 in PC platform.

In 2010, *MetaQuotes* released a successor, MetaTrader 5. However, most traders and brokers keep on using MT4.

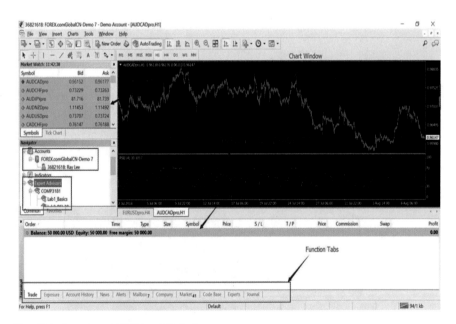

Fig. 10.7 Snapshot of MT4 in a PC platform

10.3.2 MQL—MetaQuotes Language

MetaQuotes Language 4 (MQL4) is a built-in language for programming trading strategies (Brown 2016; Walker 2018; Young 2015).

This language is developed by MetaQuotes Software Corp. based on their long experience in the creation of online trading platforms.

Using this language, we can create our own expert advisors that make trading management automated and are perfectly suitable for implementing our own trading strategies.

Besides, using MQL4 we can create our own technical indicators (custom indicators), scripts, and libraries.

MQL4 contains a large number of functions necessary to analyze current and previously received quotes, along with built-in basic indicators and functions to manage and control trade orders.

The *MetaEditor* that highlights different constructions of MQL4 is used for writing the program code. It helps users to orientate themselves in the expert system text quite easily.

Although MQL5 had launched for years, most traders and brokers keep on using MQL4 over MT4 platform. As any MQL4 program can be compiled and run in MT4/MT5 platforms (but not the other way around). In this book, we focus on MT4 programming.

Programs written in MQL4 have different features and purposes:

1. expert advisor (EA)—is a mechanical trading system linked up to a certain chart. An expert advisor starts to operate when an event happens that can be handled by it: events of initialization and deinitialization, event of a new tick receipt, a timer event, depth of market-changing event, chart event, and custom events.
2. custom indicator—is a technical indicator written independently in addition to those already integrated into the client terminal. Like built-in indicators, they cannot trade automatically and intend to implement analytical functions only.
3. scripts—is a program that intends to a single execution of some actions. Unlike expert advisors, scripts do not process any actions, except for the start event.
4. library—is a set of custom functions that intend to store and distribute used blocks of custom programs frequently. Libraries cannot start executing by themselves.
5. include File—is a source text of the most frequently used blocks of custom programs like C++.

10.3.3 MQL—A Brief Overview

MetaQuotes Language 4 (MQL4) is a built-in language for programming trading strategies.

MQL4 is an OO-programming language intended to write automated trading strategies, custom technical indicators for various financial markets analyses.

It allows not only to write a variety of expert systems, designed to operate in real time, but also to create their own graphical tools to help users make trade decisions.

MQL4 is based on the concept of the popular programming language C++.

Like C++, MQL4 provides the OO-programming basics which include the following:

- Syntax,
- data types,
- operations and expressions,
- operators,
- functions,
- variables,
- preprocessor, and
- object-oriented programming.

(Details please refer to https://docs.mql4.com/basis)

In addition to all C++-style programming fundamental tools and functions, MQL4 provides predefined variables, classes, and functions tailored for program trading in online financial markets which include the following:

- account information,
- checkup,
- market info,
- time series and indicators access
- chart operations,
- trade functions.
- trade signals,
- global variables of the terminal, and
- technical indicators.

10.3.4 MQL—Time Series and Indicators Access Functions

These are functions for working with time series and indicators.

A time series differs from usual data array by its reverse ordering—elements of time series are indexed from the end of an array to its start (from the most recent data to the oldest ones).

To copy the time series values and indicator data, it is recommended to use dynamic arrays only, because copying functions are designed to allocate the necessary size of arrays that receive values.

Commonly used functions include the following:

- iOpen
- iHigh
- iLow

iOpen

Returns the Open price of the bar (indicated by the 'shift' parameter) on the corresponding chart.

```
double  iOpen(
   const string          symbol,        // Symbol
   ENUM_TIMEFRAMES       timeframe,     // Period
   int                   shift          // Shift
   );
```

Example:

```
input int shift=0;
//+-------------------------------------------------------------+
//| Function-event handler "tick"                               |
//+-------------------------------------------------------------+
void OnTick()
   {
   datetime time  = iTime(Symbol(),Period(),shift);
   double   open  = iOpen(Symbol(),Period(),shift);
   double   high  = iHigh(Symbol(),Period(),shift);
   double   low   = iLow(Symbol(),Period(),shift);
   double   close = iClose(NULL,PERIOD_CURRENT,shift);
   long     volume= iVolume(Symbol(),0,shift);
   int      bars  = iBars(NULL,0);

   Comment(Symbol(),",",EnumToString(Period()),"\n",
           "Time:  "  ,TimeToString(time,TIME_DATE|TIME_SECONDS),"\n",
           "Open:  "  ,DoubleToString(open,Digits()),"\n",
           "High:  "  ,DoubleToString(high,Digits()),"\n",
           "Low:  "   ,DoubleToString(low,Digits()),"\n",
           "Close: "  ,DoubleToString(close,Digits()),"\n",
           "Volume: " ,IntegerToString(volume),"\n",
           "Bars:  "  ,IntegerToString(bars),"\n"
           );
   }
```

Fig. 10.8 Syntax of iOpen() function and program example of using time series and indicators access functions

- iClose
- iVolume
- iTime

Figure 10.8 shows the syntax and example of using time series and indicators access functions.

10.3.5 MQL—Trade Functions

This is the group of functions that intend to manage trading activities.

Major trade functions need to be used in EA are the following:

- OrderSend(),
- OrderClose(),
- OrderModify(), and
- OrderDelete().

A complete list of functions can be found at https://docs.mql4.com/trading.

Figure 10.9a shows the syntax and program example of OrderSend(), one of the

OrderSend

The main function used to open market or place a pending order.

```
int  OrderSend(
    string   symbol,              // symbol
    int      cmd,                 // operation
    double   volume,              // volume
    double   price,               // price
    int      slippage,            // slippage
    double   stoploss,            // stop loss
    double   takeprofit,          // take profit
    string   comment=NULL,        // comment
    int      magic=0,             // magic number
    datetime expiration=0,        // pending order expiration
    color    arrow_color=clrNONE  // color
    );
```

Example:

```
//+-------------------------------------------------------------+
//| Script program start function                               |
//+-------------------------------------------------------------+
void OnStart()
  {
//--- get minimum stop level
   double minstoplevel=MarketInfo(Symbol(),MODE_STOPLEVEL);
   Print("Minimum Stop Level=",minstoplevel," points");
   double price=Ask;
//--- calculated SL and TP prices must be normalized
   double stoploss=NormalizeDouble(Bid-minstoplevel*Point,Digits);
   double takeprofit=NormalizeDouble(Bid+minstoplevel*Point,Digits);
//--- place market order to buy 1 lot
   int ticket=OrderSend(Symbol(),OP_BUY,1,price,3,stoploss,takeprofit,"My order",16384,0,clrGreen);
   if(ticket<0)
     {
      Print("OrderSend failed with error #",GetLastError());
     }
   else
      Print("OrderSend placed successfully");
//---
  }
```

Fig. 10.9 Syntax and program example of OrderSend() function

most important and frequently used in program trading. Besides, MT4 also provides six types of trading orders: Buy, Sell, Buy Limit, Sell Limit, Buy Stop, and Sell Stop (Kaufman 2013).

10.3.6 MQL—Technical Indicators

A group of functions that intend to standard and custom indicators' calculations. Totally 39 technical indicator functions are provided.

Commonly used technical indicators include:

iBands	Bollinger Bands
iMomentum	Momentum
iMA	Moving Average
iMACD	Moving Averages Convergence-Divergence
iRSI	Relative Strength Index
iStochastic	Stochastic Oscillator

iMA

Calculates the Moving Average indicator and returns its value.

```
double  iMA(
    string      symbol,          // symbol
    int         timeframe,       // timeframe
    int         ma_period,       // MA averaging period
    int         ma_shift,        // MA shift
    int         ma_method,       // averaging method
    int         applied_price,   // applied price
    int         shift            // shift
    );
```

iRSI

Calculates the Relative Strength Index indicator and returns its value.

```
double  iRSI(
    string      symbol,          // symbol
    int         timeframe,       // timeframe
    int         period,          // period
    int         applied_price,   // applied price
    int         shift            // shift
    );
```

Fig. 10.10 Technical indicator functions in MQL of iMA and iRSI

iFractals   Fractals

Figure 10.10 shows the syntax of two most commonly used technical indicators, iMA (moving average), and iRSI (relative strength index). Appendix B shows the list of 39 technical indicators functions implemented by MQL.

10.3.7   MQL Programming Tips

If the user is an experienced C++/Java programmer, it should be easy to pick up the skill set of writing EA programs using MQL.

All we need to do is the following:

- go through MT4 documentation site,
- refresh C++/Java program skills,
- learn about MT4 program trading with technical indicator functions and how they can be used, and
- understand file I/O techniques.

In learning new program languages, nothing is more useful than actual lab practice (Fig. 10.11).

Fig. 10.11 Quantum finance MQL programming workshop in QFFC.org

10.4   System Implementation—QPL Evaluation for Worldwide Financial Products

From the implementation perspective, 129 financial products from Forex.com (one of the biggest forex MT platform providers) and AvaTrade.com (major cryptocurrency MT platform providers) are used for QPLs evaluation.

They include the following:

- 9 Cryptocurrency,
- 84 Forex,
- 19 Financial Indices, and
- 17 Major Commodities.

Fig. 10.12 A snapshot of MT4 platform of cryptocurrency provided by AvaTrade.com

Except cryptocurrency which has around 300–500 trading days information, all other financial products contain 2048 time series trading day information for QPLs evaluation.

Figure 10.12 shows a snapshot of the MT4 platform of cryptocurrency provided by AvaTrade.com.

Appendix A shows the list of 129 financial products provided by Forex.com and AvaTrade.com for their MT4 platforms.

10.5   MQL Program for QPLs Evaluation of Worldwide Financial Products (QPL2019.mq4)

10.5.1   System Flowchart and Main Program Logic

The main program logic for the evaluation of the first 21 QPLs of all 120+ worldwide financial product is shown in Fig. 10.13.

For each financial product, do the following:

- Read the daily time series and extract (Date, O, H, L, C, V)
- Calculate daily price return r(t)
- Calculate quantum price return wavefunction Q(r) (size 100)
- Evaluate λ value for the wavefunction Q(r) using F.D.M. and Eq. (5.21) & evaluate other related parameters:

  – sigma (std dev of Q)
  – maxQPR (max quantum price return—for normalization).

Fig. 10.13 Flowchart for the determination of the first 21 QFELs and QPLs

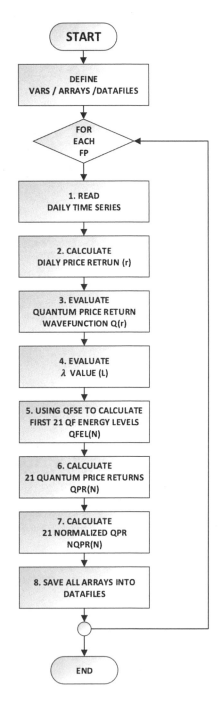

- Once λ is found, using QFSE (numerical solution) by solving the depressed cubic equation using Cardano's method Eq. (5.32a–c) to calculate first 21 quantum finance energy level, QFEL(n), n = [1 ... 20]
- Calculate quantum price return, QPR(n)

$$p = -(2n + 1)^2 \tag{10.1}$$

$$QPR(n) = \frac{QFEL(n)}{QFEL(0)} \tag{10.2}$$

where n = [1 ... 20]
- Calculate normalized QPR(n)

$$NQPR(n) = 1 + 0.21 * sigma * QPR(n) \tag{10.3}$$

where n = [1 ... 20]
- Save two level of data files:

  - For each financial product, save the QPL table contains QPE, QPR, NQPR for the first 21 energy levels
  - For all financial product, create a QPL summary table contains NQPR for all FP, which will be used for financial prediction using recurrent neural networks.

Figure 10.14 shows the program header with program logic listed for the QPL2019.mq4 program for QPLs evaluation of 120+ worldwide financial products.

```
//********************************************************************************************************************
// Date        :  6 Jan 2019
// Created by  :  Dr. Raymond LEE
// Name        :  QPL Evaluation Program of worldwide financial products
// VERSION NO: 1.1
// Objective   :  This program calculates the N(QPR) of ALL 120 Financial Products
//                For Each Financial Product:
//                1) Read the Daily Time Series and extract (Date, O, H, L, C, V) m
//                2) Calculate Dally Price Return r(t)
//                3) Calculate quantum price return wavefunction Q(r)(size 100)
//                4) Evaluate lambda (L) value for the wavefunction Q(r)using F.D.M. at ground state
//                   L = abs((r0^2*Q0 - r1^2*Q1)/(r0^4*Q0 - r1^4*Q1))
//                5) Evaluate other related parameters:
//                   - sigma  (std dev of Q)
//                   - maxQPR (max Quantum Price Return - for normalization)
//                6) Once L found, using Quartic Schrodinger Equation of Quantum Finance to find
//                   all the 21 Quantum Price Energies (QFEL0 .. QFEL20).
//                   Given by:
//                   (E(n)/(2n+1))^3 - (E(n)/(2n+1)) - K(n)^3 * L = 0
//                   where
//                   K(n) = ((1.1924+33.2383n+56.2169n^2)/(1+43.6106n))^(1/3)
//                7) Solve the 21 Cubic Eqts in (6) and extract the +ve real roots as QFEL0 .. QFEL20.
//                8) Calculate QPR(n) = QFEL(n)/QFEL(0) n = [1 .. 20]
//                9) Calculate NQPR(n) = 1 + 0.21*sigma*QPR(n);
//                10) Save TWO Level of datafiles
//                    1) For each financial product, save the QPL Table contains
//                       QFEL, QPR, NQPR for the first 21 energy levels
//                    2) For all financial product, create a QPL Summary table contains NQPR for all FP
//
//********************************************************************************************************************
```

Fig. 10.14 Main program logic of QPL evaluation program

10.5.2 Program Declaration

Program Declaration consists of the following three sections:

- declaration of program directories,
- declaration of global variables,
- declaration of timing variables, and
- declaration of financial products and related variables.

According to the numerical formulation revealed in Chap. 5, daily returns of the time series for each financial product are evaluated to calculate the wavefunction values Qf.

The application of the finite difference method (F.D.M.) (Abreu et al. 2018; Guo 2018) is to evaluate the Lambda (λ) value.

Once the Lambda value is found, apply the quartic Schrödinger equation (Singh et al. 2009; Tichy et al. 2012) of quantum finance to find all 21 quantum price energies (QFEL0 .. QFEL20) and store in QFEL datafiles. Figure 10.15 shows the program declaration of QPL evaluation program.

10.5.3 Program Initialization: init()

As MQL is a real-time program trading development platform, init() is used to perform *one-off process* such as the system initialization or QPL calculation in our case, while start() is a *time-step looping process* to implement real-time function such as real time trading and hedging.

So, in our case, all QPL calculation process in this system are placed in init() class. In other words, to implement a real-time QPL evaluation and QPL-based trading program, one will implement QPL evaluation program in the init() class real time trading algorithm in the start() class. Figure 10.16 shows the program, data, and data file initialization process.

10.5.4 Process 1: Calculate All K Values

The calculation is used in accordance with quantum anharmonic oscillation function from quantum finance revealed in Chap. 5:

$$\left(\frac{E(n)}{2n+1}\right)^3 - \left(\frac{E(n)}{2n+1}\right) - (K_0(n))^3 \lambda = 0 \qquad (10.4)$$

where

```
#property copyright "Copyright © 2019, DR. RAYMOND LEE"
#property link       "http://QFFC.ORG"

// DEFINE DIRECTORIES
string   QP_Directory   = "QPL";                    // QPL Directory
string   QPData_Directory = "QPL_Data";             // QPL Directory
string   TS_FileName = "";                          // File name for FC_DL
int      TS_FileHandle;                             // File Handle for FC_DL
string   Qf_FileName = "";                          // File name for Qfunction
int      Qf_FileHandle;                             // File Handle for Qfunction
string   ALL_QFEL_FileName = "";                    // File name for ALL_QFEL
int      ALL_QFEL_FileHandle;                       // File Handle for ALL_QFEL
string   ALL_QPR_FileName = "";                     // File name for ALL_QPR
int      ALL_QPR_FileHandle;                        // File Handle for ALL_QPR
string   ALL_NQPR_FileName = "";                    // File name for ALL_NQPR
int      ALL_NQPR_FileHandle;                       // File Handle for ALL_NQPR
string   QPD_FileName = "";                         // File name for QPL Details
int      QPD_FileHandle;                            // File Handle for QPL Details
string   QPLog_FileName = "";                       // File name for QPL Details
int      QPLog_FileHandle;                          // File Handle for QPL Details
string   Lambda_FileName = "";                      // File name for QPL Details
int      Lambda_FileHandle;                         // File Handle for QPL Details

// DEFINE GLOBAL VARIABLES
int      maxELevel = 21;                            // Max Energy Level
int      maxTP     = 120;                           // Max no of Financial Product
int      maxTS     = 2048;                          // Max no of Time Series Record
int      nTP=0;                                     // Trading Product no
double   p3=1.0/3.0;                                // Set 1/3 for MathPow

//DEFINE TIMING VARIABLES
uint     stime=0;
uint     etime=0;
uint     Gstime=0;
uint     Getime=0;
uint     tlapse=0;
uint     Gtlapse=0;

// DEFINE FINANCIAL PRODUCT RELATED VARIABLES
// Current Trading Symbol
string   TPSymbol  = "";

// Array of 120 Trading Products for Forex.com MT Platform
string   TP_Code[120]=
            {"XAGUSD","CORN","US30","AUDUSD","EURCHF","GBPCAD","NZDJPY","USDCNH","XAUAUD","XAUCHF",
             "XAUEUR","XAUGBP","XAUJPY","XAUUSD","COPPER","PALLAD","PLAT","UK_OIL","US_OIL","US_NATG",
             "HTG_OIL","COTTON","SOYBEAN","SUGAR","WHEAT","IT40","AUS200","CHINAA50","ESP35","ESTX50",
             "FRA40","GER30","HK50","JPN225","N25","NAS100","SIGI","SPX500","SWISS20","UK100","US2000",
             "AUDCAD","AUDCHF","AUDCNH","AUDJPY","AUDNOK","AUDNZD","AUDPLN","AUDSGD","CADCHF",
             "CADJPY","CADNOK","CADPLN","CHFHUF","CHFJPY","CHFNOK","CHFPLN","CNHJPY","EURAUD",
             "EURCAD","EURCNH","EURCZK","EURDKK","EURGBP","EURHKD","EURHUF","EURJPY","EURMXN",
             "EURNOK","EURNZD","EURPLN","EURRON","EURRUB","EURSEK","EURSGD","EURTRY","EURUSD",
             "EURZAR","GBPAUD","GBPCHF","GBPDKK","GBPHKD","GBPJPY","GBPMXN","GBPNOK","GBPNZD",
             "GBPPLN","GBPSEK","GBPSGD","GBPUSD","GBPZAR","HKDJPY","NOKDKK","NOKJPY","NOKSEK",
             "NZDCAD","NZDCHF","NZDUSD","SGDHKD","SGDJPY","TRYJPY","USDCAD","USDCHF","USDCZK",
             "USDDKK","USDHKD","USDHUF","USDILS","USDJPY","USDMXN","USDNOK","USDPLN","USDRON",
             "USDRUB","USDSEK","USDSGD","USDTHB","USDTRY","USDZAR","ZARJPY"};

// Trading Product No
int      TP_No[120]={1,2,3,4,5,6,7,8,9,10,11,12,13,14,15,16,17,18,19,20,21,22,23,24,25,26,27,28,29,30,
              31,32,33,34,35,36,37,38,39,40,41,42,43,44,45,46,47,48,49,50,51,52,53,54,55,56,57,58,59,60,
              61,62,63,64,65,66,67,68,69,70,71,72,73,74,75,76,77,78,79,80,81,82,83,84,85,86,87,88,89,90,
              91,92,93,94,95,96,97,98,99,100,101,102,103,104,105,106,107,108,109,110,111,112,113,114,115,
              116,117,118,119,120};

// Trading Product Decimal Digits
int      TP_nD[120]={3,2,1,5,5,5,3,5,2,2,2,2,0,2,2,2,2,0,2,3,0,2,2,2,2,0,0,0,0,0,1,1,0,0,2,1,1,1,0,1,1,5,5,4,
              3,4,5,5,5,5,3,5,5,3,3,5,5,4,5,5,4,4,5,5,4,3,3,5,5,5,5,4,3,5,4,5,5,5,5,5,5,4,3,4,5,5,5,5,
              5,5,4,4,5,4,5,5,5,5,4,3,3,5,5,3,5,5,3,5,3,5,5,5,5,5,3,5,5,3,5,5,3};
```

Fig. 10.15 Program declaration of QPL evaluation program

```
int init() {
    int         eL, d, TSsize, nQ, maxQno, nR, maxRno, tQno;
    double      auxR, maxQ, r0, r1, m1;
    double      dr, Lup, Ldw, L, mu, sigma;
    bool        bFound;

    // Variables used in Cardano's Method
    double      p, q, u, v;

    // Declare Time Series Array
    int         DT_YY[2048];
    int         DT_MM[2048];
    int         DT_DD[2048];
    double      DT_OP[2048];
    double      DT_HI[2048];
    double      DT_LO[2048];
    double      DT_CL[2048];
    double      DT_VL[2048];
    double      DT_RT[2048];

    // Declare Array for Quantum Price Wavefunction
    double      Q[100];                         // Quantum Price Wavefunction
    double      NQ[100];                        // Normalized Q[]
    double      r[100];                         // r no

    // Declare ARRAY for QPL related arrays
    double      ALL_QFEL[120][21];              // Array contains QFEL  for all FPs
    double      ALL_QPR[120][21];              // Array contains QPR  for all FPs
    double      ALL_NQPR[120][21];             // Array contains NQPR for all FPs
    double      QFEL[21];                       // QFEL for each FP
    double      QPR[21];                        // QPR for each FP
    double      NQPR[21];                       // NQPR for each FP
    double      K[21];                          // K values in QP Schrodinger Eqt

    // Set Global Start Time
    Gstime    = GetTickCount();

    // CREATE QPL SUMMARY DATA FILE
    ALL_QFEL_FileName     = "FX_QFEL.csv";
    FileDelete(QP_Directory+"//"+ALL_QFEL_FileName,FILE_COMMON);
    ResetLastError();
    ALL_QFEL_FileHandle   = FileOpen(QP_Directory+"//"+ALL_QFEL_FileName,
                                FILE_COMMON|FILE_READ|FILE_WRITE|FILE_CSV,',');

    // CREATE QPL SUMMARY DATA FILE
    ALL_QPR_FileName      = "FX_QPR.csv";
    FileDelete(QP_Directory+"//"+ALL_QPR_FileName,FILE_COMMON);
    ResetLastError();
    ALL_QPR_FileHandle    = FileOpen(QP_Directory+"//"+ALL_QPR_FileName,
                                FILE_COMMON|FILE_READ|FILE_WRITE|FILE_CSV,',');

    // CREATE QPL SUMMARY DATA FILE
    ALL_NQPR_FileName     = "FX_NQPR.csv";
    FileDelete(QP_Directory+"//"+ALL_NQPR_FileName,FILE_COMMON);
    ResetLastError();
    ALL_NQPR_FileHandle   = Filepen(QP_Directory+"//"+
                                ALL_NQPR_FileName,FILE_COMMON|FILE_READ|FILE_WRITE|FILE_CSV,',');

    // Write Header Line for ALL_QFEL DataFile
    FileWrite(ALL_QFEL_FileHandle,"CODE","QFEL[0]","QFEL[1]", "QFEL[2]", "QFEL[3]", "QFEL[4]", "QFEL[5]",
        "QFEL[6]", "QFEL[7]", "QFEL[8]", "QFEL[9]", "QFEL[10]",|QFEL[11]","QFEL[12]","QFEL[13]","QFEL[14]",
        "QFEL[15]","QFEL[16]","QFEL[17]","QFEL[18]","QFEL[19]","QFEL[20]");

    // Write Header Line for ALL_QPR DataFile
    FileWrite(ALL_QPR_FileHandle,"CODE","QPR[0]","QPR[1]", "QPR[2]", "QPR[3]", "QPR[4]", "QPR[5]",
        "QPR[6]", "QPR[7]", "QPR[8]","QPR[9]","QPR[10]","QPR[11]","QPR[12]","QPR[13]","QPR[14]","QPR[15]",
        "QPR[16]","QPR[17]","QPR[18]","QPR[19]","QPR[20]");

    // CREATE QP Log Filel
    QPLog_FileName    = "QPL_120Log.csv";
    FileDelete(QP_Directory+"//"+QPLog_FileName,FILE_COMMON);
    ResetLastError();
    QPLog_FileHandle  = FileOpen(QP_Directory+"//"+QPLog_FileName,
                            FILE_COMMON|FILE_READ|FILE_WRITE|FILE_CSV,',');

    // CREATE Lambda FILE
    Lambda_FileName   = "Lambda_FX.csv";
    FileDelete(QP_Directory+"//"+Lambda_FileName,FILE_COMMON);
    ResetLastError();
    Lambda_FileHandle = FileOpen(QP_Directory+"//"+Lambda_FileName,
                            FILE_COMMON|FILE_READ|FILE_WRITE|FILE_CSV,',');

    // Write Header Line for Lambda File
    FileWrite(Lambda_FileHandle,"CODE","Lambda");
```

Fig. 10.16 Program initialization of QPL evaluation program

```
//***********************************************************************************************************
//
//    1. Cacluate All K values [K0 .. K20] using the following formula:
//
//        K[eL] = pow((1.1924 + 33.2383*eL + 56.2169*eL*eL)/(1 + 43.6106 *eL),p3);
//
//***********************************************************************************************************
// Printout K List Header
Print("Printout ALL K values K0 .. K20 for first 20 Energy Levels");
FileWrite(QPLog_FileHandle,"Printout ALL K values K0 .. K20 for first 20 Energy Levels");

for (eL=0;eL<21;eL++)
{
  K[eL] = MathPow((1.1924 + (33.2383*eL) + (56.2169*eL*eL))/(1 + (43.6106 *eL)),p3);
  Print("Energy Level ",eL," K",eL," = ",K[eL]);
  FileWrite(QPLog_FileHandle,"Energy Level ",eL," K",eL," = ",K[eL]);
}

// LOOP OVER ALL TP
for (nTP=0;nTP<maxTP;nTP++)
{
  TPSymbol   = TP_Code[nTP];      // Get TP Symbol
  stime      = GetTickCount();    // Get timer

  // CREATE QPL Detail DATA FILE
  QPD_FileName    = TP_No[nTP]+" "+TPSymbol+"_QPR.csv";
  FileDelete(QP_Directory+"//"+QPD_FileName,FILE_COMMON);
  ResetLastError();
  QPD_FileHandle  = FileOpen(QPData_Directory+"//"+QPD_FileName,
                    FILE_COMMON|FILE_READ|FILE_WRITE|FILE_CSV,',',');

  // Write Header Line
  FileWrite(QPD_FileHandle,"Year","Month","Day","Open","High","Low","Close","Volumn","Return");

  // CREATE Qf Wavefunction Distribution DataFile
  Qf_FileName     = TP_No[nTP]+" "+TPSymbol+"_Qf.csv";
  FileDelete(QP_Directory+"//"+Qf_FileName,FILE_COMMON);
  ResetLastError();
  Qf_FileHandle  = FileOpen(QP_Directory+"//"+Qf_FileName,
                   FILE_COMMON|FILE_READ|FILE_WRITE|FILE_CSV,',',');

  // Write Header Line of Qfile
  FileWrite(Qf_FileHandle,"r","Q(r)","NQ(r)");
```

Fig. 10.17 Process 1 calculate all K values of QPL evaluation program

$$K_0(n) = \left[\frac{1.1924 + 33.2383n + 56.2169n^2}{1 + 43.6196n} \right]^{1/3} \qquad (10.5)$$

First, evaluate all $K_0(n)$ for the 20 energy levels (n).

Then open the file handles of the data files for the storage of the QPR, NQPR, QF values. Program codes of process 1 are shown in Fig. 10.17.

10.5.5 Process 2: READ ALL Daily Time Series

For each financial product (FP), this process uses MQL time series data functions iOpen, iHigh, iLow, iClose to read the daily time series data and calculate the daily return by

$$DT\_RT[d] = DT\_CL[d]/DT\_CL[d+1] \qquad (10.6)$$

```
//********************************************************************************************
//
// 2. READ ALL Daily Time Series
//
//********************************************************************************************
// Since iBars/Bars sometimes doesn't work, manually check TSsize
TSsize = 0;
while (iTime(TPSymbol,PERIOD_D1,TSsize)>0 && (TSsize<maxTS))
{
   TSsize++;
}

// Using For LOOP to get all the time series data
for (d=1;d<TSsize;d++)
{
   DT_YY[d-1] = TimeYear(iTime(TPSymbol,PERIOD_D1,d));
   DT_MM[d-1] = TimeMonth(iTime(TPSymbol,PERIOD_D1,d));
   DT_DD[d-1] = TimeDay(iTime(TPSymbol,PERIOD_D1,d));
   DT_OP[d-1] = iOpen(TPSymbol,PERIOD_D1,d);
   DT_HI[d-1] = iHigh(TPSymbol,PERIOD_D1,d);
   DT_LO[d-1] = iLow(TPSymbol,PERIOD_D1,d);
   DT_CL[d-1] = iClose(TPSymbol,PERIOD_D1,d);
   DT_VL[d-1] = iVolume(TPSymbol,PERIOD_D1,d);
   DT_RT[d-1] = 1;
}

// Cacluate DT_RT[d]
for (d=0;d<(TSsize-2);d++)
{
   if (DT_CL[d+1] > 0)
   {
      DT_RT[d] = DT_CL[d]/DT_CL[d+1];
   }else{
      DT_RT[d] = 1;
   }

   // Write out the QPD data file
   FileWrite(QPD_FileHandle,DT_YY[d],DT_MM[d],DT_DD[d],DoubleToString(DT_OP[d],TP_nD[nTP]),
            DoubleToString(DT_HI[d],TP_nD[nTP]),DoubleToString(DT_LO[d],TP_nD[nTP]),
            DoubleToString(DT_CL[d],TP_nD[nTP]),DT_VL[d], DoubleToString(DT_RT[d],8));
}

// Close QP Detail Data File
FileClose(QPD_FileHandle);
```

Fig. 10.18 Process 2 read time series of QPL evaluation program

where DT_CL[d] and DT_CL[d+1] are the closing prices for d and d+1 day, respectively. As mentioned, except cryptocurrencies which have 300–500 trading day records, the other 120 financial products (forex, commodities and financial indices) have 2048-trading day records, which are sufficient for the evaluation on distribution function (wavefunction) of the price returns (r). Program codes of process 2 are shown in Fig. 10.18.

10.5.6   Process 3: Calculate Mean (mu) and Standard Deviation (sigma) of Return Array

After the evaluation of daily returns DT_RT[d] of the financial time series, this process evaluates the mean (mu, μ) and standard deviation (sigma, σ) of the return distribution function (φ).

```
//*************************************************************************************************
//
// 3. Calculate Mean (mu) and Standard Deviation (sigma) of return array
//
//*************************************************************************************************

maxRno = TSsize - 2;

// Calculate mean mu first
mu = 0;
for (d=0;d<maxRno;d++)
{
  mu = mu + DT_RT[d];
}
mu = mu/maxRno;

// Calculate STDEV sigma
sigma = 0;
for (d=0;d<maxRno;d++)
{
  sigma = sigma + (DT_RT[d]-mu)*(DT_RT[d]-mu);
}
sigma = sqrt((sigma / maxRno));

// Calculate dr where dr = 3*sigma/50
dr = 3 * sigma / 50;

Print("TP",nTP+1," ",TP_Code[nTP]," No of r = ",maxRno," mu = ",mu," sigma = ",sigma," dr=",dr);
FileWrite(QPLog_FileHandle, "TP",nTP+1," ",TP_Code[nTP]," No of r = ",
          maxRno," mu = ",mu," sigma = ",sigma," dr=",dr);
```

Fig. 10.19 Process 3 calculate mean (mu) and standard deviation (sigma) of return array

The purpose for mu evaluation is to locate the central point of the distribution function, and sigma is used to calculate "*dr*" term—the width of each slice of returns for the distribution function is an important factor to calculate Lambda using F.D.M., which is given by

$$dr = 3 * \frac{sigma}{50};$$
(10.7)

For each QP distribution chart, 100 equal spacing slices with width *dr* are used for to evaluate the distribution function values. Program codes of process 3 are shown in Fig. 10.19.

10.5.7   Process 4: Generate the Returns Wavefunction Distribution

Based on dr value calculated in process 3, this process calculates the price returns wavefunction values Q[nQ], NQ[nQ] where nQ is the number of slice in distribution function, Q[nQ] is the pdf (probability density function values) within the particular slice nQ; and NQ[nQ] is the normalized pdf values of the returns wavefunction.

The NQ[nQ] is given by:

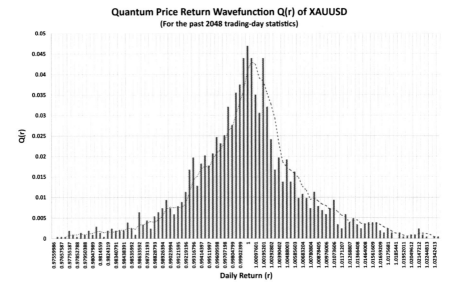

Fig. 10.20 Process 4 quantum price return wavefunction of Q(r) of gold (XAUUSD)

$$NQ[nQ] = \frac{no\ of\ trading\ days\ within\ the\ range\ of\ return\ value\ in\ nQ}{Total\ number\ trading\ day\ return\ values} \qquad (10.8)$$

Figure 10.20 shows the return wavefunction for Gold (XAUUSD) in 2048 financial time series. Program codes of process 4 are shown in Fig. 10.21.

10.5.8   Process 5: Evaluate Lambda Value for the Returns Wavefunction

Once all wavefunction values in each slice are evaluated, we make use of F.D.M., together with QFSE (Quantum Finance Schrödinger Equation) to evaluate the Lambda value (λ), which is given by:

$$\lambda = \left| \frac{r_{-1}^2 \varphi_{r-1} - r_{+1}^2 \varphi_{r+1}}{r_{+1}^4 \varphi_{r+1} - r_{-1}^4 \varphi_{r-1}} \right| \qquad (10.9)$$

where $\varphi_{r_{-1}}$ and $\varphi_{r_{+1}}$ are wavefunction values (normalized) for the 1st left and right-hand side slice from the slice with max wavefunction value, normally is the central point with returns value = 1 as in Fig. 10.22.

Program codes of process 5 are shown in Fig. 10.23.

```
//************************************************************************************************
//
// 4. Generate the QP Wavefunction distribution
//
//************************************************************************************************
auxR   = 0;

// Loop over all r from r - 50*dr to r + 50*dr and get the distribution function
// Reset all the Q[] first
for (nQ=0;nQ<100;nQ++)
{
  Q[nQ] = 0;
}

// Loop over the maxRno to get the distribution
tQno = 0;
for (nR=0;nR<maxRno;nR++)
{
  bFound = False;
  nQ = 0;
  auxR = 1 - (dr * 50);
  while (!bFound && (nQ < 100))
  {
    if ((DT_RT[nR] > auxR) && (DT_RT[nR] <= (auxR + dr)))
    {
      Q[nQ]++;
      tQno++;
      bFound = True;
    }else
    {
      nQ++;
      auxR = auxR + dr;
    }
  }
}

// Write out the Qfile for Record
auxR = 1 - (dr * 50);
for (nQ=0;nQ<100;nQ++)
{
  r[nQ]  = auxR;
  NQ[nQ] = Q[nQ]/tQno;
  FileWrite(Qf_FileHandle,auxR,Q[nQ],NQ[nQ]);
  auxR = auxR + dr;
}

// Find maxQ and maxQno
maxQ   = 0;
maxQno = 0;
for (nQ=0;nQ<100;nQ++)
{
  if (NQ[nQ] > maxQ)
  {
    maxQ   = NQ[nQ];
    maxQno = nQ;
  }
}

// Printout the maxQ, maxQno
Print("TP",nTP+1," ",TP_Code[nTP]," MaxQ= ",maxQ," maxQno=",maxQno," Total Qno =",tQno);
FileWrite(QPLog_FileHandle,"TP",nTP+1," ",TP_Code[nTP],
                " MaxQ= ",maxQ," maxQno=",maxQno," Total Qno =",tQno);
```

Fig. 10.21   Process 4 generate the returns wavefunction distribution

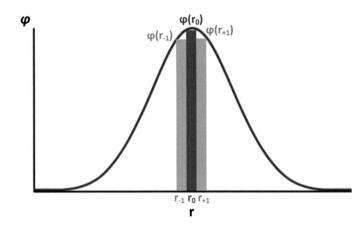

Fig. 10.22 Illustration of F.D.M. scheme for evaluation of lambda value

```
//**********************************************************************************************
//
// 5. Evaluate Lambda L for the QP Wavefuntion
//
//**********************************************************************************************
//
//  Given:
//          maxQno - i.e. ground state Q[0], r[0] = r[maxQno-dr]
//  We have:
//          Q[+1] = NQ[maxQno+1], r[+1] = r[maxQno]+(dr/2)
//          Q[-1] = NQ[maxQno-1], r[-1] = r[maxQno]-(dr*1.5)
//  Apply F.D.M. into Quantum Finance Schrodinger Equation
//          L = abs((r[-1]^2*Q[-1]-(r[+1]^2*Q[+1]))/(r[-1]^4*Q[-1]-(r[+1]^4*Q[+1])))

r0  = r[maxQno] - (dr/2);
r1  = r0 + dr;
rn1 = r0 - dr;
Lup = (pow(rn1,2)*NQ[maxQno-1])-(pow(r1,2)*NQ[maxQno+1]);
Ldw = (pow(rn1,4)*NQ[maxQno-1])-(pow(r1,4)*NQ[maxQno+1]);
L   = MathAbs(Lup/Ldw);

// Printout r0,Q0, r1, Q1, r-1 Q-1
Print("TP",nTP+1," ",TP_Code[nTP]," r0 = ",r0," r1 = ",r1," r-1 = ",rn1," Q0 = ",
          NQ[maxQno]," Q1 = ",NQ[maxQno+1]," Q-1 = ",NQ[maxQno-1]," L = Lup/Ldw = ",Lup,"/",Ldw," = ",L);

FileWrite(QPLog_FileHandle," r0 = ",r0," r1 = ",r1," r-1 = ",rn1," Q0 = ",NQ[maxQno]," Q1 = ",
          NQ[maxQno+1]," Q-1 = ",NQ[maxQno-1]," L = Lup/Ldw = ",Lup,"/",Ldw," = ",L);

FileWrite(Lambda_FileHandle,TPSymbol,L);

Print(TPSymbol,L);
```

Fig. 10.23 Process 5 Evaluate lambda value for the returns wavefunction

10.5.9 Process 6: Using QP Schrodinger Equation to FIND the First 21 Energy Levels

Once Lambda value is evaluated, we can apply the Cardano's formula to solve the quantum price return Schrödinger equation, with formulation given by:

$$E(n) = \sqrt[3]{-\frac{q}{2} + \sqrt{\frac{q^2}{4} + \frac{p^3}{27}}} + \sqrt[3]{-\frac{q}{2} - \sqrt{\frac{q^2}{4} + \frac{p^3}{27}}} \qquad (10.10)$$

where:

$$p = -(2n + 1)^2, \, q = -\lambda(2n + 1)^3 [K_0(n)]^3$$

Please refer to Chap. 5 on detailed mathematical proof and derivations for the numerical evaluation of quantum price level.

In this program, we calculate the first 21 *quantum finance energy levels (QFEL)* of every financial product (ground state 0 and 20 Energy Levels), which is sufficient for daily financial trading and analysis.

Figure 10.24 shows MQL codes for 21 energy levels evaluation.

10.5.10 Process 7: Program End

This process closes all datafile handles, calculate total CPU time spent and report the total time taken to calculate QPLs of all financial products.

As shown in Fig. 10.25, MQL also provides start() class for developers to implement real time trading/hedging programs.

In other words, one can make use of the MQL learnt in this chapter to evaluate the QPLs for any financial product, and make use of it, say, as S&R lines to implement real time trading and hedging programs learnt in Chap. 6.

Figure 10.26 shows a snapshot of implementation result of QPL evaluation program on Forex.com MT4 platform for QPL evaluation of 120 financial products.

10.5.11 Implementation Results

Figure 10.26 shows snapshot of the implementation results of QPL2019.mq4 MQL program for QPLs evaluation of 120 financial products at the Forex.com MT platform. As shown, it took only 1594 ms (i.e., less than 2 s) to evaluate the first 20 QPLs for 120 financial products, which was much faster and effective than using traditional Path Integral Techniques.

```
//*********************************************************************************************************************
// 6. Using QP Schrödinger Eqt to FIND first 21 Energy Levels
//    By solving the Quartic Anharmonic Oscillator as cubic polynomial equation of the form
//          a*x^3 + b*x^2 + c*x + d = 0
// Using (Dasqupta et. al. 2007) QAHO solving equation:
//        (E(n)/(2n+1))^3 - (E(n)/(2n+1)) - K(n)^3*L = 0
// Solving the above Depressed Cubic Eqt using Cardano's Method
// Given    t^3 + p*t + q = 0
// Let      t = u + v
// Cardano's Method deduced that:
//          u^3 = -q/2 + sqrt(q^2/4 + p^3/27)
//          v^3 = -q/2 - sqrt(q^2/4 + p^3/27)
// The first cubic root (real root) will be:
//          t = u + v
// So, combining Cardano's Method into our QF Sch Eqt.
// Substitue p = -(2n+1)^2;  q = -L(2n+1)^3*(K(n)^3) into the above equations to get the real root
//
//*********************************************************************************************************************
for (eL=0;eL<21;eL++)
{
  p = -1 * pow((2*eL+1),2);
  q = -1 * L * pow((2*eL+1),3) * pow(K[eL],3);

  // Apply Cardano's Method to find the real root of the depressed cubic equation
  u = MathPow((-0.5*q + MathSqrt(((q*q/4.0) + (p*p*p/27.0)))),p3);
  v = MathPow((-0.5*q - MathSqrt(((q*q/4.0) + (p*p*p/27.0)))),p3);

  // Store the QFEL
  QFEL[eL] = u + v;

  // Printout the QF Energy Levels
  Print("TP",nTP+1," ",TP_Code[nTP]," Energy Level",eL," QFEL = ",QFEL[eL]);
  FileWrite(QPLog_FileHandle," Energy Level",eL," QFEL = ",QFEL[eL]);
}

// Evaluate ALL QPR values
for (eL=0;eL<21;eL++)
{
  QPR[eL]  = QFEL[eL]/QFEL[0];
  NQPR[eL] = 1 + 0.21*sigma*QPR[eL];

  // Store into ALL QFEL, QPR, NQPR into array
  ALL_QFEL[nTP,eL] = QFEL[eL];
  ALL_QPR[nTP,eL]  = QPR[eL];
  ALL_NQPR[nTP,eL] = NQPR[eL];
}

// Close Qfile
FileClose(Qf_FileHandle);

// Write out ALL_QFEL into QFEL Datafile
FileWrite(ALL_QFEL_FileHandle,TPSymbol,ALL_QFEL[nTP,0],
        ALL_QFEL[nTP,1], ALL_QFEL[nTP,2], ALL_QFEL[nTP,3], ALL_QFEL[nTP,4], ALL_QFEL[nTP,5],
        ALL_QFEL[nTP,6], ALL_QFEL[nTP,7], ALL_QFEL[nTP,8], ALL_QFEL[nTP,9], ALL_QFEL[nTP,10],
        ALL_QFEL[nTP,11],ALL_QFEL[nTP,12],ALL_QFEL[nTP,13],ALL_QFEL[nTP,14],ALL_QFEL[nTP,15],
        ALL_QFEL[nTP,16],ALL_QFEL[nTP,17],ALL_QFEL[nTP,18],ALL_QFEL[nTP,19],ALL_QFEL[nTP,20]);

// Write out ALL_QPR into QPR Datafile
FileWrite(ALL_QPR_FileHandle,TPSymbol,ALL_QPR[nTP,0],
        ALL_QPR[nTP,1], ALL_QPR[nTP,2], ALL_QPR[nTP,3], ALL_QPR[nTP,4], ALL_QPR[nTP,5],
        ALL_QPR[nTP,6], ALL_QPR[nTP,7], ALL_QPR[nTP,8], ALL_QPR[nTP,9], ALL_QPR[nTP,10],
        ALL_QPR[nTP,11],ALL_QPR[nTP,12],ALL_QPR[nTP,13],ALL_QPR[nTP,14],ALL_QPR[nTP,15],
        ALL_QPR[nTP,16],ALL_QPR[nTP,17],ALL_QPR[nTP,18],ALL_QPR[nTP,19],ALL_QPR[nTP,20]);

// Write out ALL_NQPR into NQPR Datafile
FileWrite(ALL_NQPR_FileHandle,ALL_NQPR[nTP,0],
        ALL_NQPR[nTP,1], ALL_NQPR[nTP,2], ALL_NQPR[nTP,3], ALL_NQPR[nTP,4], ALL_NQPR[nTP,5],
        ALL_NQPR[nTP,6], ALL_NQPR[nTP,7], ALL_NQPR[nTP,8], ALL_NQPR[nTP,9], ALL_NQPR[nTP,10],
        ALL_NQPR[nTP,11],ALL_NQPR[nTP,12],ALL_NQPR[nTP,13],ALL_NQPR[nTP,14],ALL_NQPR[nTP,15],
        ALL_NQPR[nTP,16],ALL_NQPR[nTP,17],ALL_NQPR[nTP,18],ALL_NQPR[nTP,19],ALL_NQPR[nTP,20]);
```

Fig. 10.24 Process 6 using QP Schrodinger equation to find the first 21 energy levels

```
//*******************************************************************************
//
//      7. Program Termination
//
//*******************************************************************************

// Close All DataFiles
FileClose(QPLog_FileHandle);
FileClose(Lambda_FileHandle);
FileClose(ALL_NQPR_FileHandle);

// Check Global Time
Getime  = GetTickCount();
Gtlapse = Getime - Gstime;

// Output time taken
Print("Total Time Taken : ",Gtlapse," msec");

return(0);
}

int start()
  {

  return(0);
  }
```

Fig. 10.25   Process 7 program end

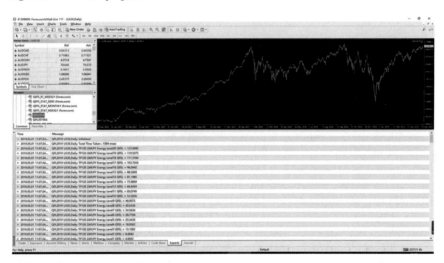

Fig. 10.26   Implementation results of QPL2019.mq4 on Forex.com MT4 platform to calculate QPLs of 120 financial products

10.6   Conclusion

Based on quantum finance concepts and theories in Chaps. 1–4, and numerical QPL (quantum price levels) evaluation learnt in Chap. 5, this chapter we studied how to implement these methods and theories to evaluate QPLs of worldwide 120+ financial products including: worldwide forex, major cryptocurrencies, international finance indices and major commodities.

From the implementation perspective, we demonstrated how to use MQL (MetaQuotes Language) on MT4 Platform—one of the world's biggest online program trading and development platform to develop real-time QPL evaluation system.

As mentioned in Chap. 6 about quantum trading and hedging strategy, besides using QPL as S&R lines in program trading; QPL can also be incorporated with chaotic neural networks (CNNs), deep neural oscillatory network (DNON) for real-time financial predictions, which will be studied in Chaps. 11 and 12, respectively.

Problems

10.1   What is the financial market? Discuss and explain why financial market prices may not indicate the true intrinsic value, say, a stock which leads to the popularity of programming trading nowadays.

10.2   What is the major difference between primary versus secondary financial markets? And what is the importance of Secondary markets in modern finance?

10.3   What are the two different categories of secondary market? Give one example for each category and discuss the major characteristics.

10.4   State and explain the five major types of financial markets. Give one example for each type of financial market and explain how it works.

10.5   What are the two major roles and functions of market makers (MM) in a financial market, such as forex market?

10.6   State and explain the five key participants in forex market, in the descending order of their importance and market shares.

10.7   State and explain the five key forex trading characteristics.

10.8   What are the major factors to affect the prices in forex market?

10.9   What is retail forex trading platform? Give one example and discuss it is different from traditional forex trading platform.

10.10 What are supports and resistance (S&R) analysis in technical analysis? Give one example in each case and discuss how it works.

10.11 Discuss and explain how S&R analysis is related to supply and demand analysis and theory in technical analysis. Give an example by using the daily chart of DJI (Dow Jones Index) and explain how such theory works.

10.12 What is the major difference between role reversal vs. trend reversal on S&R analysis by using simple charts to illustrate your explanation?

10.13 How S&R analysis is related and interpreted by QPL (quantum price level) modeling in quantum finance?

10.14 Discuss and explain why QPL (quantum price level) in quantum finance are critical for nowadays HFPT (High-Frequency-Program-Trading).

10.15 What are the major functions and characteristics of MetaTrader system as compared with traditional financial trading platforms?

10.16 State and explain the five categories of programs/systems generated by MQL of MT platform.

10.17 State and explain six major components in MQL as compared with traditional programming languages which are critical for program trading in finance.

10.18 State six types of OrderSend() in MQL and explain how they work for the design of trading and hedging systems.

10.19 Give five commonly used technical indicators in technical analysis and explain how they works in MQL programming.

10.20 MQL programming task 10.1 Design and implement a real-time trading program using MQL with the implementation of classical RSI (30-70) trading algorithm. Financial products: Using at least five forex products for trading (e.g., USDCAD, AUDJPY, etc.) Trading criteria: Using H1 as trading timeframe, trigger a sell order when RSI value hit 70, trigger a buy order when it hit 30.

Using (1) MQL built-in simulator; (2) Live test for at least 10 trading days using demo account and compared their results.

10.21 MQL Programming Task 10.2

Similar to Task 10.1, but this time using two more technical indicators to generate automatic trading rules. Commonly used technical indicators such as: iBand (Bollinger Bands); iMomentum (Momentum); iMA (Moving Average); iMACD (MACD) and iStochastic (Stochastic Oscillator).

Compare the trading results with the one generated by Task 10.1.

10.22 MQL Programming Task 10.3

Based on Workshop #2 (QPL2019.mq4), design and implement a real-time MQL trading program by combining QPLs (as S&R lines) and RSI trading program implemented in Task 10.1 to implement a simple quantum finance trading system.

Compare trading results with trading program implemented in Task 10.1.

References

Abreu, R. et al. (2018) On the accuracy of the Complex-Step-Finite-Difference method. *Journal of Computational and Applied Mathematics* 340: 390–403.

Arnett, G. W. (2011) *Global Securities Markets: Navigating the World's Exchanges and OTC Markets.* Wiley.

Atkeson, A. G. et al. (2015) Entry and Exit in OTC Derivatives Markets. *Econometrica* 83(6): 2231–2292.

Brown, J. (2016) *MT4/MT5 High Probability Forex Trading Method.* CreateSpace Independent Publishing Platform.

Bulkowski, T. N. (2005) *Encyclopedia of Chart Patterns.* Wiley 2nd edition.

De Angelis, T. and Peskir, G. (2016) Optimal prediction of resistance and support levels. *Applied Mathematical Finance* 23(6): 465–483.

Drayer, J. et al. (2008) Tradition vs. Trend: A Case Study of Team Response to the Secondary Ticket Market. *Sport Marketing Quarterly* 17(4): 235.

Durbin, M. (2010) *All About High-Frequency Trading.* McGraw-Hill Education.

Gomes, C. and Waelbroeck, H. (2010) An empirical study of liquidity dynamics and resistance and support levels. *Quantitative Finance* 10(10): 1099–1107.

Guo, G. (2005) Finite Difference Methods for the BSDEs in Finance. *International Journal of Financial Studies* 6(1): 26, 2018.

Huber, M. et al. (2012) Which factors drive product sales in OTC markets? *International Journal of Pharmaceutical and Healthcare Marketing* 6(4): 291–309.

Kaufman, P. J. (2013) *Trading Systems and Methods.* Springer.

Kirkpatrick, C. D. and Dahlquist, J. R. (2015) *Technical Analysis: The Complete Resource for Financial Market Technicians.* FT Press 3rd edition.

Kuznetsov, A. (2006) *The Complete Guide to Capital Markets for Quantitative Professionals (McGraw-Hill Library of Investment and Finance).* McGraw-Hill Education 1st edition.

Murphy, J. J. (1999) *Technical Analysis of the Financial Markets: A Comprehensive Guide to Trading Methods and Applications.* New York Institute of Finance.

Singh, R.M. et al. (2009) The solution of the Schrödinger equation for coupled quadratic and quartic potentials in two dimensions. *Pramana* 72(4): 647–654.

Tichy, V. et al. (2012) Analytic energies and wave functions of the two-dimensional Schrödinger equation: ground state of two-dimensional quartic potential and classification of solutions. *Canadian Journal of Physics* 90(6): 503–513.

Walker, W. (2018) *Expert Advisor Programming and Advanced Forex Strategies.* Independently published.

Young, A. R. (2015) *Expert Advisor Programming for MetaTrader 4: Creating automated trading systems in the MQL4 language.* Edgehill Publishing.

Zapranis, A. and Tsinaslanidis, P.E. (2012) Identifying and evaluating horizontal support and resistance levels: an empirical study on US stock markets. Applied Financial Economics 22(19): 1571–1585.

Chapter 11
Time Series Chaotic Neural Oscillatory Networks for Financial Prediction

*A computer would deserve to be called intelligent if it could
deceive a human into believing that it was human.*
Alan Turing (1912–1954)

Financial prediction, ranging from stocks, commodities to forex forecast poses real challenges to researchers and quantitative analysts (a.k.a. *quants*) due to its highly chaotic and unpredictable nature. With the blooming of cryptocurrency trading which provides 24×7 online currency trading, the financial markets, especially the worldwide currency markets are more chaotic and fluctuating than before. An effective and reliable financial prediction system is profoundly required.

With the integration of *quantum price levels* (QPL) evaluated by quantum finance technology, a *Time Series Chaotic Neural Oscillatory Network* (*TSCNON*) for worldwide financial market prediction is devised to effectively resolve the system over-training and deadlock problems imposed by traditional recurrent neural networks using classical sigmoid-based activation functions.

In terms of system implementation, TSCNON coalesces into 2048-trading day time series financial data and 39 major financial signals as input signals for the real-time prediction of 129 worldwide financial products which includes: 9 major cryptocurrencies, 84 forex, 19 major commodities, and 17 worldwide financial indices.

In terms of system performance, past 500-trading day of average system performance attained less than 1% forecast percentage errors.

11.1 Introduction

With the integration of *quantum price levels* (*QPL*) discussed in Chap. 10 and chaotic neural networks learnt in Chap. 9, in this chapter, TSCNON (time series chaotic neural oscillatory networks) is devised to effectively resolve the system over-training and deadlock problems imposed by traditional recurrent neural networks using classical sigmoid-based activation functions (Lee 2004, 2006a; Li and Lee 2005;

© Springer Nature Singapore Pte Ltd. 2020
R. S. T. Lee, *Quantum Finance*,
https://doi.org/10.1007/978-981-32-9796-8_11

Wong et al. 2008). TSCNON coalesces 2048-trading day time series financial data with *quantum finance signals* (*QFS*) based on QPL derived in Chap. 10 into input signals for the real-time prediction of 129 worldwide financial products which includes: 9 major cryptocurrencies, 84 forex, 19 major commodities, and 17 worldwide financial indices.

The chapter is comprised of (1) TSCNON system architecture and its chaotic neural network training algorithm in Sect. 11.2; (2) TSCNON system implementation framework for the real-time prediction of worldwide 129 financial products in Sect. 11.3; (3) Workshop #2—quantum finance forecast system (TSCNON.mq4) will be discussed in details in Sect. 11.4; (4) TSCNON system performance in Sect. 11.6; (5) TSCNON—system implementation results in Sect. 11.6 and followed by (6) conclusion in Sect. 11.7.

11.2 TSCNON—System Architecture

TSCNON (time series chaotic neural oscillatory networks) is the integration of multilayer feedforward backpropagation networks (FFBPNs) (Fausett 1994; Patterson 1996) with Lee-oscillators (Lee 2004, 2005, 2006a, b) by replacing all simple neurons with chaotic neural oscillators. The network training algorithm of TSCNON is shown in Fig. 11.1. Figure 11.2 shows the system architecture of TSCNON for financial prediction.

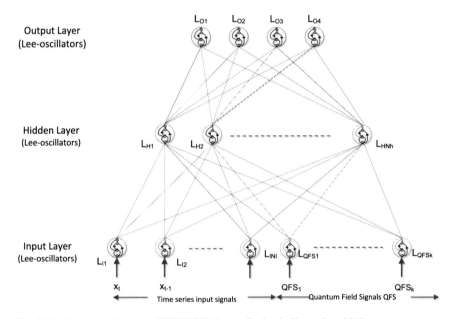

Fig. 11.1 System architecture of TSCNON (*Source* Dr. Lee's illustration, 2019)

TSCNON NETWORK LEARNING ALGORITHM

1   TSCNON CSLN Initialization Phase

    1.1 Initialization all the network weights ω by a random number generator to values between 1 and 0.

2   TSCNON CSLN Checking Stop Training Criteria

    IF MSE < Training Threshold δ (say 1×10^{-6}) STOP,

    Else CONTINUE

3   TSCNON CSLN Forward Propagation Phase

    3.1 Evaluate the total inputs for all hidden Lee-oscillators (L_H)

$$\overrightarrow{L_{Hinput}} = \sum_{n=0}^{N_I} \overrightarrow{L_{In}}\,\overrightarrow{\omega_n}$$

    Noted that $N_I = T \times S$ is the total number of input Lee-oscillators, where T is the forecast horizon and S is the dimension of the input signal vector.

    3.2 Evaluate the TCAF values of all $\overrightarrow{L_{Hinput}}$ vectors using chaotic Lee_operator given by equations (5) to (8)

$$\overrightarrow{L_H} = \overline{Lee_H\left(\overrightarrow{L_{Hinput}}\right)}$$

    3.3 Evaluate the total input vectors all output Lee-oscillators (L_O)

$$\overrightarrow{L_{Oinput}} = \sum_{n}^{N_H} \overrightarrow{L_{Hn}}\,\overrightarrow{\omega_n}$$

    Noted that N_H is the total number of hidden Lee-oscillators.

    3.4 Evaluate the TCAF values of all $\overrightarrow{L_{Oinput}}$ vectors

$$\overrightarrow{L_O} = \overline{Lee_O\left(\overrightarrow{L_{Oinput}}\right)}$$

4   TSCNON CSLN Backward Propagation Phase

    4.1 Evaluate the $\vec{\delta}_o$ (Correction Error Vector) and $\overrightarrow{\Delta\omega_{HO}}$ (weight adjustment vectors between hidden and output layer) of all $\vec{L}_o$ against the target output vectors $\vec{L}'_O$ with network learning rate β.

$$\overrightarrow{\delta_{HO}} = \left(\vec{L}'_o - \vec{L}_o\right)f'_{L_O}\left(\overrightarrow{L_{Oinput}}\right)$$

$$\overrightarrow{\Delta\omega_{HO}} = \beta\overrightarrow{\delta_{HO}}\,\overrightarrow{L_H}$$

    4.2 Evaluate the $\vec{\delta}_H$ and $\overrightarrow{\Delta\omega_{IH}}$ of all $\vec{L}_H$

$$\overrightarrow{\delta_{IH}} = \left[\sum \overrightarrow{\delta_{HO}}\cdot\overrightarrow{\omega_{HO}}\right]\left(\vec{L}'_o - \vec{L}_o\right)f'_{L_O}\left(\overrightarrow{L_{Oinput}}\right)$$

$$\overrightarrow{\Delta\omega_{IH}} = \beta\overrightarrow{\delta_{IH}}\,\overrightarrow{L_I}$$

    4.3 Evaluate the all the weight vectors at the same time.

$$\overrightarrow{\omega(t+1)} = \overrightarrow{\omega(t)} + \overrightarrow{\Delta\omega(t)}$$

5   TSCNON CSLN STEP 2 to Check for Stopping Criteria.

Note:

1. L_I , L_H and L_O are the Lee-oscillator in the input, hidden and output layer.

2.   The four output Lee oscillators correspond to the next-day forecasts of Open, High, Low and Close.

3. ω are the network weights.

4.   TCAF – Transient Chaotic Activation Function.

5. δ are the correction error vectors.

Fig. 11.2 System training algorithm of TSCNON (*Source* Dr. Lee's illustration, 2019)

As shown in Fig. 11.2, TSCNON consists of three neural network layers:

1. Input layer: consists of (1) 5-day time series input signal vector contains Open, High, Low, and Closing prices; (2) *quantum field signals (QFS)* contain the 20 closest QPLs revealed in Chap. 10. For each input node given by Lee-oscillator, totally we have 40 Lee-oscillators in the input layer (20 Lee-oscillators for time series signals, and 20 Lee-oscillators of QFS).
2. Hidden layer: consists of 40 Lee-oscillators as hidden nodes.
3. Output layer: consists of 4 Lee-oscillators which model the next-day forecasts of Open, High, Low, and Close, respectively.

Different from traditional neural networks (Fausett 1994; Patterson 1996), every neuron in the TSCNON is replaced by Lee-oscillator with neural dynamics given by Eqs. (11.1)–(11.4) mentioned in Chap. 9, which effectively converts the recurrent neural network into a chaotic recurrent neural network to resolve the over-training and deadlock problems, which is usually occurred in training of massive data such as financial data we are conducted in this chapter.

The formulations of Lee-oscillator are given by

$$E(t+1) = Sig[e_1 \cdot E(t) - e_2 \cdot I(t) + S(t) - \xi_E] \tag{11.1}$$

$$I(t+1) = Sig[i_1 \cdot E(t) - i_2 \cdot I(t) - \xi_I] \tag{11.2}$$

$$\Omega(t+1) = Sig[S(t)] \tag{11.3}$$

$$L(t) = [E(t) - I(t)] \cdot e^{-kS^2(t)} + \Omega(t) \tag{11.4}$$

where e_1, e_2, i_1, and i_2 are the weights; ξ_E and ξ_I are the threshold values and $S(t)$ is the external input.

Detail system performance analysis will be studied in the next section.

11.3 TSCNON—System Implementation Framework

Quantum finance forecast center (http://www.QFFC.org) is a nonprofit, self-funded AI-fintech R&D and worldwide financial forecast center aims at the R&D and provision of a fair and open platform for worldwide traders and individual investors to acquire free knowledge of worldwide 129 financial product forecasts based on state-of-the-art AI, intelligent agents, chaotic neural networks, and quantum field theory technologies.

With the adoption of TSCNON technology and the real-time data provided by Forex.com (one of the major international forex trading platforms) and AvaTrade.com (one of the biggest worldwide cryptocurrency trading platforms) (Narayanan et al. 2016; Vigna and Casey 2016), QFFC launched the 129 financial products' daily

Fig. 11.3 Quantum finance forecast center official site for TSCNON daily financial forecast on May 3, 2019 (*Source* Dr. Lee's illustration, 2019)

and weekly forecast services from January 1, 2018 for over 1000 worldwide traders and individual investors for testing and evaluation. Figure 11.3 shows the official site of *Quantum Finance Forecast Center*.

From the system implementation perspective, real-time and historical data of the worldwide 129 financial products provided by Forex.com and AvaTrade.com are adopted in TSCNON for chaotic neural network training and prediction.

They include the following:

– major cryptocurrencies (9);
– major worldwide forex (84);
– major commodities (19);
– major worldwide financial indices (17).

Appendix A depicts the list of 129 financial products under these four categories.

As shown in Appendix A, owing to the short trading history of cryptocurrencies (300-trading day records are provided by AvaTrade.com), all other financial products consist of 2048 past trading day records for each financial product (data provided by Forex.com), which provide sufficient training and test sets for TSCNON system testing and evaluation.

To provide a fully coherent and automation of TSCNON with both Forex.com and AvaTrade.com trading platforms for the automatic acquisition of real-time and historical data, the whole TSCNON system is developed in MT platform (Walker 2018; Young 2015) using MetaQuotes Language (MQL) and Expert Advisor (EA) system for daily financial forecast. Figure 11.4 shows the system framework of TSCNON.

As shown in Fig. 11.4 each financial product has 2048-trading day data (except cryptocurrency which only have 300-trading day data) are automatically generated by the MT4 engines of Forex.com and AvaTrade.com on a daily basis. Through the QPL (quantum price level) Generator revealed in Chap. 10, 21 closed QPL signals are generated by TSCNON together with the previous 5-day time series patterns; they are fed into TSCNON for chaotic neural network training and testing.

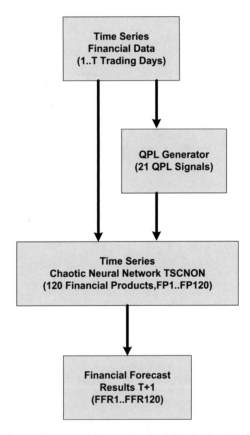

Fig. 11.4 TSCNON System Framework (*Source* Dr. Lee's illustration, 2019)

11.4 Workshop #2—Quantum Finance Forecast System (TSCNON.Mq4)

11.4.1 Introduction

In this section, we will explore the programming details of quantum finance forecast system (TSCNON) using MQL for real-time implementation on any MT4 platforms. For the *time series chaotic neural oscillatory network* (*TSCNON*), we are using the QFFC SDK developed by QFFC.org.

This section will study the complete MQL programming details for the design and development of *quantum finance forecast system* by using TSCNON technology in this chapter. Users can base on the method and program learnt in this chapter to design and implement their own quantum finance prediction systems.

Since the entire TSCNON.mq4 consists of over 1000 program lines of coding, this section will focus on the main module implementation process. Complete program coding can be found in QFFC.org Workshop #2.

11.4.2 Main Program Logic

TSCNON.mq4 is a MQL4 program with the implementation of TSCNON system in this chapter for the daily forecast of 120 financial products provided by Forex.com MT4 platform. To use this program for any other MT4 platforms, simply replace the product code array (TP_Code[]) with the ones adopted by users' own MT4 system. The entire system flowchart is shown in Fig. 11.5.

As shown in Fig. 11.6, quantum finance forecast center (a.k.a. QFFC) has implemented TSCNON system training and forecast functions and packaged into QFSDK software library. To use QFSDK software library, users should do the following:

```
//*********************************************************************************************************
// Date      : 3 May 2019
// Created by     : Dr. Raymond LEE
// Objective     : Quantum Finance Forecast System using TSCNON Technology
// Program Logic  :
//  1. Program initialization
//  2. Check for Forecast day
//  3. If Forecast_Day Then For Each Financial Product, do the following:
//  4. Read FC_DT/FC_HL/FC_PF Data Files
//  5. Create TSCNON and initialize all Lee-oscillators
//  6. Register Time Series Data
//  7. TSCNON TRAINING with Time Series data and QPLs input data using TSCNON Training Alg
//  8. TSCNON FORECAST, using the past 5 trading days time series + QPLs to  predict today
//  9. Check Forecast Result within Range or NOT
//  10.FORECAST COMPLETED - Store Forecast Results and Total Time Taken
//
//*********************************************************************************************************
#include <QFSDKv1.mqh>
#property copyright "Copyright © 2019, DR. RAYMOND LEE"
#property link     http://QFFC.ORG
```

Fig. 11.5 Main program logic of TSCNON.mq4 (*Source* Dr. Lee's illustration, 2019)

Fig. 11.6 System flowchart
of TSCNON system (*Source*
Dr. Lee's illustration, 2019)

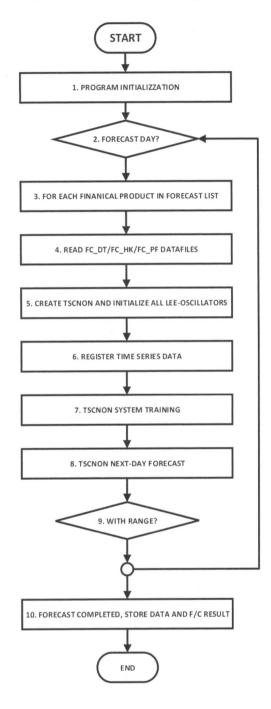

1. Visit QFFC.org official site to download QFSDK;
2. Copy QFSDKv1.dll (or latest version) library file into your MT4 Libraries folder such as "MetaQuotes\Terminal\..\MQL4\Libraries";
3. Copy QFSDKv1.mqh (or latest version) header file into your MT4 Include directory such as "\MetaQuotes\Terminal\..\MQL4\Include";
4. In users' individual TSCNON.mq4 program, add #include <QFSDKv1.mqh>, as shown in Fig. 11.5.

For the latest information about QFSDK, please visit QFFC.org official site.

In summary, TSCNON.mq4 is a real-time MQL-based multiple financial product daily forecast using TSCNON revealed in this chapter. It integrates QPL evaluation system studied in Chap. 10 with TSCNON network construction, training, and forecast algorithm into QFSDK function library. Quantum finance developers only need to follow this system flowchart and call these QFSDK functions to build their own real-time quantum finance forecast systems.

Figure 11.6 shows TSCNON system flowchart. The entire TSCNON consists of 15 steps of programming process. As the entire program consists of over 1000 program codes, we will only discuss the core concepts and program fragment of each function process. For full and complete MQL source code, please visit QFFC.org official site—Workshop #2 quantum finance forecast system.

11.4.3 Process #1—Program Initialization

11.4.3.1 Declaration of Data Files and Directories

The program initialization process consists of the following main modules:

1. declaration of data files and directories;
2. declaration of system parameters;
3. declaration of TSCNON forecast elements and variables;
4. creation of init() function class.

Actually, TSCNON is the first generation of quantum finance forecast system used in QFFC.org for real-time daily forecast and forecast performance analysis system of worldwide 129 financial products started from 2017. In other words, TSCNON not only performs the daily forecast but also analyses the previous trading day forecast performance, stored into the forecast performance data files for archive and upload to QFFC.org for public access. As shown in Fig. 11.7, four directories are used for data storage:

1. 120FC_DL stores the daily forecast results of 120 financial products that include forecasts high/low and the closest 8 QPLs, which will also upload to QFFC.org official site every morning. This directory contains two data files: FC_DAILY and PF_DAILY. FC_DAILY contains the daily forecast table and PF_DAILY contains the daily forecast performance table. They are used for daily forecast

```
//******************************************************************************
//
//       1.1 Declaration - datafiles and directory
//
//******************************************************************************
     string    DL_Directory   = "120FC_DL";        // FC DAILY Directory
     string    DT_Directory   = "120FC_DT";        // FC DATA  Directory
     string    HL_Directory   = "120FC_HL";        // FC HI/L0 Directory
     string    PF_Directory   = "120FC_PF";        // FC PERF  Directory
     string    NQPR_Directory = "QPL";             // NQPR Directory
     string    DL_FileName = "";                    // File name for FC_DL
     int       DL_FileHandle;                       // File Handle for FC_DL
     string    DT_FileName = "";                    // File name for FC_DT
     int       DT_FileHandle;                       // File Handle for FC_DT
     string    HL_FileName = "";                    // File name for FC_HL
     int       HL_FileHandle;                       // File Handle for FC_HL
     string    PF_FileName = "";                    // File name for FC_PF
     int       DPF_FileHandle;                      // File Handle for Daily FC_PF
     string    DPF_FileName = "";                   // File name for Daily FC_PF
     int       PF_FileHandle;                       // File Handle for FC_PF
     string    NQPR_FileName = "";                  // File name for NQPR
     int       NQPR_FileHandle;                     / File Handle for NQPR
```

Fig. 11.7 Declaration of data files and directories in TSCNON (*Source* Dr. Lee's illustration, 2019)

archive and upload onto the QFFC.org official site. Figure 11.8 shows the daily forecast table for the first 20 financial products on May 3, 2019. Figure 11.9 shows the daily forecast performance table for the top 20 financial products on May 2, 2019 (sorted by the % errors).

2. 120FC_DT stores the normalized daily time series (DT) data (i.e., Open, High, Low, and Close) of the past 500-trading days. Totally 120 data files with each containing the time series of one trading product. For example, "1 XAGUSD_DT" contains the (normalized) time series of Silver/USD, where "1" is the product code number.

Figure 11.10 shows the (normalized) daily time series of XAGUSD.

TP_No	TP_Code2	FC_LOW	TP_LOW	SL_LOW	FC_HIGH	TP_HIGH	SL_HIGH	QPL#1	QPL#2	QPL#3	QPL#4	QPL#5	QPL#6	QPL#7	QPL#8
1 XAGUSD	14.341	14.641	14.041	15.252	14.952	15.552	14.242	14.367	14.48	14.573	14.653	14.747	14.863	14.994
2 CORN	364.17	367.17	361.17	368.78	365.78	371.78	358.26	361.13	363.72	365.84	367.66	369.8	372.46	375.44
3 US30	25472.7	25592.7	25352.7	26320.8	26200.8	26440.8	25843.4	25969.6	26083.1	26175.3	26254.2	26346.9	26462.1	26591.3
4 AUDUSD	0.67173	0.67473	0.66873	0.72421	0.72121	0.72721	0.69259	0.69516	0.69746	0.69932	0.70092	0.70279	0.70512	0.70773
5 EURCHF	1.13385	1.13685	1.13085	1.14357	1.14057	1.14657	1.12908	1.1323	1.13519	1.13753	1.13953	1.14188	1.14479	1.14806
6 GBPCAD	1.72646	1.72946	1.72346	1.78604	1.78304	1.78904	1.73691	1.74302	1.7485	1.75294	1.75676	1.76123	1.76676	1.77298
7 NZDJPY	71.254	71.554	70.954	74.947	74.647	75.247	72.722	73.055	73.354	73.597	73.805	74.049	74.352	74.693
8 USDCNH	6.73425	6.73725	6.73125	6.77611	6.77311	6.77911	6.71674	6.7267	6.7356	6.74279	6.74893	6.75613	6.76507	6.77511
9 XAUAUD	1795.92	1798.92	1792.92	1827.68	1824.68	1830.68	1788.21	1797.33	1805.63	1812.61	1818.8	1825.82	1834.25	1843.62
10 XAUCHF	1277.3	1280.3	1274.3	1295.38	1292.38	1298.38	1273.1	1280.29	1286.75	1291.98	1296.44	1301.72	1308.28	1315.67
11 XAUEUR	1133.98	1136.98	1130.98	1152.12	1149.12	1155.12	1118.84	1124.85	1130.26	1134.66	1138.43	1142.85	1148.35	1154.52
12 XAUGBP	968.81	971.81	965.81	979.77	976.77	982.77	959.02	964.61	969.64	973.72	977.22	981.33	986.45	992.2
13 XAUJPY	140912	141212	140612	141776	141476	142076	139514	140231	140878	141404	141856	142386	143043	143778
14 XAUUSD	1262.6	1265.6	1259.6	1273.31	1270.31	1276.31	1252.23	1258.57	1264.33	1269.17	1273.47	1278.35	1284.2	1290.7
15 COPPER	271.42	274.42	268.42	278.44	275.44	281.44	272.13	274.14	275.94	277.42	278.68	280.17	282.02	284.09
16 PALLAD	1323.66	1326.66	1320.66	1353.58	1350.58	1356.58	1303.03	1315.66	1327.13	1336.54	1344.66	1354.18	1365.99	1379.23
17 PLAT	853.06	856.06	850.06	858.14	855.14	861.14	820.65	832.06	842.51	851.15	858.67	867.47	878.36	890.58
18 UK_OIL	6911	6971	6851	7014	6954	7074	6758	6834	6903	6960	7010	7067	7139	7220
19 US_OIL	58.56	58.86	58.26	63.99	63.69	64.29	59.06	59.73	60.35	60.85	61.29	61.8	62.44	63.15
20 US_NATG	2.502	2.532	2.472	2.606	2.576	2.636	2.406	2.458	2.506	2.546	2.58	2.621	2.673	2.73

Fig. 11.8 FC_DAILY—TSCNON daily forecast table of first 20 financial products on May 3, 2019 (*Source* Dr. Lee's illustration, 2019)

Fig. 11.9 PF_DAILY—TSCNON daily forecast performance table of the top 20 financial products on May 2, 2019 (*Source* Dr. Lee's illustration, 2019)

Fig. 11.10 Normalized daily time series of XAGUSD (*Source* Dr. Lee's illustration, 2019)

	A	B	C	D	E	F	G
1	2019	5	2	0.24532	0.24682	0.24253	0.24355
2	2019	5	1	0.24875	0.24943	0.24453	0.24528
3	2019	4	30	0.24802	0.25	0.24673	0.24875
4	2019	4	29	0.25085	0.25095	0.24762	0.24803
5	2019	4	26	0.24883	0.25117	0.24815	0.25077
6	2019	4	25	0.2494	0.25002	0.24747	0.24883
7	2019	4	24	0.24658	0.24957	0.24623	0.2494
8	2019	4	23	0.24978	0.2504	0.24555	0.24658
9	2019	4	22	0.24965	0.25113	0.24965	0.24977
10	2019	4	19	0.24968	0.24987	0.24927	0.24967
11	2019	4	18	0.24907	0.25028	0.24855	0.2497
12	2019	4	17	0.24948	0.25143	0.24897	0.24908
13	2019	4	16	0.24968	0.2504	0.2478	0.24947
14	2019	4	15	0.24962	0.24992	0.24732	0.2497
15	2019	4	12	0.2501	0.25172	0.24807	0.2499
16	2019	4	11	0.254	0.2547	0.24973	0.25013
17	2019	4	10	0.25472	0.25482	0.2523	0.25402
18	2019	4	9	0.25348	0.2555	0.25293	0.25472
19	2019	4	8	0.25168	0.2548	0.25165	0.25347
20	2019	4	5	0.2512	0.2533	0.25085	0.25202

3. 120FC_HL contains previous 500-trading days forecast High/Low of each financial product. They are used for forecast archive and daily performance analysis. Figure 11.11 shows the normalized forecast H/L of XAGUSD.

4. 120FC_PF contains daily forecast performance data files for each of the 120 financial products. As mentioned, TSCNON not only performs the daily forecast, but also performs the previous-day forecast performance analysis by calculating the forecast errors between previous-day forecast H/L with the actual H/L, stores into the 120FC_PF directory for the archive. Figure 11.12 shows the daily forecast performance table of XAGUSD. In which the first three columns are year, month, and day, columns D–G contain the actual Open, High, Low, and Close, columns H and I contain the forecast High and Low, columns J & K contain the forecast errors for High/Low and columns L and M contain their % errors.

	A	B	C	D	E
1	2019	5	3	15.252	14.341
2	2019	5	2	15.091	13.926
3	2019	5	1	15.827	14.479
4	2019	4	30	15.988	14.481
5	2019	4	29	16.159	14.656
6	2019	4	26	15.56	14.195
7	2019	4	25	15.691	14.657
8	2019	4	24	14.868	14.397
9	2019	4	23	15.486	14.424
10	2019	4	22	15.243	14.945
11	2019	4	19	15.684	14.427
12	2019	4	18	16.015	14.518
13	2019	4	17	16.466	14.514
14	2019	4	16	15.931	14.903
15	2019	4	15	15.359	14.615
16	2019	4	12	16.045	14.622
17	2019	4	11	15.782	14.778
18	2019	4	10	16.348	15.017
19	2019	4	9	15.509	14.334
20	2019	4	8	15.617	13.999

Fig. 11.11 Normalized forecast H/L of XAGUSD (*Source* Dr. Lee's illustration, 2019)

	A	B	C	D	E	F	G	H	I	J	K	L	M
1	2019	5	2	14.719	14.809	14.552	14.613	15.091	13.926	0.282	0.626	0.454	0.03107
2	2019	5	1	14.925	14.966	14.672	14.717	15.827	14.479	0.861	0.193	0.527	0.03581
3	2019	4	30	14.881	15	14.804	14.925	15.988	14.481	0.988	0.323	0.6555	0.04392
4	2019	4	29	15.051	15.057	14.857	14.882	16.159	14.656	1.102	0.201	0.6515	0.04378
5	2019	4	26	14.93	15.07	14.889	15.046	15.56	14.195	0.49	0.694	0.592	0.03935
6	2019	4	25	14.964	15.001	14.848	14.93	15.691	14.657	0.69	0.191	0.4405	0.0295
7	2019	4	24	14.795	14.974	14.774	14.964	14.868	14.397	0.106	0.377	0.2415	0.01614
8	2019	4	23	14.987	15.024	14.733	14.795	15.486	14.424	0.462	0.309	0.3855	0.02606
9	2019	4	22	14.979	15.068	14.979	14.986	15.243	14.945	0.175	0.034	0.1045	0.00697
10	2019	4	19	14.981	14.992	14.956	14.98	15.684	14.427	0.692	0.529	0.6105	0.04075
11	2019	4	18	14.944	15.017	14.913	14.982	16.015	14.518	0.998	0.395	0.6965	0.04649
12	2019	4	17	14.969	15.086	14.938	14.945	16.466	14.514	1.38	0.424	0.902	0.06035
13	2019	4	16	14.981	15.024	14.868	14.968	15.931	14.903	0.907	0.035	0.471	0.03147
14	2019	4	15	14.977	14.995	14.839	14.982	15.359	14.615	0.364	0.224	0.294	0.01962
15	2019	4	12	15.006	15.103	14.884	14.994	16.045	14.622	0.942	0.262	0.602	0.04015
16	2019	4	11	15.24	15.282	14.984	15.008	15.782	14.778	0.5	0.206	0.353	0.02352
17	2019	4	10	15.283	15.289	15.138	15.241	16.348	15.017	1.059	0.121	0.59	0.03871
18	2019	4	9	15.209	15.33	15.176	15.283	15.509	14.334	0.179	0.842	0.5105	0.0334
19	2019	4	8	15.101	15.288	15.099	15.208	15.617	13.999	0.329	11	0.7145	0.04698
20	2019	4	5	15.072	15.198	15.051	15.121	15.691	13.885	0.493	1.166	0.8295	0.05486

Fig. 11.12 Daily forecast performance table of XAGUSD (*Source* Dr. Lee's illustration, 2019)

11.4.3.2 Declaration of System Parameters

This section declares the key parameters for the TSCNON system, they include the following:

- number of network layers;
- number of Lee-oscillators input, hidden and output layers;
- maximum training iterations (in case of deadlock);

- training and time series sizes;
- target RMSE (root-mean-square error);
- trading Product (TP) parameters include: TP code (only shown 5 for illustration purpose); TP code number (TP_No); TP number of digits (TP_nD); TP normalization value (TP_Norm); TP daily range between daily H/L (TP_range); stop loss (TP_SL), and target_price (TP_TP).

Figure 11.13 shows the declaration section of system parameters of TSCNON program.

```
//********************************************************************************
//
//              1.2 Declaration - TSCNON System Parameters
//
//********************************************************************************
int     input_node_num = 41;                    // Number of input neurons
int     hidden_layer_num = 4;                   // Number of hidden layer(s)
int     hidden_node_num = 3;                    // Number of  neurons for each hidden layer
int     output_node_num = 4;                    // Number of outputs
double  learning_rate = 0.001;                  // Learning rate
double  neuron_bias = 0.1;                       // Neuron_bias
int     epoch = 50;                             // Training times of each training input
int     data_len = 744;                         // The length of each training input
string  trainFile = "";                         // Training file
neuron input_layer[input_node_num];             //  Create input layer
neuron hidden_layer[hidden_layer_num*hidden_node_num];  //  Create hidden layer
neuron output_layer[output_node_num];           //  Create output layer

double  trainingData[][45];                     // IMPORTANT! size = nn_input + nn_output
int     maxTraining = 2000;                     // maximum number of time we will train
double  targetRMSE   = 0.000001                 // Set the MSE to 1x10-6
int     TRAIN_SIZE  = 400;                      // USE 500 TRAINING RECORDS
int     TSDATA_SIZE = 505;                      // USE PAST 505 TIME SERIES DATA RECORDS
uint    stime=0;                                // Forecast Start Time
uint    etime=0;                                // Forecast End Time
uint    Gstime=0;                               // System Start Time
uint    Getime=0;                               // System End Time
uint    tlapse=0;                               // Forecast Time Lapse
uint    Gtlapse=0;                              // System Time Lapse
int     nTraining=0;                            // Number of training
int     nTP=0;                                  // Product number
string  TPSymbol  = "";                         // Current Trading duct Symbol
string  TP_Code[5]={ "XAGUSD",  "CORN",   "US30","AUDUSD","EURCHF"}; // Product codes
int     TP_No[5]={1,2,3,4,5};                   // Product code no
int     TP_nD[5]={3,2,1,5,5};                   // Product decimal digits
double  TP_Norm[5]={60,650,50000,2,2.5};        // Product normalization value
double  TP_Range[5]={1.2,25.00,900,0.04,0.04};  // Product daily range
double  TP_SL[5]={0.3,3,120,0.003,0.003};       // Stop-loss
double  TP_TP[5]={0.3,3,120,0.003,0.003};       // Target-profit
int     TSCNON[120];                            // Identifier of the TSCNON
int     MAX_TP=5;                               // Max NO OF Forecasting TPs
bool    bPass=false;                            // Check for forecast day
int     nRound=0;                               // Number of round
int     today;                                  // Today
int     DoW;                                    // Date-of-Week
bool    bWorkDay;                               // Check working day
bool    b_Forecast;                             // Check forecast status
bool    b_PF;                                   // Check Perforance Analysis status
bool    b_Skip;                                 // Check day skip
```

Fig. 11.13 Declaration of system parameters (*Source* Dr. Lee's illustration, 2019)

11.4.3.3 Declaration of TSCNON Forecast Elements and Variables

This section declares the forecast elements and variables include: previous-day record elements for Open, High, Low, and Close (OHLC); today forecast elements for OHLC and QPL variables for the 8 closest QPLs as in Fig. 11.14.

11.4.3.4 Declaration of Variables in Init() Function Class

Declaration of variables in init() include the creation of init() and the declaration of all variables and arrays used in the program. As mentioned previously, init() instead of start() class is used as the forecast system is a "one-off" real-time system being

```
//*************************************************************************************
//
//              1.3 Declare Daily Forecast elements
//
//*************************************************************************************
int     D1_YEAR;
int     D1_MONTH;
int     D1_DAY;
double  D0_OPEN;
double  D1_OPEN;
double  D1_HIGH;
double  D1_LOW;
double  D1_CLOSE;
double  FC_AOPEN;
double  FC_AHIGH;
double  FC_ALOW;
double  FC_ACLOSE;
double  FC_AJOPEN;
double  FC_AJHIGH;
double  FC_AJLOW;
double  FC_AJCLOSE;
double  FC_Diff;
double  FC_NOPEN;
double  FC_NHIGH;
double  FC_NLOW;
double  FC_NCLOSE;
double  FC_RANGE;
double  MSE;
bool    bMatch;

// Daily Forecast Display Items
string  PName_EN;
string  PName_CH;
string  PCode;
string  PCat_EN;
string  PCat_CH;
double  TP_Low;
double  SL_Low;
double  TP_High;
double  SL_High;
double  FC_MID;
double  QPL1;
double  QPL2;
double  QPL3;
double  QPL4;
double  QPL5;
double  QPL6;
double  QPL7;
double  QPL8;
bool    bQPLFound;
```

Fig. 11.14 Declaration of forecast elements and variables (*Source* Dr. Lee's illustration, 2019)

used only once every morning of trading day; instead of trading program, which will be operated continuously throughout the entire trading day. Figure 11.15 shows the variables and arrays declaration section of the init() class.

11.4.4 Process #2—Check for Forecast Day

As daily forecast will only perform during trading days from Monday to Friday (even though there are half-day trading in Saturday, but normally will not perform daily forecast as the half-day time period is too short for actual prediction).

This process will check whether today is a weekday (i.e., Mon–Fri) or not. If it is a weekday, b_Forecast will be set to *TRUE* and the system will read FX_NQPR.csv data file. Actually, this file is the implementation results of the QPL evaluation program (QPL2019.mq4) revealed in Chap. 10 for the QPLs evaluation of worldwide financial products, and NQPR.csv contains 21 NQPRs of 120 financial products.

By multiplying the *opening price* for any trading day, users can calculate the 21 closest QPLs on both sides of the *opening price* for TSCNON network training and the closest 8 QPLs evaluation for daily information update in the QFFC.org official site.

In addition, this section also prepares the file header for daily forecast file (FC_DAILY) and daily performance file (PF_DAILY), respectively. Figure 11.16 shows the program section of the checking for forecast day.

11.4.5 Process #3—Loop over ALL Trading Products

This section will start the looping of ALL trading product data acquisition and forecasting. For every financial market, two cases exist. Either today *day bar* exists or not. If it does not exist, we set yesterday trading day closing as today opening to calculate the 8 closest QPLs and forecast result adjustment. In either case, the system will read and store the previous-day time series (i.e., O, H, L, C) for today forecast, archive, and calculation of yesterday performance (Fig. 11.17).

11.4.6 Process #4—Read FC_DT/HL/PF Data Files

This section will read three data source of previous time series, which are as follows:

- FC_DT—to read the previous 500-trading day normalized time series data, together with the previous-day OHLC for today forecast;

```
//*********************************************************************************************
//
//              1.4 Program Init() Class
//
//********************************************************************************************int
init() {
    int     i,d,t,k,q,nL,nt,nD,ToW;

    // Declare Read DT Data
    int     RDT_YY;
    int     RDT_MM;
    int     RDT_DD;
    double  RDT_OP;
    double  RDT_HI;
    double  RDT_LO;
    double  RDT_CL;

    // Declare Read HL Data
    int     RHL_YY;
    int     RHL_MM;
    int     RHL_DD;
    double  RHL_HI;
    double  RHL_LO;

    // Declare Read PF Data
    int     RPF_YY;
    int     RPF_MM;
    int     RPF_DD;
    double  RPF_OP;
    double  RPF_HI;
    double  RPF_LO;
    double  RPF_CL;
    double  RPF_FH;
    double  RPF_FL;
    double  RPF_EH;
    double  RPF_EL;
    double  RPF_AE;
    double  RPF_PE;

    // Declare FC_DT Arrays
    int     DT_YY[505];
    int     DT_MM[505];
    int     DT_DD[505];
    double  DT_OP[505];
    double  DT_HI[505];
    double  DT_LO[505];
    double  DT_CL[505];

    // Declare FC_HL Arrays
    int     HL_YY[505];
    int     HL_MM[505];
    int     HL_DD[505];
    double  HL_HI[505];
    double  HL_LO[505];

    // Declare FC_PF Arrays
    int     PF_YY[505];
    int     PF_MM[505];
    int     PF_DD[505];
    double  PF_OP[505];
    double  PF_HI[505];
    double  PF_LO[505];
    double  PF_CL[505];
    double  PF_FH[505];
    double  PF_FL[505];
    double  PF_EH[505];
    double  PF_EL[505];
    double  PF_AE[505];
    double  PF_PE[505] ;

    // Declare 21 QPLs for network training
    double  T_QPL[21];
```

Fig. 11.15 Declaration of variables in the init() (*Source* Dr. Lee's illustration, 2019)

```
//***************************************************************************************
//
//    2. Check for Forecast day
//
//***************************************************************************************
// Declare NQPR AND QPL ARRAYS
double    NQPR[5][21];  // Array to store ALL NQPR (each 21 energy levels)

// Set Global Start Time
Gstime     = GetTickCount();

// Get Current Day
today      = TimeDay(TimeLocal());
DoW        = TimeDayOfWeek(TimeLocal());

// Check Whether Need to Forecast or NOT
b_Forecast   = false;
if ((DoW>0) && (DoW<6))
{
   // Mon - Fri -> Forecast
   b_Forecast = true;
}

// Prepare for Foreast if needed
if (b_Forecast)
{
   // Read ALL NQPR into array NQPR[][]
   ResetLastError();
   NQPR_FileName    = "FX_NQPR.csv";
   NQPR_FileHandle  = FileO-
pen(NQPR_Directory+"//"+NQPR_FileName,FILE_COMMON|FILE_READ|FILE_CSV,',');

   if(NQPR_FileHandle!=INVALID_HANDLE)
   {
      for (nTP=0;nTP<MAX_TP;nTP++)
      {
         // Get ALL the 21 NQPR for the first 20 Energy Levels
         for (nL=0;nL<21;nL++)
         {
            NQPR[nTP][nL] = StringToDouble (FileReadString(NQPR_FileHandle));
         }
      } // Loop over ALL products
   } // Valid NQPR File Handle
   // Close NQPR File
   FileClose(NQPR_FileHandle);

   // OPEN THE DAILY FOECAST DATAFILE
   DL_FileName    = "FC_DAILY.csv";
   FileDelete(DL_Directory+"//"+DL_FileName,FILE_COMMON);
   ResetLastError();

   // CREATE FC_DL HANDLE
   DL_FileHandle   = FileO-
pen(DL_Directory+"//"+DL_FileName,FILE_COMMON|FILE_READ|FILE_WRITE|FILE_CSV,',');

   // Write Header Line
   File-
Write(DL_FileHandle,"TP_No","TP_Code2","FC_LOW","TP_LOW","SL_LOW","FC_HIGH","TP_HIGH","SL_HIGH",
      "QPL#1","QPL#2","QPL#3","QPL#4","QPL#5","QPL#6","QPL#7","QPL#8");
} // b_Forecast is TRUE

// CREATE Daily DPF_HANDLE
DPF_FileName    = "PF_DAILY.csv";
FileDelete(DL_Directory+"//"+DPF_FileName,FILE_COMMON);
ResetLastError();
DPF_FileHandle  = FileO-
pen(DL_Directory+"//"+DPF_FileName,FILE_COMMON|FILE_READ|FILE_WRITE|FILE_CSV,',');

// Write Header Line
FileWrite(DPF_FileHandle,"YEAR","MONTH","DAY",
               "TP_No","TP_CODE2","TP_CCat","TP_ECat","TP_CNAME","TP_ESNAME","TP_ELNAME","Digit",
               "OPEN","HIGH","LOW","CLOSE",
               "FC_HIGH","FC_LOW",
               "ERR_HIGH","ERR_LOW","AV_ERR","%ERROR");
```

Fig. 11.16 Process #2—Check for forecast day (*Source* Dr. Lee's illustration, 2019)

```
//*************************************************************************************
//
//  3. LOOP OVER ALL TP
//
//*************************************************************************************
for (nTP=0;nTP<MAX_TP;nTP++)
{
  TPSymbol   = TP_Code[nTP];
  stime      = GetTickCount();
  nTraining  = 0;
  nt         = 0;
  b_Skip     = false;
  // Check whether the current D1 Bar is today
  if (TimeDay(iTime(TPSymbol,PERIOD_D1,0)) == today)
  {  // Check previous D1 bar is Mon-Fri or NOT
    nD  = 1;
    bWorkDay = false;
    while (!bWorkDay)
    {
      ToW =TimeDayOfWeek(iTime(TPSymbol,PERIOD_D1,nD));
      if ((ToW>0) && (ToW<6))
      {
        bWorkDay = true;
      }else{
        b_Skip = true;
        nD++;
      }
    }
    // Today D1 bar exist, get Today OPEN, D-1 OPEN HIGH LOW CLOSE VOL
    D0_OPEN   = iOpen(TPSymbol,PERIOD_D1,0);
    D1_YEAR   = TimeYear(iTime(TPSymbol,PERIOD_D1,nD));
    D1_MONTH  = TimeMonth(iTime(TPSymbol,PERIOD_D1,nD));
    D1_DAY    = TimeDay(iTime(TPSymbol,PERIOD_D1,nD));
    D1_OPEN   = iOpen(TPSymbol,PERIOD_D1,nD);
    D1_HIGH   = iHigh(TPSymbol,PERIOD_D1,nD);
    D1_LOW    = iLow(TPSymbol,PERIOD_D1,nD);
    D1_CLOSE  = iClose(TPSymbol,PERIOD_D1,nD);
  }else
  {
    // Check previous D1 bar is Mon-Fri or NOT
    nD  = 0;
    bWorkDay = false;
    while (!bWorkDay)
    {
      ToW =TimeDayOfWeek(iTime(TPSymbol,PERIOD_D1,nD));
      if ((ToW>0) && (ToW<6))
      {
        bWorkDay = true;
      }else{
        b_Skip = true;
        nD++;
      }
    }
    // Today D1 Bar does not exist, get yesterday data and set today open as yesterday close
    D1_YEAR   = TimeYear(iTime(TPSymbol,PERIOD_D1,nD));
    D1_MONTH  = TimeMonth(iTime(TPSymbol,PERIOD_D1,nD));
    D1_DAY    = TimeDay(iTime(TPSymbol,PERIOD_D1,nD));
    D1_OPEN   = iOpen(TPSymbol,PERIOD_D1,nD);
    D1_HIGH   = iHigh(TPSymbol,PERIOD_D1,nD);
    D1_LOW    = iLow(TPSymbol,PERIOD_D1,nD);
    D1_CLOSE  = iClose(TPSymbol,PERIOD_D1,nD);
    D0_OPEN   = D1_CLOSE;
  }
```

Fig. 11.17 Process #3—Loop over all trading products (*Source* Dr. Lee's illustration, 2019)

- FC_HL—to read the previous 500-trading day forecast H/L data, together with the previous-day actual H/L results to calculate yesterday forecast performance and stored into FC_PF file. The system will also store today's forecast results after daily forecast;
- FC_PF—to open the previous 500-trading day daily performance data, and store yesterday forecast performance results.

11.4.6.1   Process #4.1—Read FC_DT Data Files

This process will check whether the first data record is the current *D1 Bar*. If yes, there is no need to store it into the data array. In either case, the system reads all 500 previous-day time series into the DT data array for network training (Fig. 11.18).

11.4.6.2   Process #4.2—Write FC_DT Data File if b_PF Is TRUE

This section checks whether b_PF is true or not. If it is true, record the previous-day time series into the FC_DT data file. Figure 11.19 shows the program code of Process #4.2.

11.4.6.3   Process #4.3—Read FC_HL Data Files

This section will read FC_HL data file, which stores the daily forecast H/L of the previous 500-trading day (Fig. 11.20).

11.4.6.4   Process #4.4—Read FC_PF Data Files

This section will read the FC_PF data file. This data file contains the previous 500-trading day daily forecast performance results (Fig. 11.21).

11.4.6.5   Process #4.5—Write FC_PF Data File if bWrite Is TRUE

This section will check whether bWrite is TRUE. If it is true, calculate the yesterday forecast H/L errors, overall % errors, and store them into the FC_PF data file (Fig.11.22).

```
//******************************************************************************************
//
//   4.1 READ FC_DT FILE
//
//******************************************************************************************
Print("#",nTP," ",TPSymbol,": 1.1 READING FC_DT FILE ...");
ResetLastError();
DT_FileName    = TP_No[nTP]+" "+TPSymbol+"_DT.csv";
DT_FileHandle  = FileOpen(DT_Directory+"//"+DT_FileName,FILE_COMMON|FILE_READ|FILE_CSV,',');
Print("#",nTP," ",TPSymbol,": 1.2 READING FC_HL FILE ...");
ResetLastError();
HL_FileName    = TP_No[nTP]+" "+TPSymbol+"_HL.csv";
HL_FileHandle  = FileOpen(HL_Directory+"//"+HL_FileName,FILE_COMMON|FILE_READ|FILE_CSV,',');
Print("#",nTP," ",TPSymbol,": 1.3 READING FC_PF FILE ...");
ResetLastError();
PF_FileName    = TP_No[nTP]+" "+TPSymbol+"_PF.csv";
PF_FileHandle  = FileOpen(PF_Directory+"//"+PF_FileName,FILE_COMMON|FILE_READ|FILE_CSV,',')
// READ FC_DT[]
if(DT_FileHandle!=INVALID_HANDLE)
{
  for (t=0;t<TSDATA_SIZE;t++)
  {
    // READ FC_DT DATA
    RDT_YY = StringToInteger(FileReadString(DT_FileHandle));
    RDT_MM = StringToInteger(FileReadString(DT_FileHandle));
    RDT_DD = StringToInteger(FileReadString(DT_FileHandle));
    RDT_OP = StringToDouble (FileReadString(DT_FileHandle));
    RDT_HI = StringToDouble (FileReadString(DT_FileHandle));
    RDT_LO = StringToDouble (FileReadString(DT_FileHandle));
    RDT_CL = StringToDouble (FileReadString(DT_FileHandle));
    // CHECK WHETHER THE FIRST RECORD IS LATEST D1 BAR VALUES
    if (t==0)
    {
      if ((D1_DAY == RDT_DD)||(DoW<2)||(b_Skip))
      {
        //FIRST DATA RECORD IS THE CURRENT D1 BAR, NO NEED TO STORE INTO ARRAY
        nt    = 0;
        b_PF  = false;
      }else
      {
        //FIRST DATA RECORD IS NOT THE CURRENT D1 BAR, STORE VALUE INTO FIRST ARRAY ELEMENT
        nt    = 1;
        b_PF  = true;

        // STORE ALL THE BAR VALUES INTO TS_[0]
        DT_YY[0]  = D1_YEAR;
        DT_MM[0]  = D1_MONTH;
        DT_DD[0]  = D1_DAY;
        DT_OP[0]  = (double)D1_OPEN/TP_Norm[nTP];
        DT_HI[0]  = (double)D1_HIGH/TP_Norm[nTP];
        DT_LO[0]  = (double)D1_LOW/TP_Norm[nTP];
        DT_CL[0]  = (double)D1_CLOSE/TP_Norm[nTP];
      }
    } // Check first record
    // Set the Array Values
    if (nt<TSDATA_SIZE)
    {
      DT_YY[nt]  = RDT_YY;
      DT_MM[nt]  = RDT_MM;
      DT_DD[nt]  = RDT_DD;
      DT_OP[nt]  = RDT_OP;
      DT_HI[nt]  = RDT_HI;
      DT_LO[nt]  = RDT_LO;
      DT_CL[nt]  = RDT_CL;
    } //if (nt<TSDATA_SIZE)

    // Upcount nt
    nt++;
  } // READING FC_DT DATA RECORDS
  // Print("Reading ",TPSymbol,"READING FC_DT DATA FILE COMPLETED !!!");
}else
{
  PrintFormat("Failed to open %s file, Error code = %d",DT_FileName,GetLastError());
  Print("Reading FC_DT DataFile ERROR!!!");
} // CHECK DT_FILEHANDLE
// CLOSE TIME SERIES DATA FILE
FileClose(DT_FileHandle);
```

Fig. 11.18 Process #4.1—read FC_DT file (*Source* Dr. Lee's illustration, 2019)

```
//***********************************************************************************************
//
//   4.2 WRITE FC_DT DATA FILE IF b_PF IS TRUE
//
//***********************************************************************************************

    if (b_PF==true)
    {
      Print("#",nTP," : ",TPSymbol,": 1.1 WRITING FC_DT DATA FILE ...");
      FileDelete(DT_Directory+"//"+DT_FileName,FILE_COMMON);
      ResetLastError();
      DT_FileHandle   = FileO-
pen(DT_Directory+"//"+DT_FileName,FILE_COMMON|FILE_READ|FILE_WRITE|FILE_CSV,',');

      if(DT_FileHandle!=INVALID_HANDLE)
      {
        for (t=0;t<TSDATA_SIZE;t++)
        {
          FileWrite(DT_FileHandle,DT_YY[t],DT_MM[t],DT_DD[t],
            DoubleToString(DT_OP[t],5), DoubleToString(DT_HI[t],5), DoubleToString(DT_LO[t],5), DoubleTo-
String(DT_CL[t],5));
        }
      } // TS_FileHandle

      // CLOSE TIME SERIES DATA FILE
      FileClose(DT_FileHandle);
    } //CHECK bWriteTS
```

Fig. 11.19 Process #4.2 write FC_DT data file if b_PF is TRUE (*Source* Dr. Lee's illustration, 2019)

11.4.7 Process #5—Create TSCNON and Initialize ALL LEE-Oscillators

By using QFSDK, this process creates the TSCNON network and initial all the Lee-oscillator in different network layers (Fig. 11.23).

11.4.8 Process #6—Register Time Series Data

This process calls prepareData() function to register the 500 time series input data one-by-one. Also, it makes use of NQPR array to generate all 21 QPLs for network inputs (Fig. 11.24).

11.4.9 Process #7—TSCNON Training

This process performs network training with target RMSE defined in the declaration section (Fig. 11.25).

```
//***********************************************************************************************
//
//   4.3 READ FC_HL DATA FILE
//
//***********************************************************************************************

if(HL_FileHandle!=INVALID_HANDLE)
{
  for (t=0;t<TSDATA_SIZE;t++)
  {
    // READ FC_HL DATA
    RHL_YY = StringToInteger(FileReadString(HL_FileHandle));
    RHL_MM = StringToInteger(FileReadString(HL_FileHandle));
    RHL_DD = StringToInteger(FileReadString(HL_FileHandle));
    RHL_HI = StringToDouble (FileReadString(HL_FileHandle));
    RHL_LO = StringToDouble (FileReadString(HL_FileHandle));

    // CHECK WHETHER IT IS WORKING DAY OR THE FIRST RECORD IS TODAY
    if (t==0)
    {
      if ((RHL_DD == today)|| (TimeDayOfWeek(TimeLocal()) == 0) || (TimeDayOfWeek(TimeLocal())== 6))
      {
        nt      = 0;
      }else
      {
        nt      = 1;
        // STORE ALL THE BAR VALUES INTO TS_[0]
        HL_YY[0]  = TimeYear(TimeLocal());
        HL_MM[0]  = TimeMonth(TimeLocal());
        HL_DD[0]  = TimeDay(TimeLocal());
      }
    } // Check first record

    // Set the Array Values
    if (nt<TSDATA_SIZE)
    {
      HL_YY[nt]  = RHL_YY;
      HL_MM[nt]  = RHL_MM;
      HL_DD[nt]  = RHL_DD;
      HL_HI[nt]  = RHL_HI;
      HL_LO[nt]  = RHL_LO;
    } //if (nt<TSDATA_SIZE)

    // Upcount nt
    nt++;
  } // READING FC_HL DATA RECORDS

  // Print("Reading ",TPSymbol,"READING FC_HL DATA FILE COMPLETED !!!");

}else
{
  PrintFormat("Failed to open %s file, Error code = %d",HL_FileName,GetLastError());
  Print("Reading FC_HL DataFile ERROR!!!");
} // CHECK HL_FILEHANDLE

// CLOSE FC_HL DATA FILE
FileClose(HL_FileHandle);
```

Fig. 11.20 Process #4.3 read FC_HL data file (*Source* Dr. Lee's illustration, 2019)

11.4.10   Process #8—TSCNON Forecast

This section registers the past 5-trading day time series and 21 QPLs to for next-day forecast (Fig. 11.26).

```
//****************************************************************************
//
//  4.4 READ FC_PF DATA FILE
//
//****************************************************************************
if(PF_FileHandle!=INVALID_HANDLE)
{
  for (t=0;t<TSDATA_SIZE;t++)
  {
    // READ FC_DT DATA
    RPF_YY = StringToInteger(FileReadString(PF_FileHandle));
    RPF_MM = StringToInteger(FileReadString(PF_FileHandle));
    RPF_DD = StringToInteger(FileReadString(PF_FileHandle));
    RPF_OP = StringToDouble (FileReadString(PF_FileHandle));
    RPF_HI = StringToDouble (FileReadString(PF_FileHandle));
    RPF_LO = StringToDouble (FileReadString(PF_FileHandle));
    RPF_CL = StringToDouble (FileReadString(PF_FileHandle));
    RPF_FH = StringToDouble (FileReadString(PF_FileHandle));
    RPF_FL = StringToDouble (FileReadString(PF_FileHandle));
    RPF_EH = StringToDouble (FileReadString(PF_FileHandle));
    RPF_EL = StringToDouble (FileReadString(PF_FileHandle));
    RPF_AE = StringToDouble (FileReadString(PF_FileHandle));
    RPF_PE = StringToDouble (FileReadString(PF_FileHandle));

    // CHECK WHETHER THE FIRST RECORD IS LATEST D1 BAR VALUES
    if (t==0)
    {
      if (!b_PF)
      {
        //FIRST DATA RECORD IS THE CURRENT D1 BAR, NO NEED TO STORE INTO ARRAY
        nt     = 0;
      }else
      {
        //FIRST DATA RECORD IS NOT THE CURRENT D1 BAR, STORE VALUE INTO FIRST ARRAY ELEMENT
        nt     = 1;
        // STORE ALL THE BAR VALUES INTO TS_[0]
        PF_YY[0]  = D1_YEAR;
        PF_MM[0]  = D1_MONTH;
        PF_DD[0]  = D1_DAY;
        PF_OP[0]  = (double)D1_OPEN;
        PF_HI[0]  = (double)D1_HIGH;
        PF_LO[0]  = (double)D1_LOW;
        PF_CL[0]  = (double)D1_CLOSE;
        // CHECK WITH HL_YY, HL_MM, HL_DD TO LOCATE HL_HI & HL_LO
        bMatch = false;
        k     = 0;
        while ((!bMatch)&&(k<10))
        {
          if ((PF_YY[0]==HL_YY[k])&&(PF_MM[0]==HL_MM[k])&&(PF_DD[0]==HL_DD[k]))
          {
            bMatch = true;
            // CALCULATE THE PF_EH, PF_EL, PF_AE, PF_PE
            PF_FH[0]  = HL_HI[k];
            PF_FL[0]  = HL_LO[k];
            PF_EH[0]   = MathAbs(PF_HI[0]-HL_HI[k]);
            PF_EL[0]   = MathAbs(PF_LO[0]-HL_LO[k]);
            PF_AE[0]   = MathAbs((PF_EH[0]+PF_EL[0])/2);
            PF_PE[0]   = MathAbs(PF_AE[0]/PF_CL[0]);
          }
          k++;
        }
      }
    } // Check first record

    // Set the Array Values
    if (nt<TSDATA_SIZE)
    {
      PF_YY[nt]  = RPF_YY;
```

Fig. 11.21 Process #4.4 read FC_PF data file (*Source* Dr. Lee's illustration, 2019)

```
                PF_MM[nt]  = RPF_MM;
                PF_DD[nt]  = RPF_DD;
                PF_OP[nt]  = RPF_OP;
                PF_HI[nt]  = RPF_HI;
                PF_LO[nt]  = RPF_LO;
                PF_CL[nt]  = RPF_CL;
                PF_FH[nt]  = RPF_FH;
                PF_FL[nt]  = RPF_FL;
                PF_EH[nt]  = RPF_EH;
                PF_EL[nt]  = RPF_EL;
                PF_AE[nt]  = RPF_AE;
                PF_PE[nt]  = RPF_PE;
              } //if (nt<TSDATA_SIZE)
              // Upcount nt
              nt++;
          } // READING FC_DT DATA RECORDS
          // Print("Reading ",TPSymbol,"READING FC_PF DATA FILE COMPLETED !!!");
       }else
       {
          PrintFormat("Failed to open %s file, Error code = %d",PF_FileName,GetLastError());
          Print("Reading FC_PF DataFile ERROR!!!");
       } // CHECK PF_FILEHANDLE

       // CLOSE FC_PF DATA FILE
       FileClose(PF_FileHandle);
```

Fig. 11.21   (continued)

```
//***************************************************************************************************
//
//   4.5 WRITE FC_PF DATA FILE IF bWriteTS IS TRUE
//
//***************************************************************************************************
if (b_PF==true)
{
    Print("#",nTP," : ",TPSymbol,": 1.1 WRITING FC_PF DATA FILE ...");
    // First Write to DPF Summary File
    FileWrite (DPF_FileHandle,PF_YY[0],PF_MM[0],PF_DD[0],TP_No[nTP],TP_Code[nTP],TP_CCat[nTP],
       TP_ECat[nTP],TP_CName[nTP],TP_ESName[nTP],TP_ELName[nTP],TP_nD[nTP],DoubleToString(PF_OP[0],
       TP_nD[nTP]),DoubleToString(PF_HI[0],TP_nD[nTP]),DoubleToString(PF_LO[0],TP_nD[nTP]),DoubleToString(
       PF_CL[0],TP_nD[nTP]),DoubleToString(PF_FH[0],TP_nD[nTP]),DoubleToString(PF_FL[0],TP_nD[nTP]),
       DoubleToString(PF_EH[0],5),DoubleToString(PF_EL[0],5),DoubleToString(PF_AE[0],5),
       DoubleToString(PF_PE[0],5));
    FileDelete(PF_Directory+"//"+PF_FileName,FILE_COMMON);
    ResetLastError();
    PF_FileHandle   = FileOpen(PF_Directory+"//"+
       PF_FileName,FILE_COMMON|FILE_READ|FILE_WRITE|FILE_CSV,',');
    if(PF_FileHandle!=INVALID_HANDLE)
    {
       for (t=0;t<TSDATA_SIZE;t++)
       {
          FileWrite(PF_FileHandle,PF_YY[t],PF_MM[t],PF_DD[t], DoubleToString(PF_OP[t],TP_nD[nTP]),
            DoubleToString(PF_HI[t],TP_nD[nTP]), DoubleToString(PF_LO[t],TP_nD[nTP]),
            DoubleToString(PF_CL[t],TP_nD[nTP]), DoubleToString(PF_FH[t],TP_nD[nTP]),
            DoubleToString(PF_FL[t],TP_nD[nTP]),  DoubleToString(PF_EH[t],5),
            DoubleToString(PF_EL[t],5), DoubleToString(PF_AE[t],5), DoubleToString(PF_PE[t],5));
       }
    } // PF_FileHandle
    // CLOSE TIME SERIES DATA FILE
    FileClose(PF_FileHandle);
} //CHECK bWriteTS
```

Fig. 11.22   Process #4.5 write FC_PF data file if bWrite is TRUE (*Source* Dr. Lee's illustration, 2019)

```
//****************************************************************************************
//
//  5. CREATE TSCNON and Initialize ALL Lee-oscillators
//
//****************************************************************************************
Print("#",nTP," : ",TPSymbol,": 2 CREATE TSCNON ...");
IndicatorBuffers(0);
IndicatorDigits(6);
// Set the network size
ArrayResize(trainingData,1);
// CREATE NEW TSCNON

//create the NN model by three kinds of layer

printf("Creating Neural Network Model...");
neuron input_layer[input_node_num];
neuron hidden_layer[hidden_layer_num*hidden_node_num];
neuron output_layer[output_node_num];
printf("Done!\n");
```

Fig. 11.23 Process #5 Create TSCNON and Initialize ALL LEE-oscillators (*Source* Dr. Lee's illustration, 2019)

11.4.11 Process #9—Check Forecast Within Range or NOT

After TSCNON daily forecast, this process evaluates the range between forecast H and L can check whether it is within the acceptable range defined in the declaration section. It is to check whether the forecast OHLC are in logical values to decide whether it needs to forecast again (Fig. 11.27).

11.4.12 Process #10.1—Forecast Completed, Store Forecast Results and Total Time Taken

After the daily forecast is completed and within an acceptable range, this process calculates the total time taken, calculates the closest 8 QPLs and records the forecast results into FC_DAILY data file, for data archive and upload to QFFC.org official site (Fig. 11.28).

11.4.13 Process #10.2—Write FC_HL Data File and Close All Files

This process writes the Forecast H/L into FC_HL data file and closes all data files (Fig. 11.29).

```
//**************************************************************************************
//
// 5. REGISTER TIME SERIES DATA
//     INSERT TIME SERIES DATA FOR TRAINING ONE-BY-ONE
//     INPUTS  : 5-DAY TIME SERIES (OPEN/HIGH/LOW/CLOSE) - TOTALLY 20 DATA
//           D/D-1/D-2/D-3/D-4 + 20 closest QPLs
//     OUTPUT  : NEXT-DAY (D+1) OPEN/HIGH/LOW/CLOSE
//     Note    : d = 0 (First record is the test set)
//
//**************************************************************************************
    Print("#",nTP," : ",TPSymbol,": 3 REGISTER TIME SERIES DATA ...");

    //load the trainData
    printf("Reading Data from csv file...");
    int index;
    double *train = loadData(trainData, input_node_num, output_node_num, dataLen, &index);
    printf("Done!\n");

    //devide the data to input and output
    printf("Dividing training data to input and output...");
    double *input = (double *)malloc(index*input_node_num * sizeof(double));
    double *output = (double *)malloc(index*output_node_num * sizeof(double));

    if(input == NULL || output == NULL)
    {
        printf("Out of memery while initialize input and output data\n");
        return;
    }

    //devide the input layer and output layer
    for(int i = 0; i < index; i++){
        for(int m = 0; m < input_node_num; m++)
        {
            input[i*input_node_num+m] = train[i*(input_node_num+output_node_num)+m];
        }

        for(int n = input_node_num; n < input_node_num+output_node_num; n++)
        {
            output[i*output_node_num+(n-input_node_num)] = train[i*(input_node_num+output_node_num)+n];
        }
    }

    //create predict value
    double *predict = (double *)malloc(input_node_num * sizeof(double));
    int predict_index = 0;
    for(int j = (index - 9) * 4; j < index * 4; j++)
    {
        predict[predict_index] = train[j];
        predict_index++;
    }

    free(train);
    printf("Done!\n");
```

Fig. 11.24 Process #6 register time series data (*Source* Dr. Lee's illustration, 2019) *Note* Training data file "Data_4_mix.csv" contains 744 records and for each record includes 40 data, the first 36 data presenting 9-days of O, H, L, C, and the last 4 data presenting next day of O, H, L, C

11.5   TSCNON—Implementation Results

Figure 11.30 shows a snapshot of TSCNON system training and forecast process of 120 financial products of Forex.com on May 6, 2019 in the server farm of quantum finance forecast center using Intel i5 CPU 2.39 GHz 32 MB RAM Dell Server.

```
//**********************************************************************************************************
//
// 7. TSCNON TRAINING
//
//**********************************************************************************************************

//start training
printf("Start Neural Network Training...\n");
double cost = 0.0;
double current_error = 0.0;
int training_num = 1;


for(int epoch = 1; epoch <= epochTime; epoch++)
{

    for(int times = 0; times < index; times++)
    {
        //training proces

        //load the data to input layer
        loadInputLayer(input_layer, input_node_num, input);
        FeedForward(input_layer, hidden_layer, output_layer, input_node_num, hidden_node_num, output_node_num,
hidden_layer_num, bias);
        cost = backpropagation(input_layer, hidden_layer, output_layer, output, input_node_num, hidden_node_num,
output_node_num, hidden_layer_num, learningRate, bias);
        current_error = cost / (training_num + 1);

        printf("\rEpoch times: %d, Training %d/%d data, MSE: %e", epoch, times+1, index, current_error);
        training_num++;
    }
    printf("\n");
}

printf("Training finish\n");

//tracking the value of the hidden layer
    //checkHidden(hidden_layer, input_node_num, hidden_node_num, hidden_layer_num);

    //save the neural network model
    printf("Prepare to save the NN model\n");
    saveModel(input_layer, hidden_layer, output_layer, input_node_num, hidden_layer_num, hidden_node_num, out-
put_node_num, learningRate, bias);
```

Fig. 11.25 Process #7 TSCNON training (*Source* Dr. Lee's illustration, 2019)

```
//**********************************************************************************************************
//
// 8. TSCNON FORECAST, USING PREVIOUS 5 DAY TIME SERIES + QPLS TO PREDICT TODAY
//
//**********************************************************************************************************
// Print(TPSymbol,": Forecast Today ...");

    //using the last record to predict new value
    printf("Using the newest data to predict tomorrow value:\n");
    loadInputLayer(input_layer, input_node_num, predict);
    FeedForward(input_layer, hidden_layer, output_layer, input_node_num, hidden_node_num, output_node_num,
hidden_layer_num, bias);
    for(int q = 0; q < output_node_num; q++)
    {
        printf("\t%f", output_layer[q].output);
    }
    printf("\n");
```

Fig. 11.26 Process #8 TSCNON forecast (*Source* Dr. Lee's illustration, 2019)

```
//*******************************************************************************************************
//
// 9. CHECK FORECAST RESULT WITHIN RANGE OR NOT
//
//*******************************************************************************************************

// De-normalize FC Results
FC_AOPEN   = FC_NOPEN * TP_Norm[nTP];
FC_AHIGH   = FC_NHIGH * TP_Norm[nTP];
FC_ALOW    = FC_NLOW * TP_Norm[nTP];
FC_ACLOSE  = FC_NCLOSE * TP_Norm[nTP];

// Calculate the FC_OPEN DIFF AND CALCULATE THE AJ FC
FC_Diff    = D0_OPEN  - FC_AOPEN;
FC_AJOPEN  = FC_AOPEN  + FC_Diff;
FC_AJHIGH  = FC_AHIGH  + FC_Diff;
FC_AJLOW   = FC_ALOW   + FC_Diff;
FC_AJCLOSE = FC_ACLOSE + FC_Diff;

// Calculate TP_Range
FC_RANGE = FC_AJHIGH - FC_AJLOW;

// Check bPASS
if ((FC_AJHIGH>FC_AJOPEN)&&(FC_AJHIGH>FC_AJLOW)&&(FC_AJHIGH>FC_AJCLOSE)&&
   (FC_AJLOW<FC_AJOPEN) &&(FC_AJLOW<FC_AJHIGH)
&&(FC_AJLOW<FC_AJCLOSE)&&(FC_RANGE<(TP_Range[nTP]*2)))
   {
   bPass = true;
   Print("#",nTP," : ",TPSymbol," : Round #",nRound," PASSED!!!");
   }else
   {
   Print("#",nTP," : ",TPSymbol," : Round #",nRound," NOT PASSED!!!");
   }

nRound++;

// Randam Weight
randomize_weights (tscnon[nTP], -0.77, 0.77);

} // WHILE LOOP ON bPASS
```

Fig. 11.27 Process #9—check forecast within range or NOT (*Source* Dr. Lee's illustration, 2019)

As shown in Fig. 11.30, in a typical daily forecast of 120 financial products on Forex.com MT4 platform, the TSCNON system only took 91,547 ms (91.547 s) to finish the training and forecast of 120 financial products.

On average, it took 0.763 s (less than 1 s) to complete the network training and forecast process of a single financial product. Figure 11.31 shows the snapshot of TSCNON system for the system training and forecast of 9 major cryptocurrencies over AvaTrade.com MT platform on May 6, 2019.

As shown in Fig. 11.31, in a typical forecast day, TSCNON took 56,578 ms (56.578 s) to finish the training and forecast of the 9 cryptocurrencies. That is, on average it took 6.275 s to train and forecast a single cryptocurrency.

As compared with all those 120 non-cryptocurrency products, TSCNON took 8.22 times to predict cryptocurrency, even though cryptocurrency only have 300-trading day records while the other 120 financial products each have 2048-trading day records for system training. It may due to the fact that cryptocurrencies, in general, are much more chaotic and fluctuant in nature, which takes more time and iterations for TSCNON to learn the market pattern.

```
//****************************************************************************************
//
// 10.1 FORECAST COMPLETED - STORE FORECAST RESULTS AND REPORT TIME TAKEN
//
//****************************************************************************************
// Check time
etime  = GetTickCount();
tlapse = etime - stime;

// Output to Console
Print(TP_No[nTP]," : ",TPSymbol,": FORECAST COMPLETED !!! Time Taken : ",tlapse," RMSEc.",
        " #Training =",nTraining+1," RMSE =",DoubleToString(get_MSE(tscnon[nTP]),10));
Print(TP_No[nTP]," : ",TPSymbol,": Today Forecast OPEN(",FC_AJOPEN,"),
        HIGH(",FC_AJHIGH,"), LOW(",FC_AJLOW,"), CLOSE(",FC_AJCLOSE,")");

// UPDATE HL_HI[0], HL_LO[0]
HL_HI[0] = DoubleToString(FC_AJHIGH,TP_nD[nTP]);
HL_LO[0] = DoubleToString(FC_AJLOW, TP_nD[nTP]);

// Update ALL Daily Forecast Items
TP_Low  = FC_AJLOW  + TP_TP[nTP];
SL_Low  = FC_AJLOW  - TP_SL[nTP];
TP_High = FC_AJHIGH - TP_TP[nTP];
SL_High = FC_AJHIGH + TP_SL[nTP];

// Cacluate the EIGHT closest QPL
QPL1 = D1_CLOSE / NQPR[nTP][3];
QPL2 = D1_CLOSE / NQPR[nTP][2];
QPL3 = D1_CLOSE / NQPR[nTP][1];
QPL4 = D1_CLOSE / NQPR[nTP][0];
QPL5 = D1_CLOSE * NQPR[nTP][0];
QPL6 = D1_CLOSE * NQPR[nTP][1];
QPL7 = D1_CLOSE * NQPR[nTP][2];
QPL8 = D1_CLOSE * NQPR[nTP][3];

// PRINTOUT to DailyForecast.csv DataFile
FileWrite(DL_FileHandle,TP_No[nTP],TP_Code[nTP], DoubleToString(FC_AJLOW,TP_nD[nTP]),
    DoubleToString(TP_Low,TP_nD[nTP]), DoubleToString(SL_Low,TP_nD[nTP]),
    DoubleToString(FC_AJHIGH,TP_nD[nTP]), DoubleToString(TP_High,TP_nD[nTP]),
    DoubleToString(SL_High,TP_nD[nTP]), DoubleToString(QPL1,TP_nD[nTP]),
    DoubleToString(QPL2,TP_nD[nTP]),DoubleToString(QPL3,TP_nD[nTP]),
    DoubleToString(QPL4,TP_nD[nTP]), DoubleToString(QPL5,TP_nD[nTP]),
    DoubleToString(QPL6,TP_nD[nTP]),DoubleToString(QPL7,TP_nD[nTP]),
    DoubleToString(QPL8,TP_nD[nTP]));
```

Fig. 11.28 Process #10.1—forecast completed, store forecast results, and total time taken (*Source* Dr. Lee's illustration, 2019)

11.6 TSCNON—System Performance

From the system performance and evaluation perspective, TSCNON system evaluated the daily forecast performance of the 129 financial products in four timeframes: daily, weekly average, monthly average, and past 500-day average. Figure 11.32 presents the past 500-day performance ranking list of the top 20 financial products.

As shown, the 500-day average forecast % error of the top 20 financial products ranging from 0.011% to 0.227%, respectively, which is somewhat promising and significant as reflected by over 1000+ members of QFFC which consist of professional forex traders, quants, and investors.

```
//*********************************************************************************************************
//
// 10.2 WRITE FC_HL DATA FILE AND CLOSE ALL FILES
//
//*********************************************************************************************************
  Print("#",nTP," : ",TPSymbol,": 4.1 WRITING FC_HL DATA FILE ...");

  FileDelete(HL_Directory+"//"+HL_FileName,FILE_COMMON);
  ResetLastError();

  HL_FileHandle  = FileOpen(HL_Directory+"//"+HL_FileName,
    FILE_COMMON|FILE_READ|FILE_WRITE|FILE_CSV,',');

  if(HL_FileHandle!=INVALID_HANDLE)
  {
    for (t=0;t<TSDATA_SIZE;t++)
    {
      FileWrite(HL_FileHandle,HL_YY[t],HL_MM[t],HL_DD[t],
        DoubleToString(HL_HI[t],TP_nD[nTP]),DoubleToString(HL_LO[t],TP_nD[nTP]));
      // Print("t=",t,":",HL_YY[t],":",HL_MM[t],":",HL_DD[t],":",
      //   DoubleToString(HL_HI[t],TP_nD[nTP]),":",DoubleToString(HL_LO[t],TP_nD[nTP]));
    }
  } // PF_FileHandle

  // CLOSE FC_HL DATA FILE
  FileClose(HL_FileHandle);

  } // IF WORKING DAY
  } // LOOP OVER ALL TP

  // Close FC_DL File
  FileClose(DL_FileHandle);
  FileClose(DPF_FileHandle);

  // Check Global Time
  Getime  = GetTickCount();
  Gtlapse = Getime - Gstime;

  // Output time taken
  Print("Total Time Taken : ",Gtlapse," RMSEc");

  return(0);
}
```

Fig. 11.29 Process #10.2—write FC_HL data file and close all files (*Source* Dr. Lee's illustration, 2019)

To differentiate system performance with traditional FFBPN (feedforward backpropagation network), four categories of worldwide 129 financial products are tested using FFBPN and TSCNON systems with network training error rates ranging from 1×10^{-5} to 1×10^{-6}, respectively.

Table 11.1 presents the system performance comparison chart of both systems.

As shown in Table 11.1, system simulations using error rate 1×10^{-5} (Case I); the average total time taken for 129 financial products training using FFBPN and TSCNON were 2,915,709 ms (48.6 min) and 40,049 ms (40.05 s), respectively. TSCNON outperformed FFBPN by 72.8 times in terms of network training time to attain the same error rate. Also, system simulations in Case II using error rate 1×10^{-6}, TSCNON completed the system training task in 147,940 ms (147.9 s) while FFBPN encounters "deadlocks" during system training for cryptocurrency and forex products; which clearly demonstrated the resolution of over-training and deadlock problems with sufficient improvement in system training efficiency by using TSCNON over traditional neural networks.

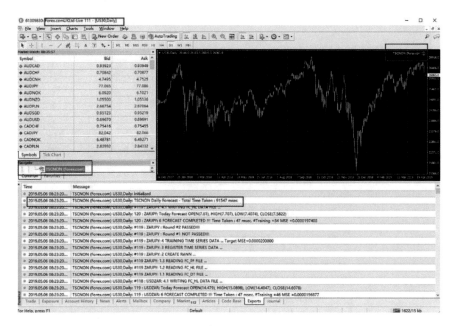

Fig. 11.30 Snapshot of TSCNON (Forex.com) for system training and forecast of 120 financial products for Forex.com MT4 platform on May 6, 2019 (*Source* Dr. Lee's illustration, 2019)

In terms of system performance across different financial products, the simulation results clearly showed that both cryptocurrency and forex are more chaotic and difficult for network training than other financial products as expected, which will be further explored in the future research of QFFC.

11.7 Conclusion

TSCNON (time series chaotic neural oscillatory network) is for real-time financial prediction incorporated into quantum finance forecast center—a nonprofit making AI-fintech R&D and worldwide financial forecast center aims at R&D and provision of a fair and open platform for worldwide traders and individual investors to acquire free knowledge of worldwide 129 financial product forecasts based on state-of-the -art AI, chaotic neural networks and quantum field theory technologies.

If TSCNON is so effective to forecast worldwide financial markets, is it the end of story?

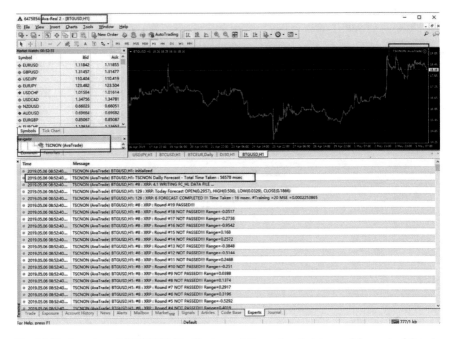

Fig. 11.31 Snapshot of TSCNON (AvaTrade.com) for system training and forecast of 9 major cryptocurrencies for AvaTrade.com MT4 platform on May 6, 2019 (*Source* Dr. Lee's illustration, 2019)

The truth is: For a professional trader and investor, a reliable and effective financial forecast system is only the beginning of the story. A good financial investment also needs: (1) good, effective trading and hedging strategies; (2) stable, logical with rational investment psychology, which are also the future R&D directions of quantum finance forecast center.

Current research of QFFC include the following:

(1) Integration of TSCNON with fractal technology for market trends/patterns mining and prediction;
(2) R&D on quantum entanglement of quantum finance system on severe financial event modeling and prediction;
(3) Design and implementation of *chaotic type-2 transient-fuzzy deep neural oscillatory network* for real-time financial prediction;
(4) Design and develop intelligent agent-based hedging and trading systems as discussed in Chap. 13 (Lee 2019).

Ranking	Product Name	Code	HIGH (Error)	LOW (Error)	Average (Error)	% Error
			Past 500-Day Forecast Performance Ranking List			
1	EUR/Danish Krone	EURDKK	0.05184	0.04977	0.05080	0.025%
2	US DollarHong Kong Dollar	USDHKD	0.056690	0.055040	0.055870	0.088%
3	EURHungarian Forint	EURHUF	0.613710	0.613490	0.613540	0.178%
4	Norwegian Krone/Swedish Krona	NOKSEK	0.052090	0.050140	0.051110	0.201%
5	US Dollar/Chinese Yuan	USDCNH	0.013080	0.013370	0.013220	0.202%
6	EURCzech Koruna	EURCZK	0.104460	0.100480	0.102460	0.209%
7	Australian Dollar/New Zealand Dollar	AUDNZD	0.052250	0.050280	0.051270	0.210%
8	Canadian Dollar/Norwegian Krone	CADNOK	0.066080	0.063300	0.064690	0.249%
9	Euro/Singapore Dollar	EURSGD	0.053760	0.052330	0.053040	0.253%
10	US Dollar/Swiss Franc	USDCHF	0.052410	0.050640	0.051530	0.261%
11	Gold/Japanese Yen	XAUJPY	361.270000	376.948000	369.037000	0.262%
12	Euro/Romanian Leu	EURRON	0.062210	0.060480	0.061340	0.266%
13	Australian Dollar/Norwegian Krone	AUDNOK	0.066760	0.065580	0.066160	0.281%
14	US Dollar/Singapore Dollar	USDSGD	0.053910	0.051710	0.052810	0.283%
15	Australian Dollar/Singapore Dollar	AUDSGD	0.053020	0.050830	0.051920	0.289%
16	Swiss Franc/Japanese Yen	CHFJPY	0.387640	0.377330	0.382410	0.295%
17	EUR/Polish Zloty	EURPLN	0.062560	0.060790	0.061670	0.300%
18	Australian Dollar/Canadian Dollar	AUDCAD	0.052940	0.051700	0.052320	0.344%
19	US Dollar/Canadian Dollar	USDCAD	0.054400	0.052470	0.053430	0.345%
20	Canadian Dollar/Swiss Franc	CADCHF	0.052870	0.050520	0.051690	0.355%

【The above information are generated by scientific computer predictions, non-profit guarantee. For reference only.】

Fig. 11.32 Past 500-day system performance ranking chart of top 20 financial products (*Source* Dr. Lee's illustration, 2019) *Note* 1. High (Error) = Abs(High$_{Forecast}$ − High$_{Actual}$) 2. Low (Error) = Abs(Low$_{Forecast}$ − Low$_{Actual}$) 3. Average (Error) = Average(High(Error), Low(Error)) 4. % Error = Average(Error)/Close$_{Actual}$

Problems

11.1 What is a recurrent neural network (RNN)? How it is different from the traditional neural network in terms of network architecture.

11.2 Explain how to integrate Lee-oscillator(s) into a recurrent neural network (RNN) to convert RNN into chaotic RNN.

11.3 In TSCNON, we use the closest, say, 20 quantum price levels (QPLs) as quantum finance signals (QFSs) for network training.

Table 11.1 System performance comparison chart (FFBPN vs. TSCNON)

Product category	No. of products	FFBPN[a]		TSCNON[a]	
		Total STT[b]	Average STT[c]	Total STT[b]	Average STT[c]
Case I (Error rate = 1×10^{-5})					
Cryptocurrency	9	1,460,000	162222.22	15,783	1753.67
Forex	84	1,235,543	14708.85	19,126	227.69
Financial index	19	111,024	5843.37	2040	107.37
Commodity	17	109,142	6420.12	3100	182.35
Overall	129	2,915,709	22602.40	40,049	310.46
Case II (Error rate = 1×10^{-6})					
Cryptocurrency	9	DL[d]	–	58,400	6488.89
Forex	84	DL[d]	–	72,679	865.23
Financial index	19	577,324	30385.47	6939	365.21
Commodity	17	687,595	40446.76	9922	583.65
Overall	129	–	–	147,940	1146.82

Note
[a]Results are generated by 500 simulations of both FFBPN and TSCNON system training (in ms)
[b]"Total STT" denotes the total average system training time for 500 simulations of network training using FFBPN and TSCNON system for each category of financial products
[c]"Average STT" denotes the average system training time for a single financial product of each category
[d]"DL" denotes deadlock during system training

(i) What is the ground state (i.e., QPL_0) of QPL? Why?
(ii) Using numerical computational method learnt in Chap. 10, work out the formulation of the closest 21 QPLs.
(i.e., $QPL_{+10} \ldots QPL_0 \ldots QPL_{-10}$)

11.4   In TSCNON, we use normalized quantum price levels (QPLs) as input vectors of QFS, can we use the normalized quantum price return (QPRs) instead of QPLs as input signals? Why?

11.5   The following figure shows the bifurcation diagram of Lee-oscillator discussed in Chap. 9.

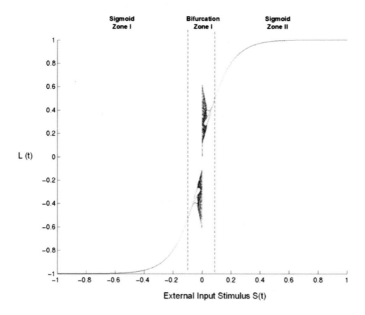

(i) Based on this bifurcation diagram and Eqs. (11.1)–(11.4), write a MAT-LAB program to generate the TCAF (transient chaotic activation function) with external input stimulus S(t) and L(t) in the normalization region [0, 1].

(ii) From the TCAF, identify the active bifurcation region.

(iii) During the normalization process of all input signals in the TSCNON, what is the optimal normalization region? Why?

11.6 In TSCNON discussed in this Sect. 11.2, 5-day timeseries are used as input signal vectors with QPLs to forecast next-day high/low Prices. Why? Try to use more trading days, says 21 days to do next-day forecast and check the difference.

11.7 In this chapter, we used TSCNON for daily financial prediction. Discuss and explain how we can use TSCNON for weekly and monthly prediction in terms of (1) time series input vectors; (2) time horizon for network training; and (3) QPLs being used. Do we need to recalculate all the QPLs? Why?

11.8 In Chap. 8, we learnt how initial conditions are critical for the prediction of complex and highly chaotic systems like weather and financial markets. Discuss why we can still use chaotic neural networks to perform financial prediction? What is the basic concept and logic behind?

11.9 In QFFC.org, four timeframes of daily system performance analysis are generated: daily, weekly, monthly, and 500-Day average daily performance. Discuss and explain physical meaning and importance of each type of performance results in terms of (1) system stability and (2) significance on different timeframes of trading. Which financial products are the best in terms of forecast performance? Why?

11.10 Programming task 11.1

(i)   Follow workshop #2 and implement the TSCNON system for at least 10 different Forex products, using 5-day time series and 21 closest QPLs to predict next-day high/low–system A.

(ii) Implement the same forecast system but this time without using QPLs as input QFS–system B.

(iii) Compare systems A and B in terms of (1) system learning performance in 1000 iterations; (2) next-day forecast performance.

11.11 Programming task 11.2

(i)   As an extension of task 11.1, this time change the number of closest QPLs from 10 to 30, with an increment of 2 QPLs (i.e.,one more QPL on each side). So totally we have 11 sets of systems for testing.

(ii) Modify TSCNON.eq4 program to automate system testing process of these 11 experiments.

(iii) Compare these 11 systems in terms of (1) System learning performance in 1000 iterations; (2) next-day forecast performance, and locate the optimal number of QPL values.

Acknowledgements The author wishes to thank Forex.com and AvaTrade.com for the provision of historical and real-time financial data. The author also wishes to thank Quantum Finance Forecast Center of UIC for the R&D supports and the provision of the channel and platform QFFC.org for worldwide system testing and evaluation.

References

Fausett, L. *Fundamentals of Neural Networks Architectures Algorithms and Applications*. Prentice Hall, 1994.

Lee, R. S. T. LEE-Associator—A Transient Chaotic Autoassociative Network for Progressive Memory Recalling. *Neural Networks*. 19(5): 644–666, 2006a.

Lee, R. S. T. Fuzzy-Neuro Approach to Agent Applications (From the AI Perspective to Modern Ontology). Springer-Verlag, Germany, 2006b.

Lee, R. S. T. *Advanced Paradigms in Artificial Intelligence From Neural Oscillators, Chaos Theory to Chaotic Neural Networks*. Advanced Knowledge International, Australia, 2005.

Lee, R. S. T. A Transient-chaotic Auto-associative Network (TCAN) based on LEE-oscillators. *IEEE Trans. Neural Networks*. 15(5): 1228–1243, 2004.

Lee, R. S. T. (2019) Chaotic Type-2 Transient-Fuzzy Deep Neuro-Oscillatory Network (CT2TFDNN) for Worldwide Financial Prediction. IEEE Transactions on Fuzzy System. https://doi.org/10.1109/tfuzz.2019.2914642.

Li, G. C. L. and Lee, R. S. T. A Real-Time Scene Segmentation System Using Solely Excitatory Oscillator Networks (SEON). *Journal of Intelligent Manufacturing*. 16(6): 669–678, 2005.

Narayanan A. et al. *Bitcoin and Cryptocurrency Technologies: A Comprehensive Introduction*. Princeton University Press, 2016.

Patterson, D. W. *Artificial Neural Networks*. Prentice Hall, 1996.

Vigna, P. and Casey, M. J. *The Age of Cryptocurrency: How Bitcoin and the Blockchain Are Challenging the Global Economic Order.* Picador, 2016.

Walker, W. *Expert Advisor Programming and Advanced Forex Strategies.* Independently published, 2018.

Wong et al. Wind Shear Forecasting by Chaotic Oscillatory-based Neural Networks (CONN) with Lee-oscillator (Retrograde Signaling) Model. *International Joint Conference on Neural Networks (IJCNN)*, 2040–2047, 2008.

Young, A. R. *Expert Advisor Programming for MetaTrader 4: Creating automated trading systems in the MQL4 language.* Edgehill Publishing, 2015.

Chapter 12
Chaotic Type-2 Transient-Fuzzy Deep Neuro-Oscillatory Network (CT2TFDNN) for Worldwide Financial Prediction

In this chapter, the author proposed a chaotic type-2 transient-fuzzy deep neuro-oscillatory network with retrograde signaling (aka CT2TFDNN) for worldwide financial prediction. With the extension of the author's original work on Lee-oscillator—a chaotic discrete-time neural oscillator with profound transient-chaotic property, CT2TFDNN provides: (1) effective modeling of Interval type-2 fuzzy logic with chaotic type-2 transient-fuzzy membership function (CT2TFMF); (2) effective time series network training and prediction using chaotic deep neuro-oscillatory network with retrograde signaling (CDNONRS). CT2TFDNN not only provides a fast chaotic fuzzy-neuro deep learning and forecast solution, but also it successfully resolves the massive data over-training and deadlock problems, which are usually imposed by traditional recurrent neural networks using classical sigmoid-based activation functions. From the implementation perspective, CT2TFDNN is integrated with 2048-trading day time series financial data and top-10 major financial signals as fuzzy financial signals (FFS) for the real-time prediction of 129 worldwide financial products which consists of: 9 major cryptocurrencies, 84 worldwide forex, 19 major commodities, and 17 worldwide financial indices.

Over the years, financial engineering ranging from the study of financial signals to the modeling of financial prediction is one of the most exciting topics for both academia and financial community. With the flourishing of AI technology in the past 20 years, various hybrid intelligent financial prediction systems with the integration of neural networks, chaos theory, fuzzy logic, and genetic algorithms were proposed. Interval type-2 fuzzy logic system (IT2FLS) with its remarkable capability for the modeling of highly uncertain events and attributes, provides a perfect tool to interpret various financial phenomena and patterns.

In this chapter, the author proposed a chaotic type-2 transient-fuzzy deep neuro-oscillatory network with retrograde signaling (aka CT2TFDNN) for worldwide financial prediction (Lee 2019a, b). With the extension of author's original work on Lee-oscillator—a chaotic discrete-time neural oscillator with profound transient-chaotic

---

© Portions of this chapter are reprinted from Lee (2019a), with permission of IEEE.

© Springer Nature Singapore Pte Ltd. 2020
R. S. T. Lee, *Quantum Finance*,
https://doi.org/10.1007/978-981-32-9796-8_12

property, CT2TFDNN provides: (1) effective modeling of Interval type-2 fuzzy logic with chaotic type-2 transient-fuzzy membership function (CT2TFMF); (2) effective time series network training and prediction using chaotic deep neuro-oscillatory network with retrograde signaling (CDNONRS). CT2TFDNN not only provides a fast chaotic fuzzy-neuro deep learning and forecast solution but also more prominently successfully resolves the massive data over-training and deadlock problems, which are usually imposed by traditional recurrent neural networks using classical sigmoid-based activation functions. From the implementation perspective, CT2TFDNN is integrated with 2048-trading day time series financial data and top-10 major financial signals as fuzzy financial signals (FFS) for the real-time prediction of 129 worldwide financial products which consists of: 9 major cryptocurrencies, 84 worldwide forex, 19 major commodities, and 17 worldwide financial indices.

12.1 Introduction

This chapter introduces an innovative chaotic type-2 transient-fuzzy deep neuro-oscillatory network with retrograde signaling (aka CT2TFDNN) for worldwide financial time series prediction.

With the extension of author's original work on Lee-oscillator—a discrete-time neural oscillator with profound transient-chaotic property on temporal information progressing (Lee 2004a, 2005, 2006a, b), together with the author's latest works on the exploration of Lee-oscillator with retrograde signaling (LORS), which generalizes into eight categories of bifurcation transfer functions for deep learning, CT2TFDNN successfully integrates discrete-time chaotic neural oscillators (for chaotic neural activity modeling), chaotic deep neuro-oscillatory network with 8-hidden layers of neuro-oscillators with various degree of retrograde signaling (for chaotic time series deep learning and prediction) and chaotic type-2 transient-fuzzy logic (for high-order uncertainty modeling of financial signals) into: (1) effective modeling of type-2 fuzzy logic with chaotic type-2 transient-fuzzy membership function (CT2TFMF); (2) effective time series network training and prediction using chaotic deep neuro-oscillatory network with retrograde signaling (CDNONRS).

CT2TFDNN not only provides a fast chaotic fuzzy-neuro deep learning and forecast solution, more prominently it successfully resolves the massive data over-training and deadlock problems, which are usually imposed by traditional recurrent neural networks and fuzzy neural networks using IT2 fuzzy membership functions and/or hybrid neural networks using classical sigmoid-based activation functions.

From the implementation perspective, CT2TFDNN is integrated with 2048-trading day time series financial data and top-10 major financial signals as input signals for the real-time prediction and agent-based trading of 129 worldwide financial products which consists of: 9 major cryptocurrencies, 84 worldwide forex, 19 major commodities, and 17 worldwide financial indices.

The main contributions and originality of this system include:

1. The introduction of an innovative type-2 fuzzy logic system (T2FLS), which is inspired by the author's original work on Lee-oscillators (Lee 2004). Different from the contemporary research on T2FLS which mainly focus on the R&D of the interval type-2 logic system (IT2FLS)—a simplified version of T2FLS due to its computational complexity, this chapter proposed a chaotic type-2 transient-fuzzy logic system (CT2TFLS)—a truly IT2LS with remarkable chaotic transient-fuzzy property to resolve the computational complexity problem.

2. The introduction of CT2TFDNN—an innovative fuzzy-neuro deep neural network, which is not only the integration of two AI technologies as two separate functional modules proposed by numerous hybrid fuzzy-neuro systems, the CT2TFDNN introduced in this chapter is constructed by LORS—chaotic neural oscillators invented by the author which serves as *transient-fuzzy input neurons* of the DNN and effectively converts it into CT2TFDNN system. In other words, the chaotic transient-fuzzification process is actually part of the neural model of CT2TFDNN.

3. The successful resolution of the massive data over-training and deadlock problems, which are usually imposed by traditional recurrent neural networks using classical sigmoid-based activation functions.

4. The successful design and implementation of real-time financial forecast system of worldwide financial products with state-of-the-art CT2TFDNN system.

This chapter is comprised of (1) literature review on (i) type-1 and interval type-2 fuzzy logic systems (FLS) on financial prediction; (ii) chaotic neural networks, Lee-oscillator, and the author's latest work on Lee-oscillator with retrograde signaling (LORS) in Sect. 12.2; (2) system framework and methodology of CT2TFDNN for financial time series prediction which include: chaotic type-2 transient-fuzzification scheme, GA-based top-10 fuzzy financial signals (FFS) selection scheme and the chaotic deep neuro-oscillatory network learning algorithm in Sect. 12.3; (3) system implementation of CT2TFDNN for real-time prediction of worldwide 129 financial products in Sect. 12.4; (4) performance analysis of proposed CTSTFDNN and comparison with five forecast systems: (i) traditional time series feedforward backpropagation networks (FFBPN), (ii) support vector machine (SVM) provided by R Project—one of the most popular financial forecasting tools used in finance industry, (iii) deep neural network (DNN) with PCA (principal component analysis) model (Singh and Srivastava 2017), (iv) classical interval type-2 fuzzy-neuro network (IT2FNN), (v) chaotic type-1 fuzzy-neuro-oscillatory network (CT1FNON) in Sect. 12.5; and (5) conclusion in Sect. 12.6.

12.2 Literature Review

In a typical financial prediction problem, there are two basic problems we must tackle with: (1) massive and highly uncertain financial patterns and signals; (2) massive time series data and information for system training and machine learning. fuzzy logic system (FLS) (Zadeh 1975; Ross 2016; Siler and Buckley 2004; Zadeh and Aliev 2019) with its intrinsic property of handling attributes of uncertainty, provides a viable solution for system modeling and data representations.

This section gives a general overview of type-1 and type-2 FLSs, their systems' characteristics and limitations, and the current research of FLS in financial predictions. To tackle with highly chaotic and massive time series data such as severe weather information and financial data, data scientists, and AI researchers are now exploring the possibility of adopting chaotic neural oscillators into traditional recurrent neural networks—chaotic neuro-oscillatory networks (CNON, or chaotic neural networks, CNN in short) for chaotic time series modeling and forecast. This section gives a general overview of chaotic neural oscillator and the author's original work on Lee-oscillator, its neural architecture, and the latest works on Lee-oscillator with retrograde signaling (LORS).

12.2.1 An Overview on Type-1 and Interval Type-2 Fuzzy Logic Systems on Financial Prediction

Like many real-world phenomena, the linguistic variables used to describe financial markets (or market patterns), ranging from worldwide financial indices and stocks, to highly volatile markets such as worldwide forex and cryptocurrencies are typical fuzzy logic systems (FLS) with high degree of uncertainty and fuzziness (Ross 2016; Siler 2004; Zadeh 1975; Zadeh and Aliev 2019).

For example, to describe a typical stock market (e.g., Nasdaq Index), financial professionals are usually using fuzzy linguistic variables such as: bearish, bullish, over-buy, or over-sell. More importantly, these fuzzy linguistic variables are usually coexisting and overlapping in nature. For instance, when a stock market (e.g., Dow Jones Index) appeared to be bearish (means short-sell), it is also having a certain degree of over-sell (means rebound—with just the opposite trading signal) simultaneously. The same case occurs in a bullish market. Besides, all these linguistic variables are *fuzzy* in the sense that the meaning of *bearish* or *bullish* are highly subjective and different for different people, even for the same person the meanings of these linguistic variables are usually different at different times and scenarios.

In order to model these highly uncertain and fuzzy linguistic variables into a computer system (such as neural networks) for system modeling, FLS is one of the most viable solutions. A typical FLS consists of three main modules:

(1) fuzzification module,
(2) inference module, and
(3) defuzzification module.

Figure 12.1 shows a typical type-1 fuzzy membership function (T1-FMF) of RSI (relative strength index)—one of the most commonly used financial indexes in financial engineering (Lee 2006b).

Zadeh (1975) in his influential paper *The Concept of a Linguistic Variable and its Application to Approximate Reasoning* introduced the concept of higher order type-n fuzzy logic system (Tn-FLS) as

> *A fuzzy set is of type n, n = 2, 3, …, if its membership function ranges over fuzzy sets of type n-1. The membership function of a fuzzy set of type 1 ranges over the interval [0, 1].*

For example, in a typical *type-2 FLS*, the membership function of each attribute of the system itself is also a fuzzy set with a fuzzy range over the interval [0, 1]. In other words, a higher order (n > 1) *type-n FLS* will increase the degree of uncertainty of each fuzzy variable substantially.

But it has an intrinsic problem of computational complexity. For a simple situation of a *T2-FLS* with k fuzzy variables, each fuzzy variable has m membership functions, and each membership function is quantized into N different states. The total number of possible fuzzy rules/states will be $(N^m)^k$, which will be an astronomical number when such method is applied to real-world financial time series prediction problems which usually have over 50 fuzzy attribute signals (from different financial indicators and oscillators) to consider and each attribute has 4 membership functions, each can span up to 100 different states, which means $(100^4)^{50}$ possible states/fuzzy rules!

In order to solve this computational problem, *interval type-2 fuzzy logic system* (IT2FLS) is introduced. In a typical IT2FLS, only certain interval(s) of the T2 membership function (usually the intervals of the *state-transition-region, STR*) are fuzzified to a second-order FMF. Figure 12.2 shows a typical interval type-2 fuzzy membership function (IT2-FMF) of relative strength index (RSI) for financial engineering.

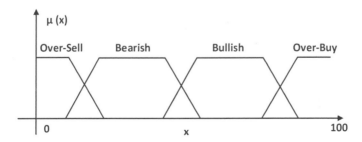

Fig. 12.1 Type-1 fuzzy membership function of relative strength index (RSI)

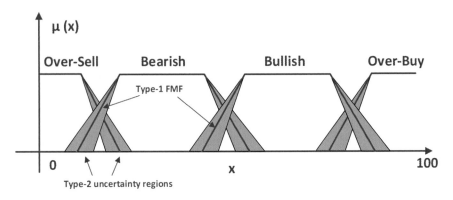

Fig. 12.2 Interval type-2 fuzzy membership function (IT2-FMF) of RSI

A typical interval type-2 fuzzy set (IT2-FS), denoted by $\tilde{A}$ is formulated by

$$\mu_{\tilde{A}}(x) = \int_{x \in X} \int_{u \in \left[\bar{\mu}_{\tilde{A}}(x), \underline{\mu}_{\tilde{A}}(x) \right]} 1/u \tag{12.1}$$

where $\bar{\mu}_{\tilde{A}}(x)$ and $\underline{\mu}_{\tilde{A}}(x)$ denote the upper and lower membership functions for the IT2-FS ($\tilde{A}$).

As shown in Fig. 12.2, the IT2-FMF is characterized by the second-order FMF (the *grey areas*) in the fuzzy membership function, which are also known as FOU (*footprints of uncertainty*). Owing to this high uncertainty modeling capability of IT2FLS as compared with its T1-FLS counterpart, extensive researches have been done in the past 20 years, especially in the fields of fuzzy financial modeling and time series prediction.

Latest works of type-2 FLS in financial time series modeling and prediction include: Castillo et al. have done extensive research of IT2FLS and fuzzy-neuro systems on time series financial predictions including Mexican Stock Exchange and Taiwan Stock Exchange (Castillo et al. 2014; Melin et al. 2012; Gaxiola et al. 2017; Ontiveros et al. 2018; Pulido et al. 2018); Bhattacharya et al. (Bhattacharya et al. 2016; Konar and Bhattacharya 2017; Bhattacharya and Konar 2018) using self-adaptive IT2FLS for stock prediction; IT2FLS for chaotic time series prediction proposed by Lee et al. (2014); Multi-order fuzzy time series forecasting model proposed by Ye et al. (2016) and Yolcu et al. (2016); Compact Evolutionary Interval-Valued Fuzzy Rule-based system for real-time financial prediction proposed by Sanz et al. (2015); Interval Type-2 Mutual Subsethood Fuzzy Neural Inference System proposed by Sumati and Patvardhan (2018) for time series prediction and function approximation; Fuzzy time series forecast of Taiwan stock market based on two-factor second-order fuzzy-trend logical relationship groups (TSFTLRGs) with particle swarm optimization technique proposed by Chen and Jian (2017); IT2FLS for stock index forecast based on fuzzy logical relationship map proposed by Jiang et al.

(2018); and type-2 neuro-fuzzy for stock price prediction based on self-constructing clustering method of fuzzy rules and particle swarm optimization technique by Liu et al. (2012).

Although IT2FLS provides a viable and computational feasible solution to model highly uncertain phenomena, the determination and formulation of FOU is still an area of interest in which the de facto usage of *triangular-shape* formulation as FOU is also queried to be a bit *artificial* in nature, which motivates the author to search for an integrated mathematical framework to model both interval type-2 fuzzy and type-1 fuzzy regions in IT2-FMF as a single continuous *transient-fuzzy* membership function.

By the extension of the author's original work on *Lee-oscillators* (Lee 2004a), this chapter proposed an innovative solution for the modeling of chaotic type-2 transient-fuzzy logic (CT2TFL) by using multiple composite *Lee-oscillators*.

The motivation for introducing CT2TFL is in three aspects, which are as follows:

1. Typical financial signals and indicators such as RSI and MACD coexist with multiple linguistic variables (over-sell, bearish, bullish, and over-buy), FLS is a viable solution for technical signal modeling;
2. These linguistic variables themselves exist with different degrees of uncertainty and fuzziness in nature, T2FLS is the best tool for handling such high-order degree of fuzziness;
3. If we can construct a chaotic neural oscillator model in which its transient-chaotic bifurcation diagram itself can directly model (simulate) a typical T2FLS with a transient-fuzzy property only appears in FOU, the computational complexity problem of T2FLS will be effectively resolved. More importantly, if such composite chaotic neural oscillators themselves are exactly input neurons of a deep neural network (DNN), the type-2 fuzzification process will become the *intrinsic property* of the DNN itself; instead of *external* fuzzification process in many contemporary fuzzy-neuro systems.

In the next section, we will study an overview of discrete-time neural oscillators and chaotic neural networks; and how it can be used to implement a type-2 chaotic fuzzy neural network for time series prediction. After that, we will explore how composite *Lee-oscillators* can be used to model *T2FMF* for financial time series financial prediction.

12.2.2 Lee-Oscillators with Retrograde Signaling (LORS)

Latest research in neuroscience revealed that *retrograde signaling (RS)* in neurons plays a vital role in temporal information processing and memory recalling in human brain (Faghihi and Moustafa 2017; Korkut et al. 2013; Naeem et al. 2015; Yoshihara et al. 2005), while abnormal retrograde signaling may lead to various types of memory disorders such as dementia, memory loss, Alzheimer disease, and Down syndrome (Sidoryk et al. 2011; Steinert et al. 2010; Verkhratsky et al. 2010; Volman et al. 2007).

Inspired by retrograde signaling in neurons, *chaotic Lee-oscillators with retrograde signaling* (aka *LORS*) is the extension of the author's previous works on *Lee-oscillator* (Lee 2004a; 2006a) with the integration of retrograde feedback signals in the neural dynamics for the improvement of temporal information processing in neural oscillatory networks. Figure 12.3 shows the neural architecture of *LORS*.

$$E(t + 1) = Sig'[a_1 LORS(t) + a_2 E(t) - a_3 I(t) + a_4 S(t) - \xi_E] \qquad (12.2)$$

$$I(t + 1) = Sig'[b_1 LORS(t) - b_2 E(t) - b_3 I(t) + b_4 S(t) - \xi_I] \qquad (12.3)$$

$$\Omega(t + 1) = Sig'[S(t)] \qquad (12.4)$$

$$LORS(t) = [E(t) - I(t)]e^{-kS^2(t)} + \Omega(t) \qquad (12.5)$$

where E, I, Ω, and LORS denote excitatory, inhibitory, input, and output neurons of LORS; a_i and b_i are the weights for excitatory and inhibitory neurons; ξ_E and ξ_I are the threshold values and S(t) is the external input stimulus.

After extensive experiments and research (2008–2018) on the bifurcation behaviors of *LORS* with different parameter settings and conditions in Table 12.1, we categorize *LORS* into 8 major categories (*LORS#0–LORS#7*), with *LORS#0* corresponds to the original *Lee-oscillator* without retrograde signals and *LORS#1–LORS#7* correspond to different degrees of bifurcations, from single to multiple bifurcation regions. Figure 12.4 illustrates the bifurcation diagrams for the 8 major categories of *LORS*.

12.3 CT2TFDNN—The System

12.3.1 CT2TFDNN—System Framework

The *chaotic type-2 transient-fuzzy deep neuro-oscillatory network with retrograde signaling* (aka *CT2TFDNN*) proposed in this chapter is a hybrid intelligent worldwide financial prediction system with the integration of the following:

1. chaotic neural oscillators (*Lee-oscillators*) as basic neural structure;
2. *chaotic type-2 transient-fuzzification* of the financial signals (FS) for signal modeling;
3. *genetic algorithms* (GA) for the selection of top-10 significant Fuzzy financial signals (FFS); and
4. *chaotic transient-fuzzy deep neuro-oscillatory networks with retrograde signaling* for the chaotic financial time series deep learning and prediction.

Figure 12.5 depicts the system framework of *CT2TFDNN* for worldwide financial prediction.

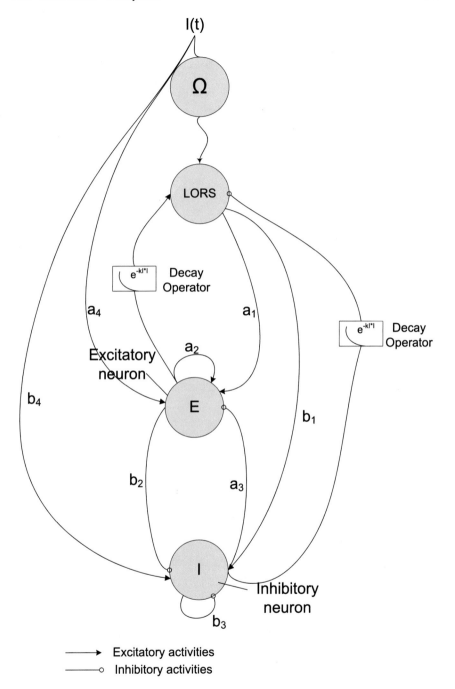

Fig. 12.3 Neural architecture of LORS

Table 12.1 Parameter settings of 8 major categories of LORS

Parameters	*Lee-oscillator* with retrograde signaling (LORS)							
	LORS #0	LORS #1	LORS #2	LORS #3	LORS #4	LORS #5	LORS #6	LORS #7
a_1	0.00	−0.50	−0.50	0.50	0.90	0.90	5.00	5.00
a_2	5.00	0.55	0.55	0.55	0.90	0.90	5.00	5.00
a_3	5.00	0.55	0.55	0.55	0.90	0.90	5.00	5.00
a_4	1.00	0.50	0.50	0.50	0.90	0.90	5.00	5.00
b_1	0.00	−0.50	0.50	0.50	−0.90	−0.90	−1.00	−1.00
b_2	−1.00	−0.55	−0.55	−0.55	−0.90	−0.90	−1.00	−1.00
b_3	1.00	−0.55	−0.55	−0.55	−0.90	−0.90	−1.00	−1.00
b_4	0.00	0.50	−0.50	−0.50	−0.90	−0.90	−1.00	−1.00
k	1.00	1.00	1.00	1.00	1.00	1.00	1.00	1.00
ξ_E	500	50	50	50	50	300	50	300
ξ_I	0.00	0.00	0.00	0.00	0.00	0.00	0.00	0.00
s	0.00	0.00	0.00	0.00	0.00	0.00	0.00	0.00

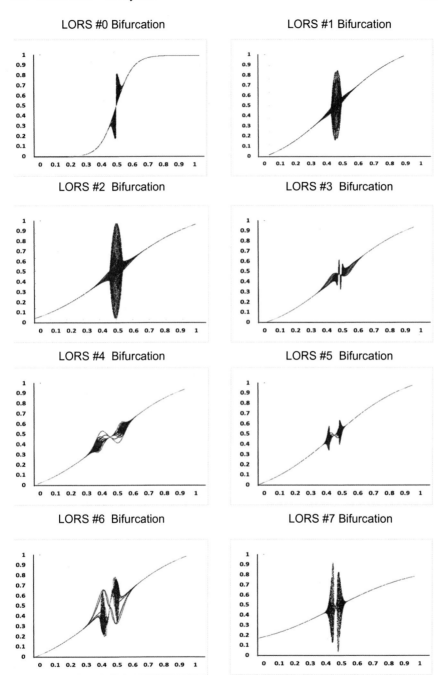

Fig. 12.4 Eight major categories of bifurcations in LORS

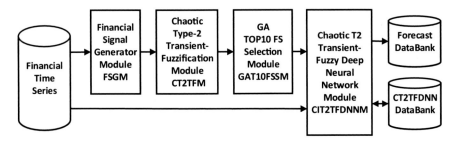

Fig. 12.5 System framework of CT2TFDNN for worldwide financial prediction

12.3.2 Financial Signal Generator Module (FSGM)

In the financial signal generator module (FSGM), 39 worldwide commonly used financial indicators/oscillators are generated. They include moving averages (MA), relative strength index (RSI), Bollinger bands (BB), MACD, Stochastic oscillators, etc. Appendix B shows the list of 39 financial trading signals generated in the proposed system.

All the daily financial time series databank will be used to generate these 39 trading signals, which are used for type-2 transient-fuzzification process performs in the next module. Also, for the ease of fuzzification and network training, all the technical indicators/oscillators being generated in this module will be converted into the technical oscillator and normalize with values between 0 and 1. For instance, the MA (Moving Average) technical indicator will be converted into MA oscillator by evaluating the signal crossing between two MA signals, such as MA5 and MA13, the two most commonly used *Fibonacci numbers* commonly used for technical signal crossing in financial engineering (Murphy 1999).

12.3.3 Chaotic Type-2 Transient-Fuzzification Module (CT2TFM)

(1) **Chaotic Type-2 Transient-Fuzzy Membership Function (CT2TFMF)**

CT2TFMF is the modeling of interval type-2 fuzzy logic by using the author's devised time-discrete chaotic neural oscillator—*LORS*. The formulation of *CT2TFMF* (normalized) is given by

$$CT2TFMF(x, t) = \begin{cases} LORS(2x, t), \ x \leq 0.5 \\ LORS(2 - 2x, t), \ x > 0.5 \geq 1 \end{cases} \quad (12.6)$$

where x is the fuzzy variable value (e.g., RSI value) which corresponds to the input stimulus S(t) in the LORS formulations Eqs. (12.2)—(12.5); t is the type-2 fuzzy logic

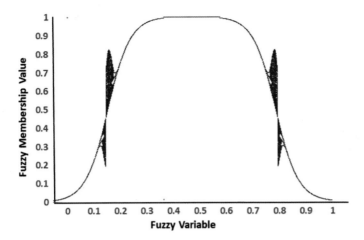

Fig. 12.6 Chaotic type-2 transient-fuzzy membership function (CT2TFMF)

index; LORS(x, t) is the normalized LORS function given by Eq. (14). Figure 12.6
shows the *CT2TFMF* by using LORS#0 with MATLAB simulation result of 1000
× 200 time-steps.

As shown, the *CT2TFMF* generated by *LORS* has certain special features:

(i)   The membership function generated is a typical type-2 fuzzy membership func-
      tion with remarkable *transient-fuzzy* FOU, which corresponds to the bifurcation
      regions of the chaotic oscillators.

(ii) The whole type-2 fuzzy membership function can be modeled and generated
      by a simple time-discrete chaotic neural oscillator, which can be easily imple-
      mented to model various Interval type-2 fuzzy logic systems (IT2FLS).

(iii) As the neural dynamics of FOUs generated by CT2TFMF is inherited from
      the Lee-oscillators, so it naturally converges to the two stable states at both
      ends, while exhibits progressive and controlled *fuzziness* (bifurcation in chaos
      theory).

(iv) By regulating various parameters (e.g., weights of inhibitory and excitatory
      neurons, threshold values ξ_E and ξ_I, decay constant k) in the *Lee-oscillator*, the
      FOUs of *CT2TFMF* can be adjusted and transformed to other chaotic patterns
      to tackle with different complex *IT2FLS* problems.

(2) **Chaotic Type-2 Composite Transient-Fuzzy Membership Function
     (CT2CTFMF)**

For the ease of modeling type-2 fuzzy logic for the 39 financial signals, chaotic type-2
composite transient-fuzzy membership function (aka CT2CTFMF) is constructed to
model the following four type-2 fuzzy financial linguistic variables: over-sell (OSell),
bearish (Bear), bullish (Bull), and over-buy (OBuy).

Figure 12.7 shows the MATLAB simulation of normalized CT2CTFMF for financial signal RSI (relative strength index) using 4 composite Lee-oscillators (LORS#0) with MATLAB simulation results of 1000×200 time-steps.

The *CT2CTFMF* of these four linguistic variables are the following:

$$CT2CTFMF_{Osell}(x, t) = \begin{cases} CT2TFMF(2x + 0.5, t), \ x \le 0.25 \\ 0, \ x > 0.25 \ge 1 \end{cases} \tag{12.7}$$

$$CT2CTFMF_{Bear}(x, t) = \begin{cases} CT2TFMF(1.6x, t), \ x \le 0.625 \\ 0, \ x > 0.625 \ge 1 \end{cases} \tag{12.8}$$

$$CT2CTFMF_{Bull}(x, t) = \begin{cases} 0, \ x \le 0.375 \\ CT2TFMF(1.6x - 0.6, t), \ x > 0.375 \ge 1 \end{cases} \tag{12.9}$$

$$CT2CTFMF_{OBuy}(x, t) = \begin{cases} 0, \ x \le 0.75 \\ CT2TFMF(2x - 1.5, t), \ x > 0.75 \ge 1 \end{cases} \tag{12.10}$$

From the fuzzy-neuro perspective, Fig. 12.7 is the bifurcation diagram of the composite Lee-oscillators with the superposition of four LORS#0 Lee-oscillators which model the membership function CT2CTFMF of the four type-2 fuzzy variables: over-sell, bearish, bullish, and over-buy with their neural dynamics governed by (12.7)–(12.10), respectively.

As shown in the above equations, the formulations of all type-2 fuzzy linguistic variables are continuous state functions with *transient-chaotic* (named as *transient-fuzzy* in *Type-n FLS*) properties within the state-transition regions, which correspond to FOUs in the type-2 FLS. Besides, since *CT2CTFMF* is generated by *Lee-oscillator*

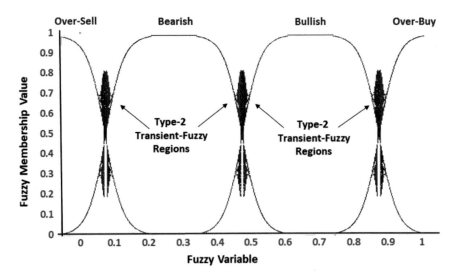

Fig. 12.7 Chaotic T2 composite transient-fuzzy membership function (CT2CTFMF) of RSI

which is a discrete-time chaotic oscillator, all type-2 fuzzy states can be easily modified, implemented, and evaluated by adjusting the total number of first-order (x) and second-order (t) CTFMV (*composite transient-fuzzy membership values*) to fit for different requirements of granularity for any IT2FLS problems.

From the theoretical perspective, the chaotic bifurcation diagram of the composite Lee-oscillators is the transient-chaotic output states of these chaotic oscillators subjected to the input stimulus. In terms of type-2 fuzzy logic, the transient-fuzzy bifurcation at *type-2 transient-fuzzy regions* shown in Fig. 12.7 are theoretically analog to the interval type-2 concept in fuzzy logic. The difference is that this chaotic transient-fuzzy membership function is not an additional or external *fuzzification engine* that usually is applied by typical IT2 FLS, but rather it is an intrinsic property of the fuzzy-neuro network itself. In other words, the type-2 fuzzification process proposed in this system is just part of the neural dynamics of the input neurons by simply replaced each of the input neuron (financial signal) by 4 LORS#0 Lee-oscillators and automatically *inherited* the transient-fuzzy capability during the system training and forecast process.

(3) **Chaotic T2 Transient-Fuzzification Module (CT2TFM)**

In this module, all 39 normalized trading signals (oscillators) will undergo the chaotic interval type-2 fuzzification process by applying the *CT2CTFMF* described in (12.7)–(12.10). The resulted 39 (each fuzzy financial signal has 4 second-order type-2 TFMV) fuzzified financial signals will enter to the GA selection module for top-10 financial fuzzy signals selection.

12.3.4 GA-Based TOP-10 Financial Signals Selection Module (GAT10FSSM)

GA-based top-10 financial signals selection module (GAT10FSSM) is a vital process in this system to screen-out noncritical signals/data, as usually there are too many signals and time series data appeared in many financial engineering problems. If all related signals and time series are used for financial prediction, it will not only slow down the whole system training and prediction process but more importantly, it will cause serious system over-training and deadlock problems by being trapped in the local minima.

Top10-FFSSM is a *GA (genetic algorithms)* based module to select the best (top 10) fuzzy financial signals (FFS) generated in the *CT2TFM*. It consists of six functional modules shown in Fig. 12.8.

(1) **Fuzzy Financial Signals Population Generation Module (FFSPGM)**

In this module, a population of 1000 *fuzzy financial signal vectors FFSV* (*s, w*) are generated, where *s* and *w* are the signal vectors of the 39 FFS and their weights with values between 0 and 1. Noted that *s* is the composite fuzzy

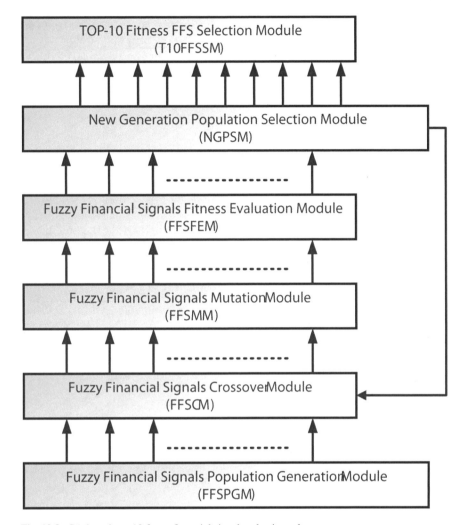

Fig. 12.8 GA-based top-10 fuzzy financial signals selection scheme

financial signal vectors with 4 dimensions correspond to the four type-2 fuzzy variables.

In the initialization stage, a random number generator is used to generate all the weights for these 1000 FFSVs. Note that these weights are also used as the initial weights used for the *CT2TFDNN* for fitness evaluation. In fact, the objective of this GA-based signal selection scheme is to evaluate the best (fitness) FFSV and retrieve the top-10 FFS with highest weights for system training and prediction in *CT2TFDNN*.

(2) **Fuzzy Financial Signals Crossover Module (FFSCM)**

The whole *FFSCM* consists of the following steps:

Step 1. Randomly selects two FFSVs from the population.
Step 2. Perform a two-point crossover operation.
Step 3. Two new FFSVs offspring are generated.
Step 4. Perform steps 1–3 for 500 times until 1000 FFSV offspring are created.

3) **Fuzzy Financial Signals Mutation Module (FFSMM)**
   In *FFSMM*, a 5% (i.e., 0.05) mutation rate is adopted.
   The whole *FFSMM* consists of the following steps:

   Step 1: For each 1000 FFSV offspring, generate a random number (between 0 and 100) for each of the 39 FFS.
   Step 2: If the random generated for a particular FFS is less than 5, then the weight (w) for that FFS will be replaced by a new random number between 0 and 1.
   Step 3: Repeat steps 1–2 until all 1000 FFSV offspring have to go through the FFS mutation operation.

(4) **Fuzzy Financial Signals Fitness Evaluation Module (FFSFEM)**
    In this module, each of the 1000 FFSV offspring will be used (in turn) with the time series financial data as input vectors for *CT2TFDNN* deep training. Detailed training algorithm will be studied in the next section.
    After 1000 iterations, the overall RMSE (Root-Mean-Square Errors) between the *target next-day close* and the *actual-day close* will be evaluated and stored as the *fitness value (FV)* for that particular FFSV offspring.

(5) **New Generation Population Selection Module (NGPSM)**
    For 1000 FFSVs of the current population and their 1000 FFSV offspring, evaluate their FVs and choose the first 1000 best FFSV with the lowest RMSE as members for the new generation.

(6) **TOP-10 Fitness FFS Selection Module (T10FFSSM)**
    Repeat the above GA modules for 1000 generations. For the last generation, select the best (fit) FFSV with the highest FV (i.e., the lowest RMSE). Inspect the weights (w) of its 39 FFS and choose the top-10 FFS with the highest weights as the target top-10 FFS and use these weights as the initial for network training in the *CT2TFDNN* module.

Table 12.2 shows the list of top-10 FFS after 1000 generations.

As shown, it is not surprising to find out that those popular financial signals such as: MACD, RSI, MA (OsMA), and Stochastic signals appeared in the top-10 list, mainly because these financial signals are so popular that many traders and investors are using them to design their trading strategies so that are more correlated with the forecast results than other financial signals.

Table 12.2 List of TOP−10 fuzzy financial signals

No	FFS code	Fuzzy financial signals
1	AO	Awesome oscillator FFS
2	Bands	Bollinger bands FFS
3	Gator	Gator oscillator FFS
4	Ichimoku	Ichimoku Kinko Hyo FFS
5	Momentum	Momentum FFS
6	OsMA	Moving average oscillator FFS
7	MACD	MACD FFS
8	RSI	Relative strength index FFS
9	RVI	Relative vigor index FFS
10	Stochastic	Stochastic oscillator FFS

12.3.5 Chaotic T2 Transient-Fuzzy Deep Neural Network Module (CT2TFDNNM)

In short, *CT2TFDNN* is a multilayer deep supervised-learning neural oscillatory network with the replacement of all neurons all layers of the networks with *LORS* with a *chaotic bifurcation transfer function (CBTF)* as activation functions. Besides, eight chaotic bifurcation hidden layers (CBHL) which correspond to the 8 different modes of bifurcation in *LORS* are used to facilitate chaotic deep learning in *CT2TFDNN*.

Figures 12.9 and 12.10 show the system architecture and the chaotic deep learning algorithm of *CT2TFDNN* for worldwide financial prediction.

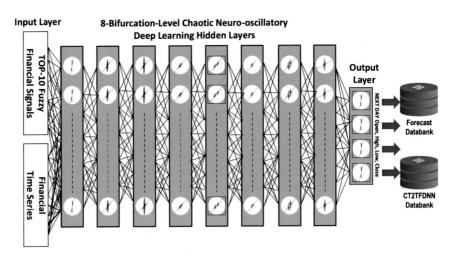

Fig. 12.9 System architecture of CT2TFDNNM

CT2TFDNN Chaotic Transient-Fuzzy Deep Learning Algorithm

1 CT2TFDNN Initialization Phase

1.1 Initialization all the network weights ω by a random number generator to values between 1 and 0.

2 CT2TFDNN Checking Stop Training Criteria

IF MSE < Training Threshold δ (say 1×10^{-6}) STOP,

Else CONTINUE

3 CT2TFDNN Forward Propagation Phase

3.1 Evaluate the total inputs for all hidden $LORS$ (L_{H0})

$$\overrightarrow{L_{H0input}} = \sum_{n=0}^{N_I} \overrightarrow{L_{In}}\,\overrightarrow{\omega_n}$$

Noted that N_I is the total number of input LORS.

3.2 Evaluate the TCAF values of all $\overrightarrow{L_{H0input}}$ vectors using chaotic Lee_operator given by equations (6) to (9)

$$\overrightarrow{L_{H0}} = Lee_{H0}\big(\overrightarrow{L_{H0input}}\big)$$

3.3 Evaluate the total input vectors for hidden $LORS$ (L_{H1})

$$\overrightarrow{L_{H1input}} = \sum_{n}^{N_{H0}} \overrightarrow{L_{H0}}\,\overrightarrow{\omega_n}$$

Noted that N_{H0} is the total number of hidden LORS at layer H0.

3.4 Evaluate the CBTF values of all $\overrightarrow{L_{H1input}}$ vectors

$$\overrightarrow{L_{H1}} = Lee_{H1}\big(\overrightarrow{L_{H1input}}\big)$$

3.5 Repeat the forward propagation to next layer until propagate to the output layer.

4 CT2TFDNN Backward Propagation Phase

4.1 Evaluate the $\overrightarrow{\delta_o}$ (Correction Error Vector) and $\overrightarrow{\Delta\omega_{H70}}$ (weight adjustment vectors between hidden layer 7 and output layer) of all $\vec{L}_o$ against the target output vectors $\vec{L}'_O$ with network learning rate β.

$$\overrightarrow{\delta_{HO}} = \big(\vec{L}'_O - \vec{L}_O\big)f'_{L_O}\big(\overrightarrow{L_{Oinput}}\big)$$

$$\overrightarrow{\Delta\omega_{H70}} = \beta\overrightarrow{\delta_{H70}}\,\overrightarrow{L_{H7}}$$

4.2 Evaluate the $\overrightarrow{\delta}_{H7}$ and $\overrightarrow{\Delta\omega_{H6H7}}$ of all $\vec{L}_{H7}$

$$\overrightarrow{\delta_{H6H7}} = \big[\sum \overrightarrow{\delta_{H70}} \cdot \overrightarrow{\omega_{H70}}\big]\big(\vec{L}'_O - \vec{L}_O\big)f'_{L_O}\big(\overrightarrow{L_{Oinput}}\big)$$

$$\overrightarrow{\Delta\omega_{H6H7}} = \beta\overrightarrow{\delta_{H6H7}}\,\overrightarrow{L_{H6}}$$

4.3 Evaluate the all the weight vectors at the same time.

$$\overrightarrow{\omega_{H6H7}(t+1)} = \overrightarrow{\omega_{H6H7}(t)} + \overrightarrow{\Delta\omega_{H6H7}(t)}$$

4.4 Backward propagate to the previous hidden layer.

5 CT2TFDNN STEP 2 to Check for Stopping Criteria.

Note:

1. L_I, L_o, L_{H0} ... L_{H7} are the *Lee-oscillator* in the input, output and the 8 hidden deep-learning layers.
2. The four output Lee oscillators correspond to the next-day forecasts of Open, High, Low and Close.
3. ω are the network weights.
4. CBTF – Chaotic Bifurcation Transfer Function.
5. δ are the correction error vectors.

Fig. 12.10 CT2TFDNN chaotic transient-fuzzy deep learning algorithm

As shown in Fig. 12.9, the input vectors of *CT2TFDNN* include the following:

(1) Top-10 T2 fuzzy financial signals (FFS), each consists of 4 fuzzy attributes indicate the T2 FMV of: *over-sell (OS), bearish (BR), bullish (BU), and over-buy (OB)*;

(2) Time series financial data, each set consists of the *open (O), high (H), low (L), close (C), and volume (V)* of 10-trading days normalized time series data, each is multiplied by time-step decay *(forgetting)* function fg() given by

$$fg(D) = e^{-kD} \qquad\qquad (12.11)$$

where D is the time-horizon for time series and k = 0.5 is the decay factor.

Figure 12.11 shows the decay chart for financial time series data (Lee 2004b).

As shown in Fig. 12.11 the decay function almost converges to 0 when *time horizon (TH)* close to day-10. So, we set the time horizon for time series financial data to 10. As reflected from previous works (Lee 2004b), the introduction of decay (forgetting) factor is vital not only because it simulates time series relationships of the historical data, but also it can avoid over-training by effectively control the data size of time series input vectors and speedup the whole training process. Also, by replacing all neurons of *CT2TFDNN* with *Lee-oscillators*, the traditional FFPBN effectively transforms into a multilayer neuro-oscillatory network with inherited transient-chaotic neuron activation properties from *Lee-oscillators*, which not only significantly speedup the whole network training process, but more importantly it

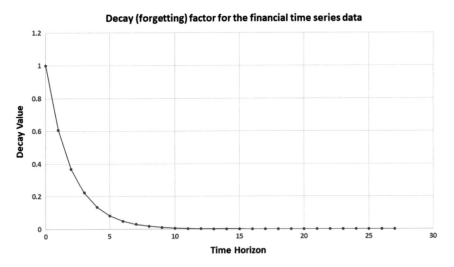

Fig. 12.11 Decay (forgetting) factor for the financial timeseries data

resolves the over-training and deadlock problems which commonly occurs in most recurrent neural networks with massive financial time series data. Detailed system performance analysis will be studied in Sect. 12.5.

12.4   CT2TFDNN—System Implementation

12.4.1   Quantum Finance Forecast Center

As shown in Fig. 12.12, CT2TFDNN forecaster is a server-side intelligent forecast agent located at the MT4 server farm of QFFC using Dell precision T5820 tower workstation with Intel 6-Core 3.6 GHz Xeon processor. For each financial product, 2048-trading day time series (except cryptocurrency which only has 300-trading day data) include: open (O), high (H), low (L), close (C), and volume (V) are automatically generated by MT4 engines of Forex.com and AvaTrade.com. Through four functional processes studied in the previous section, the next-day forecasts of these 129 financial products are stored at the CT2TFDNN forecast database for CT2TFDNN trading agents to access and release for public access.

12.4.2   129 Worldwide Financial Products

With the successful cooperation with AvaTrade.com (major cryptocurrency MT4 service provider) and Forex.com (major forex MT4 service provider), QFFC launched

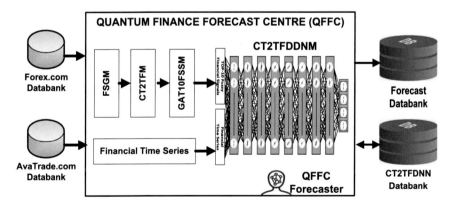

Fig. 12.12 System architecture of quantum finance forecast center (QFFC)

the 129 financial products free daily and weekly forecast services from January, 2018 for worldwide traders and individual investors. They include the following:

(1) 9 major cryptocurrencies;
(2) 84 worldwide forex;
(3) 19 major commodities;
(4) 17 worldwide financial indices.

Appendix A shows the list of 129 financial products under these four categories.

As shown in Appendix A, owing to the short trading history of cryptocurrency, 300-trading day of historical time series are provided by AvaTrade.com, while all other financial products consist of a 2048-trading day of historical time series for each financial product provided by Forex.com which provide sufficient training and test sets for system implementation and performance analysis. Also, each time series record consists of daily information: open (O), high (H), low (L), close (C), and volume (V).

12.4.3 CT2TFDNN Network Parameter Settings

Table 12.3 shows parameter settings of CT2TFDNN for network training and forecast. As shown, 10 trading day time series are input into CT2TFDNN for network training, so totally we have 50 input nodes (LORS#0) for time series inputs and 100 fuzzy financial signals (FFS) inputs with totally 400 FFS input nodes (LORS#0). For each hidden layer, 400 LORS (LORS#0–LORS#7) hidden nodes are used for network training. For financial products with 2048-trading day time series data, 1228 (60%) time series dataset is used for training, 401 (20%) time series dataset is used for network testing and validation. For cryptocurrency with only 300 trading day time

Table 12.3 Parameter settings of CT2FDNN

Parameters	Values
No. of trading days for network input	10
No. of timeseries LORS in input layer	50
No. of input Fuzzy Financial Signals (FFS)	100
No. of FFS LORS in input layer	400
No. of LORS in each hidden layer	400
No. of output LORS	4
Network training stop criteria (RMSE)	1×10^{-6}
Deadlock criteria	10,000 epochs
LORS parameter settings in input and output layers	LORS#0 settings (Table I)
LORS parameter settings in 8 hidden layers	LORS#0-LORS#7 (Table I)

series data, same proportion of datasets are employed for the next training, testing, and validation.

12.4.4 System Implementation and Pilot Run

From the implementation perspective, the whole CT2TFDNN system are implemented over MT4 (MetaTrader4) platform (Walker 2018; Young 2015)—the world's biggest online financial program trading and development platform with over hundreds of participating financial institutions for the provision of online program trading and development services of all common worldwide financial products.

Since January 1, 2018, QFFC released daily financial forecast using CT1FNON (type-1 fuzzy chaotic neuro-oscillatory networks) as pilot run and open testing for worldwide traders and investors.

From October 1, 2018, after completing the system implementation of *CT2TFDNN*, QFFC daily worldwide financial predictions are conducted by *CT2TFDNN* system.

12.5 CT2TFDNN—Performance Analysis

12.5.1 Introduction

From the system performance perspective, three types of system performance analysis were conducted. They were the following:

1. System training performance analysis;
2. System forecast simulation performance analysis;
3. Actual daily forecast performance analysis.

In all these performance analyses, the system performance of CT2TFDNN was compared with five forecast models:

1. Traditional time series feedforward backpropagation network (FFBPN);
2. Support vector machine (SVM) forecasting tool provided by R Project—one of the most popular financial forecasting tools used in the finance industry;
3. Deep neural network (DNN) with PCA (principal component analysis) model (Singh and Srivastava 2017);
4. Classical interval type-2 fuzzy-neuro network (IT2FNN);
5. Chaotic type-1 fuzzy-neuro-oscillatory network (CT1FNON).

12.5.2   System Training Performance Analysis

In CT2TFDNN training performance analysis, 60% of time series data of 129 financial products are employed for system training in two aspects. Figure 12.13 shows system performances of the six forecast models over 500 epochs of network training

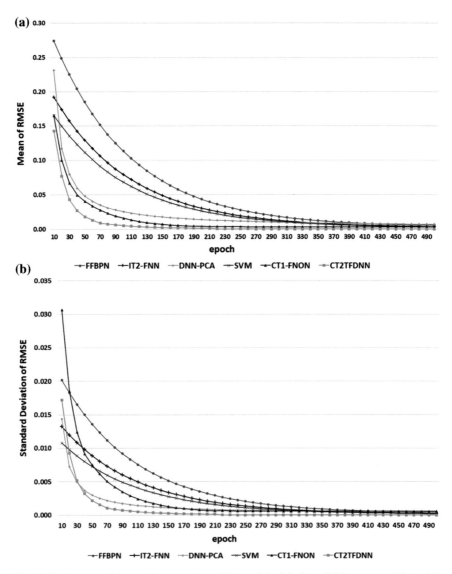

Fig. 12.13 System training performance (over 500 epochs) of six financial forecast models for 129 financial products. (TOP) Mean of RMSE. (BOTTOM) Standard deviation of RMSE

of the 129 financial products in terms of mean and standard deviations of RMSE (root-mean-square errors).

As shown in Fig. 12.13, two observations can be found: (1) CT2FDNN outperformed the other five models in terms of both mean and standard deviation of RMSE; (2) CT2FDNN attained the promisingly low RMSE in 70 epochs while the RMSE of other networks especially FFBPN, SVM, and IT2FNN were still *half-way* of their lowest RMSE levels.

12.5.3 System Forecast Simulation Performance Analysis

In the system forecast simulation performance analysis, four categories of worldwide 129 financial products were tested with target RMSE (Root-Mean-Square Error) of the forecast next-day closing price ranging from 1×10^{-4} to 1×10^{-7}, respectively. The test was done by applying 500 forecast simulations for each system. Table 12.4 presents the system forecast simulation performance test of these six systems. Certain interesting findings are revealed in Table 12.4:

(1) For case 1 simulation (RMSE 1×10^{-4}), *CT2TFDNN forecaster* outperformed FFBPN (453.94), SVM (298.92), DNN-PCA (227.40), IT2FNN (246.81), and CT1FNON (5.58) times. Similar findings can be found in case II simulation results. It clearly reflected the improvement of network learning rate achieved by the hybrid type-2 transient-fuzzy with deep chaotic neuro-oscillatory network system and GA-based top-10 FFS selection scheme.

(2) Across the 3 cases with decreasing RMSE from 1×10^{-4}(case 1), 1×10^{-5} (case 2), 1×10^{-6} (case 3) to 1×10^{-7} (case 4). All forecast systems achieved the target RMSE in cases 1 and 2. However, for cases 3 and 4 simulations using target RMSE 1×10^{-6} and 1×10^{-7}, both FFBPN and IT2FNN (which were using sigmoid-based FFBPN for machine learning) encountered deadlock problems during the network training of cryptocurrency and forex products; while *CT2TFDNN forecaster* could still complete the network training with promising training speeds. It clearly demonstrated the resolution of over-training and deadlock problems and improvement of system training efficiency by CT2TFDNN forecaster over traditional recurrent neural networks using classical sigmoid-based activation function.

(3) Comparing CT2TFDNN against CT1FNON across the three cases, it was interested to reveal that CT2TFDNN outperformed its counterpart by 5.58 times (case 1), 5.65 times (case 2), 17.48 (case 3), and 28.50 (case 4) respectively. As shown in Table 12.4, the overall performance of the CT1FNON forecast system deteriorated substantially when the target RMSE was set to 1×10^{-6} and beyond, especially on cryptocurrency and forex, while CT2TFDNN still performed stable result. It clearly reflected the merits for the integration of type-2 transient-fuzzification scheme with chaotic neural oscillator technology with retrograde signaling for time series chaotic deep learning.

Table 12.4 System forecast simulation performance analysis

Product category	FFBPN		SVM		DNN-PCA		IT2-FNN		CT1-FNON		CT2TFDNN	
	Total STT	Av. STT	Total STT	Av. STT	Total STT	Av. STT	Total STT	Av. STT	Total STT	Av. STT	Total STT	Av. STT
Case 1 (RMSE = 1 × 10^{-4})												
Cryptocurrency	557,251	61,916.78	372,244	41,360.41	244,633	27,181.47	305,240	33,915.56	5720	635.56	1023	113.67
Forex	508,453	6053.01	331,511	3946.56	291,344	3468.38	275,154	3275.64	6323	75.27	1128	13.43
Financial Index	41,120	2164.21	26,152	1376.44	19,738	1038.82	21,183	1114.89	923	48.58	159	8.37
Commodity	46,641	2743.59	29,384	1728.46	22,108	1300.46	25,564	1503.76	1223	71.94	231	13.59
Overall	1,153,465	8941.59	759,291	5885.98	577,823	4479.25	627,141	4861.56	14,189	109.99	2541	19.70
Case 2 (RMSE = 1 × 10^{-5})												
Cryptocurrency	1,460,000	162,222.22	937,320	104,146.67	823,440	91,493.33	862,334	95,814.89	15,121	1680.11	2673	297.00
Forex	1,235,543	14,708.85	841,405	10,016.72	720,322	8575.26	782,507	9315.56	17,231	205.13	2894	34.45
Financial Index	111,024	5843.37	68,391	3599.51	50,627	2664.58	62,236	3275.58	2312	121.68	481	25.32
Commodity	109,142	6420.12	71,925	4230.86	44,967	2645.09	64,733	3807.82	2640	155.29	551	32.41
Overall	2,915,709	22,602.40	1,919,041	14,876.29	1,639,356	12,708.19	1,771,810	13,734.96	37,304	289.18	6599	51.16
Case 3 (RMSE = 1 × 10^{-6})												
Cryptocurrency	DL	–	6,102,255	678,028.38	3,601,307	400,145.17	DL	–	234,663	26,073.67	10,324	1147.11
Forex	DL	–	5,477,816	65,212.10	4,184,833	49,819.44	DL	–	218,661	2603.11	13,234	157.55
Financial Index	577,324	30,385.47	373,529	19,659.40	318,683	16,772.78	459,441	24,181.11	6623	348.58	1292	68.00
Commodity	687,595	40,446.76	468,252	27,544.25	385,741	22,690.64	622,775	36,633.82	9126	536.82	1983	116.65
Overall	–	–	12,421,852	96,293.43	8,490,564	65,818.33	–	–	469,073	3636.22	26,833	208.01
Case 4 (RMSE = 1 × 10^{-7})												
Cryptocurrency	DL	–	25,141,291	2,793,476.73	14,405,228	1,600,580.89	DL	–	1,928,929	214,325.44	51,213	5690.33
Forex	DL	–	26,896,077	320191.39	17,785,540	211732.62	DL	–	1,600,598	19054.74	62,256	741.14
Financial Index	DL	–	1,714,498	90,236.74	1,446,821	76,148.46	2,451,732	129,038.53	37,088	1952.00	6231	327.95

(continued)

Table 12.4 (continued)

Product category	FFBPN		SVM		DNN-PCA		IT2-FNN		CT1-FNON		CT2TFDNN	
	Total STT	Av. STT	Total STT	Av. STT	Total STT	Av. STT	Total STT	Av. STT	Total STT	Av. STT	Total STT	Av. STT
Commodity	DL	–	2,088,404	122,847.29	1,917,133	112,772.52	2,756,693	162,158.41	48,641	2861.24	7132	419.53
Overall	–	–	55,840,269	432,870.30	35,554,721.84	275,618.00	–	–	3,615,256	28,025.24	126,832	983.19

Note

1. Results are generated by 500 simulations of each neural network system (in msec)
2. "Total STT" denotes the total average system training time for 500 simulations
3. "Av. STT" denotes the average system training time for a single financial product
4. "DL" denotes deadlock during system training

(4) In terms of the system performance across different financial products, the simulation results clearly reflected that both cryptocurrency and forex were more chaotic and difficult for network training than other financial products as expected, which will be one of the future R&D directions of quantum finance forecast center.

12.5.4   Actual Daily Forecast Performance Analysis

Started from October 1, 2018, CT2TFDNN provides an official daily forecast of these 129 financial products via QFFC official site on every trading day at 08:00 HKT (00:00 UTC) and reports the previous day forecast performance. Figure 12.14 shows the actual daily forecast performance chart of the six forecast models between October 1, 2018 and March 15, 2019 (120 trading days).

The daily forecast % error (DFPE) is defined by

$$DFPE = \frac{Average(DFErr(H), DFErr(L))}{DClose} \times 100\% \qquad (12.12)$$

where $DFErr(H), DFErr(L)$ are the daily errors of *daily forecast high (low)* with the *daily actual high (low),* respectively, and *DClose* is the daily closing price.

As shown in Fig. 12.14, the 120-trading day actual forecast performance of six forecast models were rather consistent with their system training results shown in previous tests, in which FFPBN performed the worst and CT2TFDNN performed the best.

But one interesting difference is that although the ranking of their performances were the same between system training performance against actual forecast performance, the *overall forecast performance gaps* between these six forecast systems were more distinct in which CT2FDNN outperformed the other five forecast models significantly and it took less than 20-trading days to attain its stable forecast performance state while other forecast systems took 40–60 trading days to attain their stable forecast performance states.

Figure 12.15 shows the daily forecast performance of CT2TFDNN across the four different categories of financial products. As shown, owing to difference levels of chaotic behavior and disturbance between different categories of financial products, there were two interesting findings: (1) the forecast performance of cryptocurrency was significantly poorer than all other categories even though their differences in training performance were not so significant. This might due to the highly chaotic property of cryptocurrency and insufficient time series history data; (2) the forecast performance of forex was better than its performance in system training, which was rather consistent with the observations from the professional traders using the

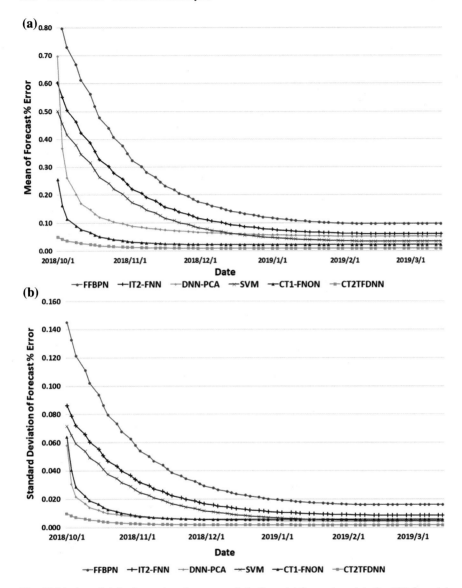

Fig. 12.14 Actual daily forecast performance of six financial forecast models for 129 financial products between 10/1/2018 and 03/15/2019. (TOP) Mean of forecast % error. (BOTTOM) Standard deviation of forecast % error

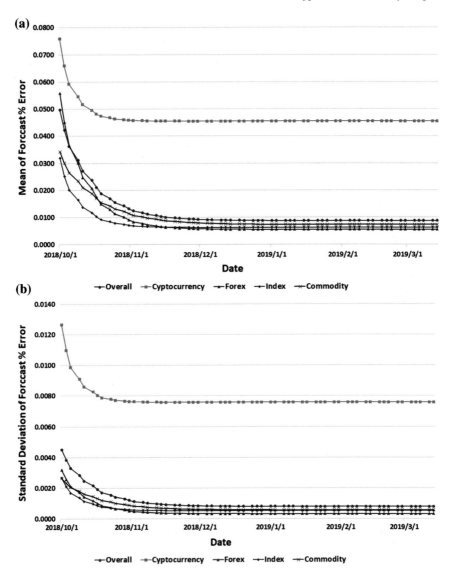

Fig. 12.15 Actual daily forecast performance of CT2TFDNN of four different categories of financial products between 10/1/2018 and 3/15/2019. (TOP) Mean of RMSE. (BOTTOM) Standard deviation of RMSE

CT2TFDNN forecast service during this period of time also concluded that although forex products overall were highly chaotic in nature, CT2TFDNN provided a reliable daily forecast with rather a stable degree of accuracy.

12.6 Conclusion

The dawn of big data epoch has driven us to face new challenges from over-flooding data and information. This chapter devised an innovative chaotic type-2 transient-fuzzy deep neuro-oscillatory network (CT2TFDNN) to address the over-training and deadlock problems, which are commonly occurred during network training of massive data such as financial, weather, and biomedical big data.

For the system architecture perspective, CT2TFDNN provided integration of four different innovative AI technologies:

(1) chaotic neural oscillator for the modeling of neural dynamics;
(2) type-2 transient-fuzzy logic for the modeling of fuzzy financial signals;
(3) genetic algorithm for the selection of the best fuzzy financial signals;
(4) deep chaotic neural network for network training and prediction.

From the implementation perspective, CT2TFDNN has been adopted for the real-time prediction of 129 worldwide financial products. In comparison with five forecast systems ranging from traditional FFBPN, SVM, DNN with PCA, IT2FNN to CT1FNON, CT2TFDNN produced promising results in terms of system training performance and actual daily forecast performance.

The future and related works include the following:

1. Further research and study of CT2TFLS for the modeling, analysis, and data mining of other big data problems such as weather, biometric and biomedical engineering.
2. Further study of CT2TFDNN for the categorization of various CT2TFMF using neural retrograde signaling techniques.
3. R&D of intelligent agent-based hedging and trading systems based on CT2TFLS.
4. Integration of CT2TFDNN with quantum price level (QPL) study using quantum anharmonic oscillatory model (QAOM) for financial trend prediction and long-term investment.

Problems

12.1 What is fuzzy logic? Discuss and explain the difference between type-1 vs. type-2 fuzzy logic. Give two examples and explain how type-2 fuzzy logic can be applied in finance engineering and quantum finance.

12.2 What are the major advantages and shortcomings of type-1 fuzzy logic? Use two applications of type-1 fuzzy logic on technical indicators in the financial market for illustration.

12.3   What is interval type-2 fuzzy logic (IT2FL)? State and explain the major difference between type-2 fuzzy logic and IT2FL. Use two examples in real-world situation for illustration.

12.4   The following figure shows the type-1 fuzzy membership function of RSI

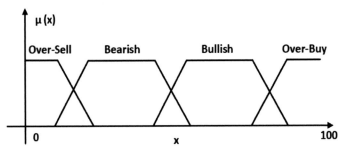

(i)   Use the same methodology, draw the type-1 fuzzy membership function to describe the fuzzy variable of *temperature*.

(ii)   Compare these two fuzzy membership functions, discuss and explain why the fuzzy membership of RSI for the description of financial market is more meaningful and critical for fuzzification than traditional fuzzy variable such as temperature and humidity for the description of weather situation.

(iii)   Based on the given type-1 membership function of RSI:
  • Draw the corresponding type-2 membership function and write the corresponding membership function formulation;
  • Draw the corresponding internal type-2 membership function and write the corresponding membership function formulation;
  • Contract their difference (i.e., pros and cons) and describe how they work;
  • Why they are more meaningful and useful than type-1 fuzzy membership?

12.5   State and describe the formulation and logic behind the type-n fuzzy logic. Use two real-time examples (weather and finance) to illustrate your explanation.

12.6   What are the major advantages and shortcomings of type-n fuzzy logic? Justify your explanation by using real-world example in finance engineering.

12.7   What is retrograde signaling in biological neuroscience? State and discuss the latest research and findings on symptoms and illness in human memory related to this effect.

12.8   What is Lee-oscillators with retrograde signaling? What is the mathematical formulation? How can it relate to biological neuroscience?

12.9   State and explain the formulation of Lee-oscillator with retrograde signaling. Discuss how it works in real-world application on weather prediction and financial engineering.

12.10 Discuss and explain the major differences between Lee-oscillator and Lee-oscillator with retrograde signaling in terms of (1) physical meaning in biological neuroscience; (2) mathematical formulation; physical meaning in chaotic neural network modeling.

12.11 Fig. 12.4 shows the bifurcation diagrams of 8 major Lee-oscillators with retrograde signaling. Discuss and explain their major characteristic in terms of (1) bifurcation; (2) chaotic transfer function capability.

12.12 Below figure shows the system framework of CT2TFDNN (chaotic type-2 transient-fuzzy deep neural network)

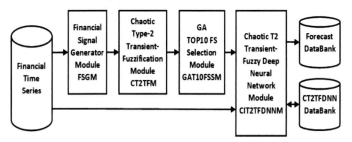

(i) Discuss and explain the major roles and functions of each functional modules of CT2TFDNN.

(ii) What is deep neural network (DNN)? What is different between DNN and ANN?

(iii) What are the major shortcomings of DNN for modeling a complex system such as real-time financial prediction systems?

(iv) What are the major advantages and characteristics of CT2TFDNN versus traditional DNN (deep neural networks)?

12.13 Below figure shows the chaotic type-2 composite transient-fuzzy membership functions (CT2CTFMF) of RSI.

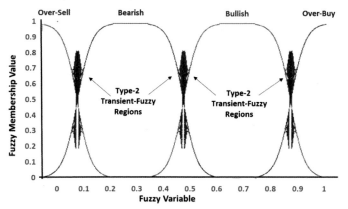

(i) Write down the formulation of CT2CTFMF of RSI.

(ii) In addition to RSI, assume we also need to model MA crossing of two MA signal lines (e.g., MA5 and MA21):

- Draw the corresponding CT2CTFMF of MA-crossing indicator.
- Write down the corresponding formulation for CT2CTFMF of MA-crossing indicator.
- What is the advantage of using CT2CTFMF of MA-crossing indicator values instead of the MA-crossing signals as an input node in the DNN for system training?

12.14 Discuss and explain why GA-based top-10 financial signals section module is critical in the CT2TFDNN system.

12.15 Programming exercise I

(i)   Draw the system flowchart for the implementation of CT2TFDNN system for real-time financial prediction.

(ii) Write the system training algorithms of CT2TFDNN.

(iii) Based on the MQL program skill learnt in QFFC.org, implement CT2TFDNN for at least five forex products using MQL on MT4 platform.

(iv) Compare their system training performance in terms of (1) RMSE (root-mean-square error) and (2) standard deviations in 1000 iterations using error rate 1×10^{-6}.

(v)   Compare their next-day forecast performance results in terms of (1) % Error and (2) Standard deviation of % error for at least 20 trading days (or 3 months MT4 simulation results).

(vi) Implement and perform the same performance test for three other categories of financial products: (1) financial indices; (2) major commodity and (3) cryptocurrency.

(vii) Compare their results in terms of different categories of financial products. Discuss and explain the experimental results.

12.16 Figure 12.15 shows the actual daily forecast performance of CT2FDNN for four different categories of financial products: cryptocurrency, forex, financial indices, and major commodity.

(i)   Discuss and explain why cryptocurrency is always the worst in terms of forecast performance. How can we improve it?

(ii) As compared with system training performance shown in Fig. 12.4, the forecast performance of forex products is significantly improved. Why?

(iii) What is a suggestion and conclusion in terms of a financial investment perspective?

Acknowledgements The author wishes to thank Forex.com and AvaTrade.com for the provision of historical and real-time financial data. The author also wishes to thank Quantum Finance Forecast Center of UIC for the R&D supports and the provision of the channel and platform qffc.org for worldwide system testing and evaluation.

References

Bhattacharya et al. (2016) Secondary Factor Induced Stock Index Timeseries Prediction using Self-Adaptive Interval Type-2 Fuzzy Sets. *Neurocomputing* 171:551–568.

Bhattacharya, D. and Konar, A. (2018) Self-adaptive type-1/type-2 hybrid fuzzy reasoning techniques for two-factored stock index timeseries prediction. *Soft Computing* 22(18): 6229–6246.

Castillo et al. (2014) Application of Interval Type-2 Fuzzy Neural Networks in Non-Linear Identification and Timeseries Prediction. *Soft Computing* 18(6): 1213–1224.

Chen, S. M. and Jian, W. S. (2017) Fuzzy Forecasting Based on Two-Factors Second-Order Fuzzy-Trend Logical Relationship Groups, Similarity Measures and PSO Techniques. *Information Sciences* 391–392: 65–79.

Faghihi, F. and Moustafa, A. A. (2017) Sparse and burst spiking in artificial neural networks inspired by synaptic retrograde signaling. *Journal of Information Science* 421: 40–42.

Gaxiola et al. (2017) Comparison of T-Norms and S-Norms for interval type-2 fuzzy numbers in weight adjustment for neural networks. *Information* 8(3): 114.

Jiang et al. (2018) An Interval Type-2 Fuzzy Logic System for Stock Index Forecasting Based on Fuzzy Timeseries and a Fuzzy Logical Relationship Map. *IEEE Access* 6: 69107–69119.

Konar, A. and Bhattacharya, D. (2017) Handling main and secondary factors in the antecedent for type-2 fuzzy stock prediction. *Intelligent Systems Reference Library* 127: 105–132.

Korkut et al. (2013) Regulation of Postsynaptic Retrograde Signaling by Presynaptic Exosome Release. *Neuron* 77: 1039–1046.

Lee, R. S. T. (2019a) Chaotic Type-2 Transient-Fuzzy Deep Neuro-Oscillatory Network (CT2TFDNN) for Worldwide Financial Prediction. IEEE Transactions on Fuzzy System. https://doi.org/10.1109/tfuzz.2019.2914642.

Lee, R. S. T. (2019b) Chaotic Interval Type-2 Fuzzy Neuro-oscillatory Network (CIT2-FNON) for Financial Prediction. International Journal of Fuzzy Systems. https://doi.org/10.1007/s40815-019-00688-w

Lee, R. S. T. (2006a) LEE-Associator – A Transient Chaotic Autoassociative Network for Progressive Memory Recalling. *Neural Networks* 19(5): 644–666.

Lee, R. S. T. (2006b) *Fuzzy-Neuro Approach to Agent Applications (From the AI Perspective to Modern Ontology)*. Springer-Verlag, Germany.

Lee, R. S. T. (2005) *Advanced Paradigms in Artificial Intelligence From Neural Oscillators, Chaos Theory to Chaotic Neural Networks*. Advanced Knowledge International, Australia.

Lee, R. S. T. A (2004a.) Transient-chaotic Auto-associative Network (TCAN) based on Lee-oscillators. *IEEE Trans. Neural Networks* 15(5): 1228–1243,

Lee, R. S. T. (2004b) iJADE Stock Advisor: An Intelligent Agent based Stock Prediction System using Hybrid RBF Recurrent Networks. *IEEE Trans. on SMC A* 34(3): 421–428.

Lee et al. (2014) An Efficient Interval Type-2 Fuzzy CMAC for Chaos Timeseries Prediction and Synchronization. *IEEE Transactions on Cybernetics* 44(3) 29–341.

Liu et al. (2012) Application of Type-2 Neuro-Fuzzy Modeling in Stock Price Prediction. *Applied Soft Computing Journal* 12(4): 1348–1358.

Melin et. al. (2012) A new approach for timeseries prediction using ensembles of ANFIS models. *Expert Systems with Applications* 39(3): 3494–3506.

Murphy, J. J. (1999) *Technical Analysis of the Financial Markets: A Comprehensive Guide to Trading Methods and Application.* New York Institute of Finance, New York.

Naeem et al. (2015) On the Role of Astroglial Syncytia in Self-Repairing Spiking Neural Networks. *IEEE Trans. On Neural Networks and Learning Systems* 26(10): 2370–2380.

Ontiveros et al. (2018) High order α-planes integration: A new approach to computational cost reduction of General Type-2 Fuzzy Systems. *Engineering Applications of Artificial Intelligence* 74: 186–197.

Pulido, M. and Melin, P. (2018) Optimization of Ensemble Neural Networks with Type-1 and Type-2 Fuzzy Integration for Prediction of the Taiwan Stock Exchange. *Studies in Fuzziness and Soft Computing* 361:151–164.

Ross, T. J. (2016) *Fuzzy Logic with Engineering Applications*. Wiley.

Sanz et al. (2015) Compact Evolutionary Interval-Valued Fuzzy Rule-Based Classification System for the Modeling and Prediction of Real-World Financial Applications with Imbalanced Data. *IEEE Trans. on Fuzzy Systems* 23(4): 973–990.

Sidoryk et al. (2011) Role of astrocytes in brain function and disease. *Toxicol. Pathol.* 39(1): 115–123.

Siler, W. and Buckley, J. J. (2004) *Fuzzy Expert Systems and Fuzzy Reasoning*. Wiley-Interscience.

Singh, R. and Srivastava, S. (2017) Stock prediction using deep learning. *Multimedia Tools and Applications* 76(18): 18569–18584.

Steinert et al. (2010) Nitric oxide signaling in brain function, dysfunction, and dementia. *Neuroscientist* 16(4): 435–452.

Sumati, V. and Patvardhan, C. (2018) Interval Type-2 Mutual Subsethood Fuzzy Neural Inference System (IT2MSFuNIS). *IEEE Trans. on Fuzzy Systems* 26(1): 203–215.

Verkhratsky et al. (2010) Astrocytes in Alzheimer's disease. *Neurotherapeutics* 7(4): 399–412.

Volman et al. (2007) The astrocyte as a gatekeeper of synaptic information transfer. *Neural Computation* 19(2): 303–326.

Walker, W. (2018) *Expert Advisor Programming and Advanced Forex Strategies*. Independently published.

Ye et al. (2016) A Novel Forecasting Method Based on Multi-Order Fuzzy Timeseries and Technical Analysis. *Information Sciences* 367–368: 41–57.

Yolcu et al. (2016) High Order Fuzzy Timeseries Forecasting Method Based on an Intersection Operation. *Applied Mathematical Modelling* 40: 8750–8765.

Yoshihara et al. (2005) Retrograde signaling by Syt 4 induces presynaptic release and synapse-specific growth. *Science* 310: 858–863.

Young, A. R. (2015) *Expert Advisor Programming for MetaTrader 4: Creating automated trading systems in the MQL4 language*. Edgehill Publishing.

Zadeh, L. A. (1975) The Concept of a Linguistic Variable and Its Application to Approximate Reasoning-1. *Information Sciences* 8: 199–249.

Zadeh, L. A. and Aliev, R. A. (2019) Fuzzy Logic Theory and Applications: Part I and Part II. World Scientific Publishing Company.

Chapter 13
Quantum Trader—A Multiagent-Based Quantum Financial Forecast and Trading System

Over the years, financial engineering ranging from the study of financial signals to the modeling of financial prediction is one of the most stimulating topics for both academia and financial community. Not only because of its importance in terms of financial and commercial values, but also it vitally poses a real challenge to worldwide researchers and quants owing to its highly chaotic and almost unpredictable nature. This chapter devises an innovative multiagent-based quantum financial forecast and trading system (a.k.a. quantum trader) for worldwide financial prediction and intelligent trading. With the adoption of author's theoretical works on Lee-oscillator with profound transient-chaotic property, quantum trader effectively integrates quantum field signals (QFS) and quantum field oscillators (OFS) studied in Part I for neural network training and prediction into: (1) quantum forecaster—chaotic FFBP-based time series supervised-learning agent for worldwide financial forecast and; (2) quantum trader—chaotic RBF-based actor-critic reinforcement-learning agents for the optimization of trading strategies. Quantum trader not only provides a fast reinforcement-learning and forecast solution, more prominently it successfully resolves the massive data over-training and deadlock problems which is usually imposed by traditional recurrent neural networks and RBF networks using classical sigmoid or Gaussian-based activation functions.

From the implementation perspective, the quantum trader is integrated with 2048-trading day time series financial data and 39 major financial signals as input signals for the real-time prediction and intelligent agent trading of 129 worldwide financial products which consists of: 9 major cryptocurrencies, 84 forex, 19 major commodities, and 17 worldwide financial indices. In terms of system performance, past 500-day average daily forecast performance of quantum trader attained less 1% forecast percentage errors and with promising results of 8–13% monthly average returns.

© Springer Nature Singapore Pte Ltd. 2020
R. S. T. Lee, *Quantum Finance*,
https://doi.org/10.1007/978-981-32-9796-8_13

13.1 Introduction

Conventional technical analysis and chart analysis methods probe various trading signals (e.g., K-lines, MAs, Bollinger-band index, KDJ and RSI indices) and chart patterns (e.g., major reversal and trend patterns, golden-ratio patterns, Fibonacci patterns, Elliot-wave patterns) that believe to affect the market price trends, trading patterns and individual reasonings to determine the best time to trigger buy or sell decisions (Borden 2018; Brown 2012; Bulkowski 2005; Murphy 1999). However, these numerous trading signals and chart-patterns are usually self-contradictory between different timeframes, let's alone with the fact that they are highly subjective to the traders' own judgment and psychological condition.

With the advancement of computational capacity in past decades, the modeling of complex financial prediction systems ranging from artificial neural networks (ANNs) (Chang et al. 2012; Dai et al. 2012) to fractal-based financial forecast systems (Ling 2013) using ordinary desktop personal computers and workstations is no more a reverie.

Current research on financial prediction include: stock prediction using deep neural network (DNN) with principal component analysis (PCA) by Singh and Srivastava (2017); fuzzy models by Hwang and Oh (2010); Postfix-GP (Genetic Programming) models by Dabhi and Chaudhary (2016); SVM (Support Vector Machine) and hybrid SVR (Support Vector Regression) models by Henrique et al. (2018), Nahil and Lyhyaoui (2018) and Ouahilal et al. (2017).

Moreover, the popularity of free and open quantitative financial system development platforms such as MetaTrader (MT) platform provides an ideal environment for worldwide researchers and quants to test their trading algorithms, strategies and financial signals with real-time financial data streams which prosper the popularity of program trading, especially the high-frequency algorithmic program trading (HFAPT) in the past 10 years (Durbin 2010; Walker 2018; Young 2015).

From the financial market perspective, the blooming of cryptocurrency originated from bitcoin in 2009 had increased to more than 4000 cryptocurrencies in the worldwide financial market (Narayanan 2016; Vigna and Casey 2016). More importantly, major international fund houses and forex trading platforms integrate 24×7 electronic trading of cryptocurrency into their forex trading platforms since 2013. Together with the flourishment of HFAPT in the past 10 years, the worldwide financial markets; especially the international currency market becomes more volatile and unpredictable. An effective, open, and reliable worldwide financial prediction, trading and advisory system is profoundly required for worldwide traders and investors, especially independent investors than ever before.

This chapter devises an innovative multiagent-based quantum financial forecast and trading system (a.k.a. *Quantum Trader*) for worldwide financial prediction and intelligent trading. With the adoption of author's theoretical works on Lee-oscillator with profound transient-chaotic property (Lee 2004), quantum trader effectively integrates quantum field signals (QFS) and quantum field oscillators (OFS) studied in Part I for neural network training and prediction into: (1) quantum finance forecaster (or *quantum forecaster* in short)—chaotic FFBP-based supervised-learning agent for worldwide financial forecast and; (2) quantum finance trader (or *quantum trader* in short)—chaotic RBF-based actor-critic reinforcement-learning agents for the optimization of trading strategies. Quantum trader not only provides a fast reinforcement-learning and forecast solution, more importantly, it successfully resolves the massive data over-training and deadlock problems, which usually imposed by traditional recurrent neural networks and RBF networks using classical sigmoid or Gaussian-based activation functions.

13.2 Quantum Forecaster—Chaotic FFBP-Based Time Series Supervised-Learning Neural Networks for Financial Prediction

13.2.1 Time Series Prediction Using Supervised-Learning-Based Artificial Neural Networks

Time series prediction, ranging from weather forecast to stock prediction has been studied for over 50 years. With the improvement of computational speed, nonlinear models such as artificial neural networks (ANN) have proven success to tackle these problems. Traditional ANN such as feedforward backpropagation neural network (FFBPN) is a typical kind of supervised-learning (SL) neural network to tackle these problems with certain success.

However, when it comes to massive input data such as financial prediction with over thousands time series financial data and signals as input vectors, these conventional ANNs using classical sigmoid-based activation function are usually "trapped" in local minima during neural network training; which not only affects the efficiency (time cost), but also the accuracy of the forecast results (Lee 2006).

13.2.2 Quantum Finance CSL Network—Chaotic FFBP-Based SL Network for Time Series Financial Prediction with Quantum Field Signals (QFF)

Quantum finance chaotic supervised-learning (CSL) network is the integration of Lee-oscillator with classical FFBPN by replacing all neurons with Lee-oscillators. Besides time series signals, the most important characteristics of quantum finance CSL network is the adoption of all related quantum field signals (QFS) for the financial market studied in Part I as quantum field oscillators (QFO) for network training and financial prediction. Figures 13.1 and 13.2 show the system architecture and

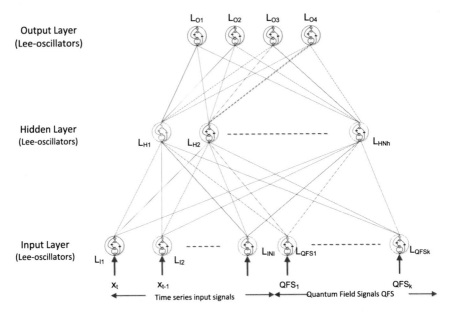

Fig. 13.1 System architecture of quantum finance CSL network for time series financial prediction

chaotic FFBP-based time series supervised-learning algorithm of quantum finance CSL network for financial prediction.

13.3 Quantum Trader—Chaotic RBF-Based Actor-Critic Reinforcement-Learning Networks—Optimization of Trading Strategy

13.3.1 An Overview of Reinforcement Learning (RL)

Different from the supervised-learning model with well-defined *input/target-output* pairs to train the network, there are many situations in which input/target-output pairs do not exist. For example, in stock investment, even though we have the best stock forecast to tell us when to set the buy/sell bid, *how much we should invest?* or *when to close the bid?* in order to get the highest returns is a typical optimization problem without an exact solution. In that case, we can make use of the reinforcement-learning (RL) method.

Reinforcement-learning (RL) theory is originated from behavior psychology. Its main concept is to train the neural network with the adoption of feedback signals, namely reinforcement signal (RS). For the right behavior, the network will respond with a positive RS to *award* the RL network; while for the wrong behavior, the

QUANTUM FINANCE CSL NETWORK LEARNING ALGORITHM

1   Quantum Finance CSLN Initialization Phase

    1.1 Initialization all the network weights ω by a random number generator to values between 1 and 0.

2   Quantum Finance CSLN Checking Stop Training Criteria

    IF MSE < Training Threshold δ (say 1×10^{-6}) STOP,

    Else CONTINUE

3   Quantum Finance CSLN Forward Propagation Phase

    3.1 Evaluate the total inputs for all hidden Lee-oscillators (L_H)

$$\overrightarrow{L_{Hinput}} = \sum_{n=0}^{N_I} \overrightarrow{L_{In}}\,\overrightarrow{\omega_n} + \sum_{j=0}^{k} \overrightarrow{L_{QFSj}}\,\overrightarrow{\omega_j} + \qquad (13.9)$$

    Noted that $N_I = T \times S$ is the total number of input Lee-oscillators, where T is the forecast horizon and S is the dimension of the input signal vector. L_{QFSj} (j = 1..k) are the k QFS for the financial product.

    3.2 Evaluate the TCAF values of all $\overrightarrow{L_{Hinput}}$ vectors using chaotic Lee_operator given by equations (13.5) to (13.8)

$$\overrightarrow{L_H} = Lee_H\left(\overrightarrow{L_{Hinput}}\right) \qquad (13.1)$$

    3.3 Evaluate the total input vectors all output Lee-oscillators (L_O)

$$\overrightarrow{L_{Oinput}} = \sum_{n}^{N_H} \overrightarrow{L_{Hn}}\,\overrightarrow{\omega_n} \qquad (13.2)$$

    Noted that N_H is the total number of hidden Lee-oscillators.

    3.4 Evaluate the TCAF values of all $\overrightarrow{L_{Oinput}}$ vectors

$$\overrightarrow{L_O} = Lee_O\left(\overrightarrow{L_{Oinput}}\right) \qquad (13.3)$$

4   Quantum Finance CSLN Backward Propagation Phase

    4.1 Evaluate the $\vec{\delta}_o$ (Correction Error Vector) and $\overrightarrow{\Delta\omega_{HO}}$ (weight adjustment vectors between hidden and output layer) of all $\vec{L}_o$ against the target output vectors $\vec{L}_O$ with network learning rate β.

$$\overrightarrow{\gamma_{HO}} = \left(\vec{L'}_O - \vec{L}_O\right) f'_{L_O}\left(\overrightarrow{L_{Oinput}}\right) \qquad (13.4)$$

$$\overrightarrow{\Delta\omega_{HO}} = \beta\overrightarrow{\gamma_{HO}}\,\overrightarrow{L_H} \qquad (13.5)$$

    4.2 Evaluate the $\vec{\delta}_H$ and $\overrightarrow{\Delta\omega_{IH}}$ of all $\vec{L}_H$

$$\overrightarrow{\gamma_{IH}} = \left[\sum \overrightarrow{\gamma_{HO}}\cdot\overrightarrow{\omega_{HO}}\right]\left(\vec{L'}_O - \vec{L}_O\right) f'_{L_O}\left(\overrightarrow{L_{Oinput}}\right) \qquad (13.6)$$

$$\overrightarrow{\Delta\omega_{IH}} = \beta\overrightarrow{\gamma_{IH}}\,\overrightarrow{L_I} \qquad (13.7)$$

    4.3 Evaluate the all the weight vectors at the same time.

$$\overrightarrow{\zeta(t+1)} = \overrightarrow{\zeta(t)} + \overrightarrow{\Delta\omega(t)} \qquad (13.8)$$

5   Quantum Finance CSLN STEP 2 to Check for Stopping Criteria.

Note:

1. L_I, L_H and L_O are the Lee-oscillators in the input, hidden and output layer.
2. L_{QFS} are the Quantum Field Oscillators (QFO) which are the Lee-oscillators incorporate with the Quantum Field Signals (QFS) in the input signal layer.
3.   The four output Lee oscillators correspond to the next-day forecasts of Open, High, Low and Close.
4. ω are the network weights.
5.   TCAF – Transient Chaotic Activation Function.
6. δ are the correction error vectors.

Fig. 13.2 Quantum finance CSL network learning algorithm

network will respond with a negative RS to *punish* the RL network. As we can see, RL networks don't need well-defined input/target-output pairs, all they need to do is to search for a set of optimal weights to minimize the negative reinforcement signals.

Classical RL model such as Markov decision process (MDP) using stochastic-based reinforcement-learning algorithms such as Q-Learning, dynamic-programming (DP) and TD-learning (TDL) with certain success (Kaelbling et al. 1996). However, when it comes with time series financial optimization problems with over thousands of input signals and possible trading strategies, these classical stochastic RL methods are either too computationally intensive or difficult to adopt for actual implementation.

With the flourishing of recurrent neural networks in the past decades, researchers began to explore how recurrent neural networks can be applied for RL on various optimization problems (Li et al. 2009; Liu et al. 2005); financial engineering include Deng et al. (2017) using deep direct reinforcement-learning network for financial signal representation and trading; Pendharkar and Cusatis (2018) applied reinforcement-learning agents for trading financial indices; Tan et al. (2011) used reinforcement-learning system for stocking trading with cycles; Chang et al. (2017) used asymmetric reinforcement-learning and conditioned responses technique to analyze global financial crisis; Carapuco et al. (2018) using reinforcement-learning system for short-term speculation in the foreign exchange market and Jeong and Kim (2019) using deep Q-Learning to improve financial trading decisions.

13.3.2 Discrete-Time Actor-Critic RL Model

DTAC-RLM is a multiagent system (Lee 2006) consists of four components: environment space (E), action space (A), actor agent (Actor), and critic agent (Critic). Different from its continuous-time AC model counterpart (such as MDP), DTAC-RLM visualizes its *world* as a collection of *discrete-time-step states* and *actions*. More vitally, all these states are related to time series events. The role of the actor is based on input signals provided at current state at time t (s_t) to respond with the *best* available actions (a_t); whereas the role of the critic is to evaluate the reinforcement signal (RS, or *reward*, r_t in short) based on action(s) taken by the actor and signals provided by the new states (s_{t+1}). Then the reward signal r_t feedbacks as input to the actor to decide the next action **a** at time t + 1 (Fig. 13.3).

The formulations are given by

$$s \in \{S\}, \ a \in \{A\}, \ r \in \{\mathbb{R}\} \tag{13.9}$$

$$\Pi = \{a_t, t = 1 \ldots T\} \tag{13.10}$$

$$V = \sum_{t=1}^{T} r_t(s_t, a_t) \tag{13.11}$$

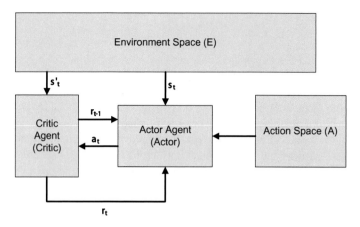

Fig. 13.3 Discrete-time actor-critic RL model

$$\mathbf{V}^* = \max_{\Pi} V^{\Pi} \tag{13.12}$$

As shown in the above formulation: S and A denote the state space and action space, s and a denote the state and action vectors; r denotes the reward vector; Π denotes the action policy from time t = 1 to t = T (dimension of policy space); V denotes the value function which represents the total returns for a particular policy Π. For DRL with maxI iterations, there will be totally maxI policies Π being generated, and $\mathbf{V}^*$ is the optimal returns obtained by the calculation of all possible return values V_k^{Π} (where k = 1... maxI) from these policies. So, the whole optimization problem is to find the best policy Π^* via DRL in order to attain the optimal returns $\mathbf{V}^*$. Next section we will explore how to adopt CRBFN (chaotic radial basis function neural network) for DRL.

13.3.3 Radial Basis Function Neural Network (RBFN) for RL

Like FFBPN, a typical multilayer radial basis function neural network (RBFN) also consists of input, hidden, and output layers. Although network architectures between FFBPN and RBFN are highly similar, the network activations in RBF networks are localized RBF functions such as Gaussian functions, resulting in a much faster training rate (Markopoulos et al. 2016).

Unlike FFBPN, input vectors in RBFN distribute values to the hidden layer neurons uniformly, without multiplying them with weights. The neurons in the hidden layer are presented by RBF neurons using radial basis function such as Gaussian function as activation function, which is given by

$$G(\mathrm{r}) = exp\left[-\frac{r}{2\sigma^2}\right] \qquad\qquad (13.13)$$

Different from FFBPN which is only tailored for supervised learning, experimental results revealed that RBFNs can basically approximate any functions. As a result, they are usually known as "universal-approximators". In other words, RBFN can be used for both SL and RL by using RBFN to approximate the V-value function.

However, like FFBPNs, when it comes with handling massive input data and/or highly chaotic nature such as time series financial forecast using over thousands of input values and financial signals as network input vectors, RBFNs also encounter "over-training" and "deadlock" problems, which hinder further improvement of network accuracy and resulted in an expensive time costs from the computational perspective.

13.3.4   Quantum Finance CRL Network—Chaotic RBF-Based Actor-Critic RL Network for the Optimization of Trading Strategy

Quantum finance chaotic reinforcement-learning (CRL) network integrates the Lee-oscillator technology with conventional RBFN by: (1) replacing all Gaussian-based radial basis functions in the hidden layer of RBFN with Lee-oscillatory RBF functions (namely, *LRbf*); and (2) replacing output neurons with Lee-oscillator to facilitate transient-chaotic neural activations. Figure 13.4 shows LRbf() which served as the chaotic RBF activation function in the quantum trader.

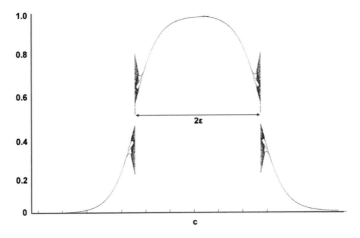

Fig. 13.4 Chaotic RBF activation function using Lee-oscillators

The formulation of LRbf is given by

$$LRbf(x) = \begin{cases} L(x), & x < c \\ L(2c - x), & x \geq c \end{cases} \tag{13.14}$$

where $L(x)$ is the Lee-oscillator function as mentioned in Chap. 9 and c is the center of LRbf().

In contrast with the Gaussian function counterpart, the LRbf exhibits a progressive-transient-chaotic property in two RBF activation regions which provide some sort of *controlled-hysteresis* to sort out the over-training and deadlock problems during RL with massive input vectors.

13.3.5 Quantum Finance CRL Network System Architecture for the Optimization of Trading Strategy

As depicted in Fig. 13.5, quantum finance CRL network is a three-layer chaotic RBF-based neural network with input, hidden, and output layers. To accomplish actor-critic-based RL, the input layer consists of two classes of input vectors: the reward signal vector **r** and input signal vector **x**. The input signal vector **x** composes

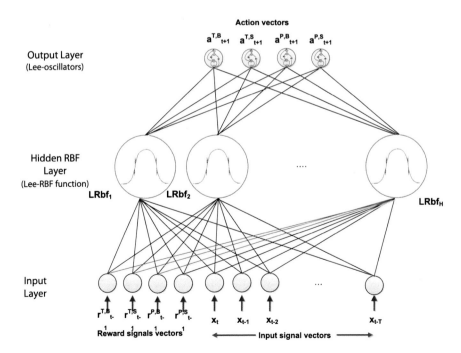

Fig. 13.5 Quantum finance CRL network system architecture

of all time series input signals together with the corresponding quantum price levels (QPLs) as quantum field signal (QFS) for reinforcement training. Reward signal vector **r** contains four reinforcement feedback signals which are generated by the critic agent in the previous time-step. These four reward signals $r^{T,B}$, $r^{T,S}$, $r^{P,B}$, $r^{P,S}$ correspond to four actions $a^{T,B}$, $a^{T,S}$, $a^{P,B}$, and $a^{P,S}$ which emulate de facto short-term trading strategies:

(1) Time-driven-buy strategy—trigger buy action $a^{T,B}$ when current price reaches forecast low, day-end harvest;
(2) Time-driven-sell strategy—trigger sell action $a^{T,S}$ when current price reaches forecast high, day-end harvest;
(3) Price-driven-buy strategy—trigger buy action $a^{P,B}$ when current price reaches forecast low, closing when reaching target price/stop-loss price;
(4) Price-driven-sell strategy—trigger sell action $a^{P,S}$ when current price reaches forecast high, closing when reaching target price/stop-loss price.

The hidden layer of quantum finance CRL network consists of H LRbf nodes, which facilitate transient-chaotic RRB-based RL.

The output layer consists of four possible actions: $a^{T,B}$, $a^{T,S}$, $a^{P,B}$, and $a^{P,S}$ as described above. Their scalar values are normalized values between 0 and 1 which represent the size (lot) of investment ("0" means *not invest*; "1" means to *invest one complete lot*).

13.3.6 Quantum Finance CRL Network Learning Algorithm

The overall quantum finance CRL network learning algorithm is presented in Fig. 13.6.

As shown in Fig. 13.6, the Lee-oscillators are adopted into the algorithm in two aspects: (1) replacement of the Gaussian RBF with chaotic RBF (LRbf) in the hidden layer; (2) replacement of sigmoid-based activation function with Lee-transient-chaotic activation function L in output layer for the determination of action **a**. From the neural dynamic perspective, the adoption of chaotic RBF and activation function can be considered as the emulation of chaotic trading behaviors in the time series events and actions. While from the financial perspective, such adoption of transient-chaotic activation and RBF can be considered as the imitation of *transient-hysteresis* investors' behavior during real-time trading.

For the value function V, the physical meaning of its optimal value V* in terms of financial investment is even more profound: while r_t is total returns (rewards) at time-step t which represents the short-term returns in day trade; its value function V plainly represents the long-term returns of the whole investment policy Π; and its optimal V* directly represents the optimal long-term returns by using the optimal quantum trader network configuration and hence the optimal policy Π*.

Quantum Finance CRL Network Learning Algorithm
1. System Initialization Phase
     1.1   Initialization all the network weights ω between hidden and output layer by a random number generator to values between 1 and 0.
2. For Iteration I = 1 to MaxI, do the following:
3. For each Time-Step t=1 to M, do the following:
4. Input Data Clustering
     4.1   For each hidden RBF oscillator $LRbf_1$ .. $LRbf_H$, cluster all the input vectors x (including all QPLs as Quantum Field Signals);
     4.2   Determine the cluster centers (c_1 .. c_H) of these LRbf oscillators by using the k-centers algorithm.
     4.3   Evaluate the LRbf amplitudes (ε) by:

$$\varepsilon_j = \sqrt{\left[\frac{1}{M}\Sigma_{k=1}^{M}\|x_j - x_k\|^2\right]} \qquad (13.15)$$

        where x_k is the M-closest neighbor of x_j
     4.4   Using the computed ε_j to rescale all the LRbf() functions in the hidden layer accordingly.
5. LRbf evaluation
     5.1   Apply LRbf() function to evaluate the chaotic-transfer-function values ($LRbf_1$ .. $LRbf_H$) for all hidden LRbf oscillators.
6. Evaluation of the four action vector elements using the Lee-chaotic Activation Function L(t)

$$\mathbf{a} = L(\ \Sigma_{y=1}^{H} LRbf_y * \mathbf{w}) \qquad (13.16)$$

     where w is the weight vector between hidden and output layer.
7. Evaluation of the four rewards ($r^{T.B}$, $r^{T.S}$, $r^{P.B}$, $r^{P.S}$) by using the Reward Function Evaluation Algorithm listed in section 4.4.3.
8. Calculate the total rewards R:
$$R = r^{T.B} + r^{T.S} + r^{P.B} + r^{P.S}$$
9. Next Time-step t
10. Evaluate the Value Function VI at I-th iteration
$$V_I = \Sigma_{t=1}^{M} R_t \qquad (13.17)$$
11. Next Iteration I
12. Evaluate the optimal Value Function V*
$$V^* = \max_{1 \leq I \leq MaxI} V_I \qquad (13.18)$$
13. Record the optimal RACTS network configuration C*.
Note:
1. M is the size of time series dataset.
2. H is the number of hidden nodes.
3. MaxI = 1000 is the maximum number of iterations.

Fig. 13.6 Quantum finance CRL network learning algorithm

13.4 System Implementation

13.4.1 Quantum Finance—TEQNA 5-Tier Architecture

From the implementation perspective, quantum finance system model adopts a 5-tier system implementation architecture, namely, *TEQNA* as shown in Fig. 13.7.

Quantum finance TEQNA system architecture consists of the following:

1. Technology layer—supports MT4/MT5 platforms and related programming technologies such as expert advisors (EAs), quantum trader agent protocol (CAP).
2. Encryption layer—supports critical cryptographical technologies include both encryption and blockchain technologies.

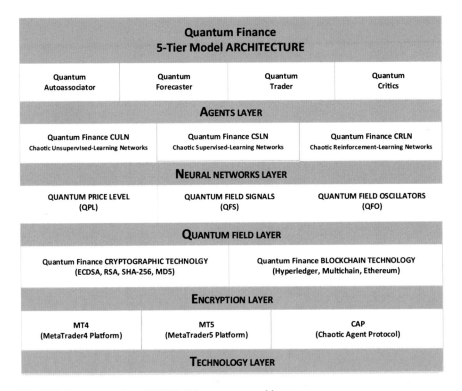

Fig. 13.7 Quantum trader—TEQNA 5-layer system architecture

3. Quantum field layer—supports three basic components of quantum field modeling of financial markets: quantum price levels (QPL), quantum field signals (QFS), and quantum field oscillators (QFO) studied in Part I.
4. Neural networks layer—supports quantum finance chaotic supervised-learning (CSL) networks, quantum finance unsupervised-learning (CUSL) networks, and quantum finance chaotic reinforcement-learning (CRL) networks.
5. Agents layer—supports intelligent chaotic agent applications include quantum auto-associator, forecaster, trading and critic agents.

13.4.2 Quantum Trader System Implementation Using 129 Worldwide Financial Products

From the application perspective, real-time and historical data of worldwide 129 financial products provided by Forex.com (major online Forex trading platform) and AvaTrade.com (the biggest cryptocurrency trading platform) are adopted for quantum trader system implementation. They include: major cryptocurrencies (9), worldwide forex (84), major commodities (19), worldwide financial indices (17). Appendix A shows the list of financial products under these four categories.

13.4.3 Quantum Trader—System Implementation

For the ease of fully integration and automation of quantum trader system with both real-time and historical financial data provided by Forex.com and AvaTrade.com, the whole intelligent agent-based system fully integrated with MT platforms. Figure 13.8 shows the system implementation framework of quantum trader.

The whole quantum trader system consists of two main modules: (1) quantum forecaster using quantum finance CSL network for time series worldwide financial prediction; and (2) quantum trader and critic agents using quantum finance CRL network for the optimization of trading strategy.

The agent activities of these quantum finance agents are described as follows:

Quantum Forecaster Agent

Quantum forecaster is a server-side forecast agent located at the server farm of quantum finance forecast center (QFFC) using Intel i5 CPU 2.39 GHz 32 MB ram Dell servers.

For each financial product, 2048-trading day data (except cryptocurrency which only have 300-trading day data) include: open (O), high (H), low (L), close (C), and volume (V) are automatically generated by MT4 engines of Forex.com and AvaTrade. com. Through the trading signal generator, 39 most common trading signals are generated, together with the 2048-trading day data, they are fed into the forecast system for training and forecast of next-day open (O), high (H), low (L), and close (C). The predicted forecasts of these 129 financial products are stored at the quantum forecast database for the quantum trader trading agents to access. Appendix B shows the 39 trading signals generated by trading signal generator.

Besides input time series data and trading signals, one of the major characteristics of a quantum forecaster is to incorporate with quantum field signals (QFS) generated by quantum field signal generator (QFSG) as shown in Fig. 13.8 that was studied in Part I. These QFS are stored in QFS databank which will also be used by quantum traders and critic agents for reinforcement learning.

Quantum Trader Agents

The quantum trader receives inputs from four sources: (1) input signals from markets (stored at the input signal databanks of the quantum finance server farm) of current and past records; (2) quantum field signals (QFS) from QFS databank; (3) current forecasts stored at quantum forecast database of the quantum finance server farm; and (4) reward signals from previous time-step generated by quantum finance critic agent. By using the quantum finance CRL algorithm, it evaluates four possible actions a. Updates its trading policy and keeps track of the market movements in order to trigger the trading actions.

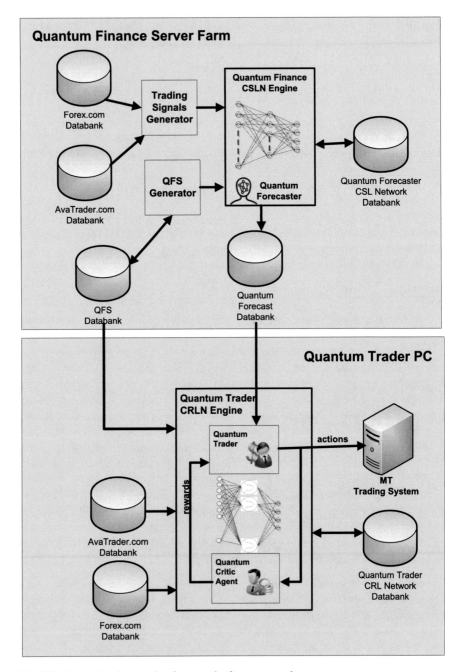

Fig. 13.8 System implementation framework of quantum trader

Quantum Finance Critic Agent

Quantum finance critic agent receives inputs from two sources: (1) input signals from the markets of current and past records; (2) quantum field signals (QFS) from QFS databank; (3) latest action a taken by the quantum trader. By using the reward evaluation algorithm, quantum finance critic agent evaluates the latest rewards r of four actions and feedback to the quantum trader as reinforcement signals for the determination of investment actions in the next time-step.

13.4.4 Quantum Finance Daily Forecast at QFFC

Figure 13.9 shows a snapshot of quantum finance forecaster (or *quantum forecaster* in short) for the training and forecast of 120 financial products of Forex.com on November 09, 2018. As shown, in a typical daily forecast, the quantum forecaster only took 62,341 ms (62.341 s) to finish the training forecast of 120 financial products. On average, it took 0.519 s (less than 1 s) to finish the network training and forecast process of a single financial product.

Figure 13.10 shows the snapshot of quantum forecaster for the system training and forecast of 9 major cryptocurrencies over AvaTrade.com MT platform on the same trading day.

As shown, quantum forecaster took 44,976 ms (44.976 s) to finish the training and forecast of 9 cryptocurrencies. On average, it took 4.99 s to train and forecast a single cryptocurrency.

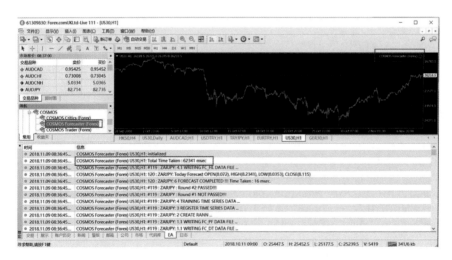

Fig. 13.9 Snapshot of quantum forecaster for the training and forecast of 120 financial products for Forex.com MT4 platform on November 09, 2018

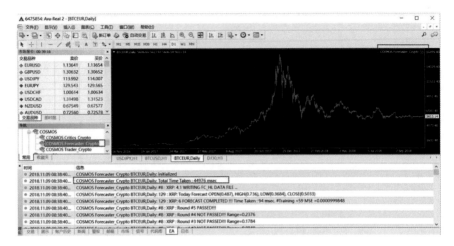

Fig. 13.10 Snapshot of quantum forecaster for training and forecast of 9 major cryptocurrencies for AvaTrade.com MT4 platform on November 09, 2018

As compared with all those 120 non-cryptocurrency products, quantum forecaster took 9.62 times of computer time to predict cryptocurrency, even though cryptocurrency only have 300-trading day records while the other 120 financial products each have 2048-trading day records for system training. It might due to cryptocurrencies, in general, are much more chaotic and fluctuant in nature, which takes more time and iterations for the quantum forecaster to learn the market patterns.

13.4.5 Quantum Forecaster Performance Test

The forecast performance of quantum forecaster is compared with traditional FFBPN by applying 500 forecast simulations for each system. For the ease of comparison, four categories of worldwide 129 financial products are tested network MSE (mean-square-error) ranging from $1 \times 10\text{-}4$ to $1 \times 10\text{-}6$, respectively.

Besides traditional FFPBN, two state-of-the-art financial forecasting methods: (i) SVM (support vector machine) forecasting tool provided by R project, one of the most popular financial forecasting tools used in finance industry and (ii) deep neural network (DNN) with principal component analysis (PCA) model (Singh and Srivastava 2017) are contrasted with quantum forecaster for forecast performance comparison. Table 13.1 presents quantum finance forecast performance comparison chart of these four systems.

Table 13.1 Quantum forecaster forecast performance comparison chart

Product category	No. of PRD	FFBPN		SVM		DNN-PCA		Quantum forecaster	
		Total STT	Av. STT	Total STT	Av. STT	Total STT	Av. STT	Total STT	Av. STT
Case 1 (MSE = 1 × 10-4)									
Cryptocurrency	9	557,251	61916.78	372,244	41360.41	244,633	27181.47	5720	635.56
Forex	84	508,453	605. 01	331,511	3948.56	291,344	3488.38	6323	75.27
Financial index	19	41,120	2184.21	26,152	1378.44	19,738	1038.82	923	48.58
Commodity	17	46,641	2743.59	29,384	1728.46	22,108	1300.46	1223	71.94
Overall	129	1,153,465	8941.59	759,291	5885.98	577,823	4479.25	14,189	109.99
Case 2 (MSE = 1 × 10-5)									
Cryptocurrency	9	1,460,000	162222.22	937,320	104148.67	823,440	91493.33	15,121	1680.11
Forex	84	1,235,543	14708.85	841,405	10016.72	720,322	8575.26	17,231	205.13
Financial index	19	111,024	5843.37	68,391	3599.51	50,627	2664.58	2312	121.68
Commodity	17	109,142	6420.12	71,925	4230.86	44,967	2645.09	2640	155.29
Overall	129	2,915,709	22602.40	1,919,041	14878.29	1,639,356	12708.19	37,304	289.18
Case 3 (MSE = 1 × 10-6)									
Cryptocurrency	9	DL		6,102,255	678028.38	3,601,307	400145.17	57,235	6359.44
Forex	84	D L		5,477,816	65212.10	4,184,833	49819.44	71,243	848.13
Financial index	19	577,324	30385.47	373,529	19659.40	318,683	16772.78	6623	348.58
Commodity	17	687,595	40446.76	468,252	27544.25	385,741	22690.64	9126	536.82
Overall	129	–		12,421,852	96293.43	8,490,564	65818.33	144,227	1118.04

Note
1. Results are generated by 500 simulations of each neural network system (measured in ms)
2. "Total STT" denotes the total average system training time for 500 simulations of network training
3. "Av. STT" denotes the average system training lime for a single financial product of each category
4. "DL" denotes deadlock during system training

Certain interesting findings were revealed:

(i)   For the case I simulation using MSE $1 \times 10\text{-}4$, quantum forecaster outperformed FFBPN, SVM, and DNN-PCA by 81.29, 53.51, and 40.72 times, respectively. Similar findings can be found in case II simulation results. It clearly reflected the improvement of network learning rate by the adoption of chaotic neural oscillators into traditional FFBPN.

(ii) Across the 3 cases with decreasing MSE from $1 \times 10\text{-}4$ (case I), $1 \times 10\text{-}5$ (case II) to $1 \times 10\text{-}6$ (case III). All forecast systems achieved MSE in case I and case II. However, case III simulations using MSE $1 \times 10\text{-}6$, FFBPN encountered "deadlock" problems during the network training of cryptocurrency and forex products; while quantum forecaster could still finish the network training with promising training speeds. It clearly demonstrated the resolution of overtraining and deadlock problems and sufficient improvement of system training efficiency by quantum forecaster over traditional neural networks using classical sigmoid-based activation function.

(iii) In terms of system performance across different financial products, the simulation results clearly reflected that both cryptocurrency and forex were more chaotic and difficult for network training than other financial products as expected, which will be one of the future R&D directions of quantum finance forecast center.

13.4.6   Quantum Trader Trading Strategy Performance Test

In order to conduct a systematic test of quantum trader on intelligent trading, two more benchmark agents were implemented for comparison purposes:

(1)   time-driven trading agent (TmAgent);
(2)   price-driven trading agent (PrAgent).

These two agents simulated real-world traders by using quantum finance forecasts together with simple short-term trading strategies of:

(1)   day-end harvest strategy (time-driven strategy); or
(2)   target-profit and stop-loss strategy (price-driven strategy) with ratio 2:1 (i.e., target-profit $= 2 \times$ stop-loss).

For the 9 cryptocurrency products, 300-trading day simulations were performed for each trading agent; while 500-trading day simulations were performed for 120 non-cryptocurrency financial products. Table 13.2 summarizes the trading performance results for three trading strategies.

Table 13.2 Quantum trader trading strategy performance comparison chart

Product category	Time-driven strategy			Price-driven strategy			Quantum trader strategy		
	BUY (%)	SELL (%)	Overall (%)	BUY (%)	SELL (%)	Overall (%)	BUY (%)	SELL (%)	Overall (%)
Cryptocurrency	2.66	3.17	2.92	3.12	3.12	3.47	6.71	9.37	8.04
Fores	5.52	7.22	6.37	6.32	1.16	7.04	12.41	13.83	13.12
Financial index	5.72	6.11	5.92	5.49	6.71	6.10	9.56	10.21	9.89
Commodity	4.39	6.01	5.20	5.12	6.43	5.7S	10.43	11.88	11.16
Average	4.57	5.63	5.10	5.01	6.18	5.60	9.78	11.32	10.55

Note
1. Results (Monthly Returns %) are generated by 300 trading day simulations of the 9 Cryptocurrencies and 500 trading day simulations of the 120 non-cyptocurrency products

Major findings can be concluded in three aspects:

(i)   Across the three trading strategies, overall speaking, quantum trader trad-ing strategy outperformed time-driven strategy and price-driven strategy by 106.86% and 88.52%, respectively. It clearly reflected the effectiveness of the combined strategy in quantum trader by using the chaotic reinforcement-learning technique. Comparing time-driven strategy with price-driven strategy, it was interested to reveal that price-driven strategy was consistently outper-formed its time-driven counterpart by around 9.73%. It can be explained by the fact that although the time-driven strategy was rather typical in automatic program trading for day trade, the closing-price normally wouldn't be the opti-mal price for bid-closing, as compared with the 2:1 target-profit and stop-loss price-driven strategy which was more sensible in terms of short-term trading.

(ii)   Across the four different categories of financial products, the overall returns of cryptocurrency products were the worst as expected, since the forecast perfor-mance of cryptocurrency products were also the lowest as compared with the other three categories. However, for forex products, it was interested to find out that although the forecast performance of forex products was not as good as financial indices and commodities, their overall returns outperformed these two categories and rank no. 1. As reflected by experienced traders, it might be owing to the fact that although the daily patterns of forex products were highly chaotic, their price patterns were mostly (over 80% of the time) oscillations between the predicted high/low without trigger the stop-loss thresholds.

(iii) As compared with the buy versus sell strategy, it was interested to find out that the overall returns from sell strategy were consistently higher than its buy strategy counterpart by around 16%, which was quite consistent with the comments from experienced traders that automatic program trading usually favors sell versus buy strategy.

13.4.7 Quantum Trader Worldwide Trading Strategy Evaluation

Started from January 1, 2018, 342 members of QFFC which consists of professional traders and quants from major fund houses are invited to join the quantum trader system evaluation focus group for a 1-year worldwide evaluation program of quantum trader trading system as compared with their own trading strategies.

They were all provided with free quantum finance daily forecasts for the 129 financial products and the two de facto agent trading programs (i.e., the time-driven and price-driven trading agents). Based on quantum finance daily forecasts, they are free to use either the de facto trading agents or their own trading strategies to do trading.

Table 13.3 Quantum trader trading performance comparison chart

Product	FFBPN (%)	SVM (%)	DNN-PCA (%)	ALL traders (%)	Top 15% traders (%)	Top 5% traders (%)	Quantum trader (%)
Cryptocurrency	2.03	5.08	5.32	2.76	6.54	8.76	7.62
Forex	2.72	7.91	8.29	5.32	12.31	16.71	13.16
Financial index	3.65	7.59	7.89	4.24	11.22	15.23	11.37
Commodity	3.87	9.52	10.01	4.12	11.81	16.71	12.67
Average	3.07	7.53	7.88	4.11	10.47	14.35	10.55

Table 13.3 presents the nine months (1.1.2018–9.30.2018) trading performance comparison chart for the worldwide evaluation scheme.

For the ease of comparison, quantum trader trading performance (overall 10.55%) was compared with the following:

(1) Traditional FFBPN with day-trade algorithm (overall 3.07%);
(2) SVM forecasting tool with day-trade algorithm (overall 7.53%);
(3) DNN-PCA forecasting model with day-trade algorithm (overall 7.88%);
(4) Overall average performance of the focus group (overall 4.11%);
(5) Top 15% traders' performance (overall 10.47%);
(6) Top 5% traders' performance (overall 14.35%).

As revealed from the performance analysis, it was clear that the overall trading performance of quantum trader outperformed FFBPN, SVM, and DNN-PCA by around 7.49%, 3.03%, and 2.67%, respectively (in terms of overall returns). On the other hand, as compared with professional traders, quantum trader trading performance was close to a typical top 15% trader, outperformed the average traders by over 6.44%, which was rather promising as feedback from professional traders.

13.5 Conclusion

This chapter devised an innovative multiagent-based quantum finance forecast and trading system—quantum trader. From the implementation perspective, the quantum trader was integrated with quantum finance forecast center for the provision of worldwide 129 financial product forecasts and intelligent trading systems.

In fact, for a professional trader and investor, a reliable and effective financial forecast system is only the beginning of the story. A good financial investment also needs: (1) good and effective trading and hedging strategies; (2) stable, logical, and rational investment psychology.

Problems

13.1 What is an intelligent agent? What are the major differences between traditional computer programs and intelligent agents? Give two real-world examples on how to apply intelligent agent technology in finance engineering.

13.2 Discuss and explain the major advantages of using intelligent agent technology for the implementation of intelligent trading systems.

13.3 Discuss and explain why program trading become so popular in nowadays financial community. What are the major differences between traditional financial trading versus program trading? Give two examples of program trading and describe how they work.

13.4 Describe and explain what is high-frequency algorithmic program trading (HFAPT). Give two examples of HFAPT in forex markets and describe how they work.

13.5 Discuss and explain why nowadays AI become so popular in building program trading systems. Give two examples of financial product trading for illustration.

13.6 Financial prediction nowadays becomes one of the hottest topics in financial engineering. Why now as AI prediction already exists for over half a century?

13.7 State and discuss three contemporary AI-based financial prediction technologies and briefly explain how they work.

13.8 As mentioned in Chap. 8, chaos theory told us that the forecast of complex systems such as weather and financial markets are bounded by the initial condition. How can AI such as artificial neural networks or recurrent neural networks get around with this intrinsic problem? Use forex prediction system as an example for explanation.

Acknowledgements The author wishes to thank Forex.com and AvaTrade.com for the provision of historical and real-time financial data. The author also wishes to thank Quantum Finance Forecast Center of UIC for the R&D supports and the provision of the channel and platform Qffc.org for worldwide system testing and evaluation.

References

Borden, C. (2018) *Fibonacci Trading: How to Master the Time and Price Advantage*. McGraw-Hill Education, UK.

Brown, C. (2012) *Elliot Wave Principle: Elementary Concepts, Wave Patterns, and Practice Exercises*. Bloomberg Press, New York.

Bulkowski, T. N. (2005) *Encyclopedia of Chart Patterns*. Wiley, UK.

Carapuço, J. et al. (2018) "Reinforcement learning applied to Forex trading," *Applied Soft Computing Journal* 73: 783–794.

Chang, P. C. et al. (2012) A novel model by evolving partially connected neural network for stock price trend forecasting. *Expert Systems with Applications* 39(1): 611–620.

Chang, C. et al. (2017) Asymmetric Reinforcement Learning and Conditioned Responses During the 2007–2009 Global Financial Crisis: Evidence from Taiwan. *Review of Pacific Basin Financial Markets and Policies* 20(2): 1750010.

Dabhi, V. K. and Chaudhary, S. (2016) Financial time series modeling and prediction using postfix-GP. *Computational Economics* 47(2): 219–253.

Dai, W. et al. (2012) Combining nonlinear independent component analysis and neural network for the prediction of Asian stock market indexes. *Expert Systems with Applications* 39(4): 4444–4452.

Deng Y. et al. (2017) Deep Direct Reinforcement Learning for Financial Signal Representation and Trading. *IEEE Transactions on Neural Networks and Learning Systems* 28(3): 653–664.

Durbin, M. (2010) *All About High-Frequency Trading.* McGraw-Hill Education.

Henrique, B. M. et al. (2018) Stock price prediction using support vector regression on daily and up to the minute prices. *The Journal of Finance and Data Science* 4(3): 183–201.

Hwang, H. and Oh, J. (2010) Fuzzy models for predicting time series stock price index. *International Journal of Control, Automation, and Systems* 8(3): 702–706.

Jeong G. and Kim, H. Y. (2019) Improving financial trading decisions using deep Q-learning: Predicting the number of shares, action strategies, and transfer learning. *Expert Systems with Applications* 117: 125–138.

Kaelbling, L. P. et al. (1996) Reinforcement Learning: A Survey. *Journal of Artificial Intelligence Research* 4: 237–285.

Lee, R. S. T. (2004) A Transient-chaotic Auto-associative Network (TCAN) based on LEE-oscillators. *IEEE Trans. Neural Networks* 15(5): 1228–1243.

Lee, R. S. T. (2006) Fuzzy-Neuro Approach to Agent Applications (From the AI Perspective to Modern Ontology). Springer-Verlag, Germany.

Lee, R. S. T. (2019) COSMOS Trader – Chaotic Neuro-oscillatory Multiagent Financial Prediction and Trading System," The Journal of Finance and Data Science 5: 61–82.

Li, C. G. et al. (2009) An Actor-Critic reinforcement learning algorithm based on adaptive RBF network. *2009 IEEE International Conference on Machine Learning and Cybernetics* 984–988.

Ling, P. W. (2013) The stock price forecasting comparative research of the use of fractal theory at Taiwan traditional industry and technology industry. *Applied Mechanics and Materials* 274: 53–56.

Liu, F. et al. (2005) Neural network model for time series prediction by reinforcement learning. *2005 IEEE International Joint Conference on Neural Networks* 2: 809–814.

Markopoulos, A. P. et al. (2016) On the use of back propagation and radial basis function neural networks in surface roughness prediction. *Journal of Industrial Engineering International* 12(3): 389–400.

Murphy, J. J. (1999) *Technical Analysis of the Financial Markets: A Comprehensive Guide to Trading Methods and Application.* New York Institute of Finance, New York.

Nahil, A. and Lyhyaoui, A. (2018) Short-term stock price forecasting using kernel principal component analysis and support vector machines: the case of Casablanca stock exchange. *Procedia Computer Science* 127: 161–169.

Narayanan A. et al. (2016) *Bitcoin and Cryptocurrency Technologies: A Comprehensive Introduction.* Princeton University Press.

Ouahilal, M. et al. (2017) A novel hybrid model based on Hodrick–Prescott filter and support vector regression algorithm for optimizing stock market price prediction. *Journal of Big Data* 4(1): 1–22.

Pendharkar P. C. and Cusatis, P. (2018) Trading financial indices with reinforcement learning agents. *Expert Systems with Applications* 103: 1–13.

Singh, R., & Srivastava, S. (2017) Stock prediction using deep learning. *Multimedia Tools and Applications* 76(18): 18569–18584.

Tan, Z. et al. (2011) Stock trading with cycles: A financial application of ANFIS and reinforcement learning. *Expert Systems with Applications* 38(5): 4741–4755.

Vigna, P. and Casey, M. J. (2016.) *The Age of Cryptocurrency: How Bitcoin and the Blockchain Are Challenging the Global Economic Order*. Picador,

Walker, W. (2018) *Expert Advisor Programming and Advanced Forex Strategies.* Independently published.

Young, A. R. (2015) *Expert Advisor Programming for MetaTrader 4: Creating automated trading systems in the MQL4 language.* Edgehill Publishing.

Chapter 14
Future Trends in Quantum Finance

> *The most important application of quantum computing in the future is likely to be a computer simulation of quantum systems, because that's an application where we know for sure that quantum systems in general cannot be efficiently simulated on a classical computer.*
> David Deutsch (Born in 1953)

Professor David Deutsch in his famous quotation pointed out an important role and function of a quantum computer—the simulation of quantum systems and related applications, which cannot be efficiently simulated by classical computers.

Is it really true?

The answer is yes and no. "*Yes*" in the sense that the major objective of the quantum computer is to model quantum dynamics and simulate quantum applications with quantum computing technology—so-called *hard quantum computing*. "*No*" in the sense that without the usage of quantum computers, technically we can still model quantum dynamics and quantum applications such as quantum finance model we have studied in this book—so-called *soft quantum computing*.

In this final chapter, we will explore the future of quantum finance and the new era of quantum computing. Similar to the history of artificial intelligence (AI) that has both hard AI and soft AI. Each of them focuses on different aspects and applications, but after over half a century; the boundary between them becomes obscure. In this chapter, we will study two disciplines of quantum computing and how they interact to shape the new age of intelligent computing.

As mentioned in previous chapters, quantum finance is a cross-discipline subject to challenge the utmost question of finance engineering—the exploration of the nature and dynamics of financial market activities in the quantum realm.

Quantum finance is one of the major research areas of quantum finance forecast center (QFFC.org). In this chapter, we will explore and introduce other aspects and active research of QFFC in terms of (1) AI-fintech—the integration of AI technology with state-of-the-art finance technology such as blockchains and digital ledgers; (2) quantum computing—the R&D of potential applications of quantum computing technology in various areas such as quantum entanglements in quantum finance, quantum cryptography, quantum holographic systems, etc.

© Springer Nature Singapore Pte Ltd. 2020
R. S. T. Lee, *Quantum Finance*,
https://doi.org/10.1007/978-981-32-9796-8_14

14.1   Introduction—The Dawn of Quantum Computing

Since the birth of AI and computing technology from over half a century ago, we are now coming to the historical moment of a new era on computing technology—quantum computing. With the official launch of IBM's System Q—the first commercially available quantum computer at CES 2019 Las Vegas USA in January 2019 signify the dawn of quantum computing technology (Moran 2019; Silva 2018).

Different from traditional computers which use bits of 0s and 1s as the fundamental binary representation in data storage and program calculation, quantum computers use *qubits* (stands for *quantum bits*)—quantum superpositions of the two states which allow for far greater flexibility than the traditional binary system in data representation and programming (Hirvensalo 2010; Scherer 2019; Zygelman 2018).

In general, a quantum computer technically would be able to perform calculations on a far greater order of magnitude than traditional computers. As mentioned by Professor David Deutsch in his famous quotation stated at the beginning of this chapter, quantum computing is the best solution tailored for the modeling of quantum systems with related highly complex and sophisticate quantum phenomena such as quantum cryptography and quantum finance in our case (Assche 2006; Bernhardt 2019; Jaeger 2018).

14.2   Two Sides of the Same Coin—Hard Quantum Computing Versus Soft Quantum Computing

The origin of quantum computing can be traced back to 1959 a speech by Professor Richard Feynman in which he spoke about the effects of miniaturization, including the idea of exploiting quantum effects to create more powerful computers. However, before the quantum effects of computing could be realized, quantum computing is still a concept and idea in the scientific community. It was not until 1985, Professor David Deutsch proposed a new idea of *quantum logic gates* in his paper *Quantum theory as a universal physical theory* as a means of harnessing the quantum realm inside a computer. In fact, his paper on the subject showed that any physical process could be modeled by a physical quantum computer—so-called hard quantum computing (Deutsch 1985; Steane 1998).

The road of hard quantum computing was not an easy path. Owing to technical complexity and difficulty, after almost decades, Professor Peter Shor devised an algorithm in 1994 that could use only 6 qubits to perform basic factorization operations. Since then, a handful of quantum computers were built. The first, a 2-qubit quantum computer in 1998, could perform trivial calculations before losing decoherence after a few nanoseconds. In 2000, quantum computing teams including IBM and AT&T successfully built both a 4-qubit and a 7-qubit quantum computers.

IBM has showcased its IBM Q System One at CES 2019, which was claimed as the world's first integrated quantum computing system designed for scientific and commercial use. The 20-qubit system by IBM, which was seen as one of the leaders in the field of quantum computing (Silva 2018; Moran 2019).

Just like hard AI versus soft AI in artificial intelligence (AI) realm, soft quantum computing refers to the modeling of quantum theory and its mathematical models by means of computing systems and programs, similar to the analog of soft AI with the focus of building computer models using artificial neural networks (ANN) to model human intelligence and learning process.

In other words, all quantum theory concepts and models we have studied in this book, ranging from Feynman's path integral for the modeling of forward interest rate to quantum anharmonic oscillator model for the modeling of quantum price levels; along with quantum financial signals are components and frameworks of soft quantum computing.

If we can model quantum theory and related phenomena by means of traditional computers, why do we need to build quantum computers?

The answer is simple. If we can model quantum theory effectively using traditional computers with binary formulation and representations, integrating these quantum models with quantum computers will become more direct and effective while the fundamental construction of quantum computer itself totally conforms to quantum model and architecture. In other words, by implementing a quantum model, say quantum finance model into a traditional computer; technically speaking, we are converting a traditional financial computation system into a quantum financial computation machine!

14.3 From Quantum Finance to AI Fintech

As mentioned earlier in this chapter, quantum finance is one of the major R&D projects of quantum finance forecast center (QFFC.org)—a worldwide financial prediction and R&D center for the design and implementation of next generation of AI-fintech standards, toolkits and applications (Nicoletti 2017).

Figure 14.1 depicts the AI-fintech 5-layer architecture of QFFC.

(1) AI-applications Layer

- This layer focuses on AI-fintech-related systems and applications in four major areas: forecasts such as quantum finance forecast, deep learning/reinforcement learning such as multiagent-based trading agents; data mining and NLP (natural language processing) such as robo-advisor (Lim et al. 2011; Sironi 2016).

(2) Business Layer

- This layer focuses on the business applications and fintech systems including cryptocurrency, blockchain systems on financial trading, payment, clearing and hedging transactions (Antonopoulos 2017; Gaur 2018; Sironi 2016; Nielsen 2011, Swan 2015).

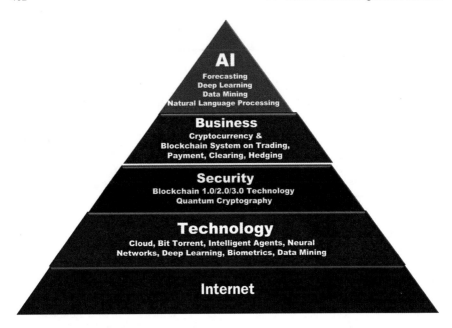

Fig. 14.1 AI-Fintech 5-layer architecture

(3) Security Layer

- This layer focuses on the implementation of fintech security systems and technology include Blockchain 1.0/2.0/3.0 systems and quantum cryptosystems on crypto-payment and transactions (Assis 2012; Balygin 2017; Elliott 2004; Fehr 2010).

(4) Technology

- This layer focuses on all the fundamental and building blocks technology and standards for the implementation of AI-fintech applications.

(5) Internet Layer

- This layer provides the basic framework and backbone for AI-fintech intelligent agent communication technology, protocols, and technology.

14.4 Quantum Finance System Development Platform

Besides the provision of various AI-fintech forecasts and applications, QFFC also actively developing a next-generation AI-fintech development kit—quantum finance development kit (QFDK), an easy to use C/C++ library with the implementation of all state-of-the-art AI-fintech functions and libraries; so that AI-fintech developers

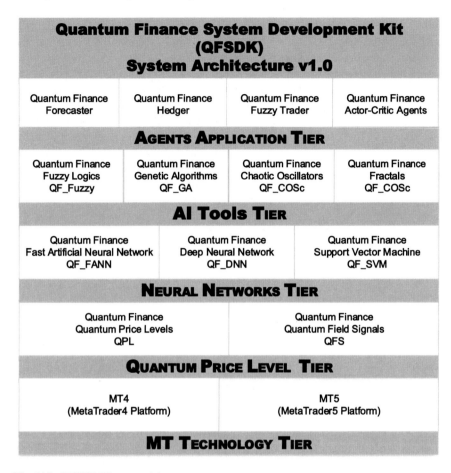

Fig. 14.2 QFSDK 5-layer model

and data scientists can base on these functional libraries to design and develop their own quantum finance and AI-fintech systems and applications.

Figure 14.2 depicts the QFSDK 5-layers model of version 1.0, which will be launched officially in late 2019 at QFFC.org official site.

14.5   Conclusion

In this final chapter, we studied the future trend in quantum finance—a cross-discipline subject that comprises quantum theory as a theoretical foundation; numerical computational theory as computer model; artificial intelligence as machine learning; prediction and trading optimization models; and finally quantum computing as the ultimate intelligence machine. In other words, the future R&D and growth of quantum finance technology not only depends on the development of quantum finance theory itself but also depends on future development and growth of related technologies such as AI and quantum computing technologies.

As mentioned in the introductory chapter, since Professor Heisenberg introduced the ground-breaking quantum mechanics' idea to nowadays, we have concrete mathematical and computational models to model quantum dynamics and QPL of financial markets, along with the interpretation of path integrals of the forward interest rate. It is both a challenging and exciting path. The contribution is not only the credit of an individual but also is the collective contribution of intelligence, innovation, and R&D from many centuries.

In view of the advancements of technology and AI in the past decades, quantum finance, together with AI and quantum computing technology should work together to provide more scientific and intelligent applications, systems and services for worldwide investors and financial professionals.

Problems

14.1   What is a quantum computer? State and explain the major difference between a quantum computer and the traditional computer systems.

14.2   State and explain the major challenges in quantum computing. Give two examples of illustrations.

14.3   State and contrasts the major differences between hard quantum computing and soft quantum computing. Give two examples in each case for illustration.

14.4   Discuss whether quantum computer in the future will or will not replace traditional computers. Why?

14.5   What is finance technology (fintech)? State three typical fintech-related technologies and briefly explain how they work.

14.6   What is a robo-advisor in fintech? State and explain the basic underlying technology. Give two examples of robo-advisor in fintech and explain briefly how they work.

14.7   What is blockchain technology? State and explain the difference between Blockchain 1.0, 2.0, and 3.0 technologies. Give one example in each case for illustration.

14.8   What is quantum cryptography? State and explain the major difference between traditional cryptosystem and quantum cryptosystem. What are the major challenges in quantum cryptography?

References

Antonopoulos, A. M. (2017) Mastering Bitcoin: Programming the Open Blockchain, O'Reilly.

Assche, G. (2006) Quantum Cryptography and Secret-Key Distillation. Cambridge University Press.

Assis et al. (2012) Improving classical authentication over a quantum channel. Entropy 14(12): 2531–2549.

Balygin et al. A. (2017) Practical quantum cryptography. JETP Letters 105(9): 606–612.

Bernhardt, C. (2019) Quantum Computing for Everyone. MIT Press.

Deutsch, D. (1985) Quantum theory as a universal physical theory. International Journal of Theoretical Physics 24(1): 1–41.

Elliott, C. (2004) Quantum cryptography," IEEE Security & Privacy, vol. 2, (4), pp. 57–61.

Fehr, S. (2010) Quantum cryptography. Foundations of Physics 40(5): 494–531.

Gaur, N. (2018) Hands-On Blockchain with Hyperledger: Building decentralized applications with Hyperledger Fabric and Composer, Packt Publishing.

Hirvensalo, M. (2010.) Quantum Computing. Springer.

Jaeger, L. (2018) The Second Quantum Revolution: From Entanglement to Quantum Computing and Other Super-Technologies. Copernicus.

Lim, E. H. Y., Liu, J. N. K. and Lee, R. S. T. (2011) Knowledge Seeker – Ontology Modelling for Information Search and Management, Springer-Verlag, Germany.

Moran, C. C. (2019) Mastering Quantum Computing with IBM QX: Explore the world of quantum computing using the Quantum Composer and Qiskit. Packt Publishing.

Nicoletti, B. (2017) The Future of Fintech: Integrating Finance and Technology in Financial Services, Palgrave Macmillan.

Nielsen, M. A. and Chuang, I. L. (2011) Quantum Computation and Quantum Information: 10th Anniversary Edition. Cambridge University Press.

Scherer, W. (2019) Mathematics of Quantum Computing: An Introduction. Springer.

Silva V. (2018) Practical Quantum Computing for Developers: Programming Quantum Rigs in the Cloud using Python, Quantum Assembly Language and IBM QExperience, APress.

Sironi, P. (2016) Fintech Innovation: From Robo-Advisors to Goal Based Investing and Gamification, Wiley.

Steane, A. (1998) Quantum computing. Reports on Progress in Physics 61(2): 117–173.

Swan, M. (2015.) Blockchain: Blueprint for a New Economy, O'Reilly.

Zygelman, B. (2018) A First Introduction to Quantum Computing and Information. Springer.

Appendix A
List of 139 Financial Products

Code	Product description	Code	Product description
Cryptocurrencies (Data provided by AvaTrade.com)			
BCHUSD	BitCoin Cash versus US Dollar	EOSUSD	EOS versus US Dollar
BTCEUR	BitCoin versus Euro	ETH	Ethereum
BTCJPY	BitCoin versus Japanese Yen	LTC	Litecoin
BTCUSD	BitCoin versus US Dollar	XRP	XRP
BTGUSD	Bitcoin Gold versus US Dollar		
Financial index (Data provided by Forex.com)			
AUS200	AUSSIE 200	N25	Netherlands 25 Index
CHINAA50	China A50 Index	NAS100	Nasdaq Index
ESP35	Spain 35 Index	SIGI	Singapore Index
ESTX50	EURO STOXX 50 Index	SPX500	SP500 Index
FRA40	CAC 40 Index	SWISS20	Switzerland 20 Index
GER30	DAX 30 Index	UK100	FTSE 100 Index
HK50	Hang Seng Index	US2000	US Small Cap 2000
IT40	Italy 40 Index	US30	Dow Jones Index
JPN225	Nikkei Index		
Commodity (Data provided by Forex.com)			
COPPER	Copper	US_OIL	WTI Crude Oil
CORN	Corn	WHEAT	Wheat
COTTON	Cotton	XAGUSD	Silver versus US Dollar
HTG_OIL	HTG Oil	XAUAUD	Gold versus Australian Dollar
PALLAD	Palladium	XAUCHF	Gold versus Swiss Franc
PLAT	Platinum	XAUEUR	Gold versus Euro
SOYBEAN	Soybean	XAUGBP	Gold versus British Pound

(continued)

© Springer Nature Singapore Pte Ltd. 2020
R. S. T. Lee, *Quantum Finance*,
https://doi.org/10.1007/978-981-32-9796-8

(continued)

Code	Product description	Code	Product description
SUGAR	Sugar	XAUJPY	Gold versus Japanese Yen
UK_OIL	Brent Crude Oil	XAUUSD	Gold versus US Dollar
US_NATG	US Natural Gas		

Forex (Data provided by Forex.com)

Code	Product description	Code	Product description
AUDCAD	Australian Dollar versus Canadian Dollar	GBPDKK	British Pound versus Danish Krone
AUDCHF	Australian Dollar versus Swiss Franc	GBPHKD	British Pound versus Hong Kong Dollar
AUDCNH	Australian Dollar versus Chinese Yuan	GBPJPY	British Pound versus Japanese Yen
AUDJPY	Australian Dollar versus Japanese Yen	GBPMXN	British Pound versus Mexican Peso
AUDNOK	Australian Dollar versus Norwegian Krone	GBPNOK	British Pound versus Norwegian Krone
AUDNZD	Australian versus New Zealand Dollar	GBPNZD	British Pound versus New Zealand Dollar
AUDPLN	Australian Dollar versus Polish Zloty	GBPPLN	British Pound versus Polish Zloty
AUDSGD	Australian Dollar versus Singapore Dollar	GBPSEK	British Pound versus Swedish Krona
AUDUSD	Australian Dollar versus US Dollar	GBPSGD	British Pound versus Singapore Dollar
CADCHF	Canadian Dollar versus Swiss Franc	GBPUSD	British Pound versus US Dollar
CADJPY	Canadian Dollar versus Japanese Yen	GBPZAR	British Pound versus South African Rand
CADNOK	Canadian Dollar versus Norwegian Krone	HKDJPY	Hong Kong Dollar versus Japanese Yen
CADPLN	Canadian Dollar versus Polish Zloty	NOKDKK	Norwegian Krone versus Danish Krone
CHFHUF	Swiss Franc versus Hungarian Forint	NOKJPY	Norwegian Krone versus Japanese Yen
CHFJPY	Swiss Franc versus Japanese Yen	NOKSEK	Norwegian Krone versus Swedish Krona
CHFNOK	Swiss Franc versus Norwegian Krone	NZDCAD	New Zealand versus Canadian Dollar
CHFPLN	Swiss Franc versus Polish Zloty	NZDCHF	New Zealand Dollar versus Swiss Franc
CNHJPY	Chinese Yuan versus Japanese Yen	NZDJPY	New Zealand Dollar versus Japanese Yen
EURAUD	Euro versus Australian Dollar	NZDUSD	New Zealand Dollar versus US Dollar

(continued)

(continued)

Code	Product description	Code	Product description
EURCAD	Euro versus Canadian Dollar	SGDHKD	Singapore versus Hong Kong Dollar
EURCHF	Euro versus Swiss Franc	SGDJPY	Singapore Dollar versus Japanese Yen
EURCNH	Euro versus Chinese Yuan	TRYJPY	Turkish Lira versus Japanese Yen
EURCZK	Euro versus Czech Koruna	USDCAD	US Dollar versus Canadian Dollar
EURDKK	Euro versus Danish Krone	USDCHF	US Dollar versus Swiss Franc
EURGBP	Euro versus British Pound	USDCNH	US Dollar versus Chinese Yuan
EURHKD	Euro versus Hong Kong Dollar	USDCZK	US Dollar versus Czech Koruna
EURHUF	Euro versus Hungarian Forint	USDDKK	US Dollar versus Danish Krone
EURJPY	Euro versus Japanese Yen	USDHKD	US Dollar versus Hong Kong Dollar
EURMXN	Euro versus Mexican Peso	USDHUF	US Dollar versus Hungarian Forint
EURNOK	Euro versus Norwegian Krone	USDILS	US Dollar versus Israeli Shekel
EURNZD	Euro versus New Zealand Dollar	USDJPY	US Dollar versus Japanese Yen
EURPLN	Euro versus Polish Zloty	USDMXN	US Dollar versus Mexican Peso
EURRON	Euro versus Romanian Leu	USDNOK	US Dollar versus Norwegian Krone
EURRUB	Euro versus Russian Ruble	USDPLN	US Dollar versus Polish Zloty
EURSEK	Euro versus Swedish Krona	USDRON	US Dollar versus Romanian Leu
EURSGD	Euro versus Singapore Dollar	USDRUB	US Dollar versus Russian Ruble
EURTRY	Euro versus Turkish Lira	USDSEK	US Dollar versus Swedish Krona
EURUSD	Euro versus US Dollar	USDSGD	US Dollar versus Singapore Dollar
EURZAR	Euro versus South African Rand	USDTHB	US Dollar versus Thai Baht
GBPAUD	British Pound versus Australian Dollar	USDTRY	US Dollar versus Turkish Lira
GBPCAD	British Pound versus Canadian Dollar	USDZAR	US Dollar versus South African Rand
GBPCHF	British Pound versus Swiss Franc	ZARJPY	South African Rand versus Jap Yen

Appendix B
List of 39 Financial Indicators

No	Function name	Signal name
1	iAC	Accelerator Oscillator
2	iAD	Accumulation/Distribution
3	iADX	Average Directional Index
4	ADXWilder	Av Dir by Welles Wilder
5	iAlligator	Alligator
6	iAO	Awesome Oscillator
7	iATR	Average True Range
8	iBearsPower	Bears Power
9	iBands	Bollinger Bands
10	iBandsOnArray	Bollinger Bands indicator, stored in array
11	iBullsPower	Bulls Power
12	iCCI	Commodity Channel Index
13	Chaikin	Chaikin Oscillator
14	iCustom	Custom indicator
15	DEMA	Double Exponential MA
16	iDeMarker	DeMarker
17	iEnvelopes	Envelopes
18	iEnvelopesOnArray	Calculation of Envelopes, stored in array
19	iForce	Force Index
20	iFractals	Fractals
21	iGator	Gator Oscillator
22	iIchimoku	Ichimoku Kinko Hyo
23	iBWMFI	Market Facilitation Index by Bill Williams
24	iMomentum	Momentum

(continued)

© Springer Nature Singapore Pte Ltd. 2020
R. S. T. Lee, *Quantum Finance*,
https://doi.org/10.1007/978-981-32-9796-8

(continued)

No	Function name	Signal name
25	iMomentumOnArray	Calculation of Momentum, stored in array
26	iMFI	Money Flow Index
27	iMA	Moving Average
28	iMAOnArray	Calculation of Moving Average indicator on data
29	iOsMA	Moving Average of Oscillator
30	iMACD	MACD indicator
31	iOBV	On Balance Volume
32	iSAR	Parabolic Stop And Reverse System
33	iRSI	Relative Strength Index
34	iRSIOnArray	Calculation of Momentum, stored in array
35	iRVI	Relative Vigor Index
36	iStdDev	Standard Deviation
37	iStdDevOnArray	Standard Deviation indicator, stored in array
38	iStochastic	Stochastic Oscillator
39	iWPR	Williams' Percent Range

Printed in the United States
By Bookmasters